Trends in Catalytic Wet Peroxide Oxidation Processes

Trends in Catalytic Wet Peroxide Oxidation Processes

Special Issue Editors

Asuncion Quintanilla
Macarena Munoz

MDPI • Basel • Beijing • Wuhan • Barcelona • Belgrade

Special Issue Editors
Asuncion Quintanilla
Universidad Autónoma de Madrid
Spain

Macarena Munoz
Universidad Autónoma de Madrid
Spain

Editorial Office
MDPI
St. Alban-Anlage 66
4052 Basel, Switzerland

This is a reprint of articles from the Special Issue published online in the open access journal *Catalysts* (ISSN 2073-4344) from 2018 to 2019 (available at: https://www.mdpi.com/journal/catalysts/special_issues/CWPO)

For citation purposes, cite each article independently as indicated on the article page online and as indicated below:

LastName, A.A.; LastName, B.B.; LastName, C.C. Article Title. *Journal Name* **Year**, *Article Number*, Page Range.

ISBN 978-3-03921-924-7 (Pbk)
ISBN 978-3-03921-925-4 (PDF)

Cover image courtesy of Freepik.com.

© 2019 by the authors. Articles in this book are Open Access and distributed under the Creative Commons Attribution (CC BY) license, which allows users to download, copy and build upon published articles, as long as the author and publisher are properly credited, which ensures maximum dissemination and a wider impact of our publications.

The book as a whole is distributed by MDPI under the terms and conditions of the Creative Commons license CC BY-NC-ND.

Contents

About the Special Issue Editors vii

Asunción Quintanilla and Macarena Munoz
Editorial Catalysts: Special Issue on Trends in Catalytic Wet Peroxide Oxidation Processes
Reprinted from: *Catalysts* **2019**, *9*, 918, doi:10.3390/catal9110918 1

Juan José Rueda Márquez, Irina Levchuk and Mika Sillanpää
Application of Catalytic Wet Peroxide Oxidation for Industrial and Urban Wastewater Treatment: A Review
Reprinted from: *Catalysts* **2018**, *8*, 673, doi:10.3390/catal8120673 3

Carmen S.D. Rodrigues, Ricardo M. Silva, Sónia A.C. Carabineiro, F.J. Maldonado-Hódar and Luís M. Madeira
Wastewater Treatment by Catalytic Wet Peroxidation Using Nano Gold-Based Catalysts: A Review
Reprinted from: *Catalysts* **2019**, *9*, 478, doi:10.3390/catal9050478 21

Asunción Quintanilla, Jose L. Diaz de Tuesta, Cristina Figueruelo, Macarena Munoz and Jose A. Casas
Condensation By-Products in Wet Peroxide Oxidation: Fouling or Catalytic Promotion? Part I. Evidences of an Autocatalytic Process
Reprinted from: *Catalysts* **2019**, *9*, 516, doi:10.3390/catal9060516 54

Asunción Quintanilla, Jose L. Diaz de Tuesta, Cristina Figueruelo, Macarena Munoz and Jose A. Casas
Condensation By-Products in Wet Peroxide Oxidation: Fouling or Catalytic Promotion? Part II: Activity, Nature and Stability
Reprinted from: *Catalysts* **2019**, *9*, 518, doi:10.3390/catal9060518 69

Zhongda Liu, Qiumiao Shen, Chunsun Zhou, Lijuan Fang, Miao Yang and Tao Xia
Kinetic and Mechanistic Study on Catalytic Decomposition of Hydrogen Peroxide on Carbon-Nanodots/Graphitic Carbon Nitride Composite
Reprinted from: *Catalysts* **2018**, *8*, 445, doi:10.3390/catal8100445 85

Sarto Sarto, Pacsal Pacsal, Irine Bellina Tanyong, William Teja Laksmana, Agus Prasetya and Teguh Ariyanto
Catalytic Degradation of Textile Wastewater Effluent by Peroxide Oxidation Assisted by UV Light Irradiation
Reprinted from: *Catalysts* **2019**, *9*, 509, doi:10.3390/catal9060509 101

Xiyan Xu, Shuming Liu, Yong Cui, Xiaoling Wang, Kate Smith and Yujue Wang
Solar-Driven Removal of 1,4-Dioxane Using $WO_3/n\gamma\text{-}Al_2O_3$ Nano-Catalyst in Water
Reprinted from: *Catalysts* **2019**, *9*, 389, doi:10.3390/catal9040389 112

Ana María Campos, Paula Fernanda Riaño, Diana Lorena Lugo, Jenny Alejandra Barriga, Crispín Astolfo Celis, Sonia Moreno and Alejandro Pérez
Degradation of Crystal Violet by Catalytic Wet Peroxide Oxidation (CWPO) with Mixed Mn/Cu Oxides
Reprinted from: *Catalysts* **2019**, *9*, 530, doi:10.3390/catal9060530 123

David Lorenzo, Carmen M. Dominguez, Arturo Romero and Aurora Santos
Wet Peroxide Oxidation of Chlorobenzenes Catalyzed by Goethite and Promoted by Hydroxylamine
Reprinted from: *Catalysts* **2019**, *9*, 553, doi:10.3390/catal9060553 137

Elena Magioglou, Zacharias Frontistis, John Vakros, Ioannis D. Manariotis and Dionissios Mantzavinos
Activation of Persulfate by Biochars from Valorized Olive Stones for the Degradation of Sulfamethoxazole
Reprinted from: *Catalysts* **2019**, *9*, 419, doi:10.3390/catal9050419 159

Selamawit Ashagre Messele, Christophe Bengoa, Frank Erich Stüber, Jaume Giralt, Agustí Fortuny, Azael Fabregat and Josep Font
Enhanced Degradation of Phenol by a Fenton-Like System (Fe/EDTA/H_2O_2) at Circumneutral pH
Reprinted from: *Catalysts* **2019**, *9*, 474, doi:10.3390/catal9050474 173

About the Special Issue Editors

Asunción Quintanilla is currently Associate Professor of the Chemical Engineering Department at the Universidad Autónoma de Madrid (Spain). She has worked as a pre-doctoral researcher at the Universidad Complutense de Madrid (2000–2002), Assistant Professor at the Universidad Autónoma de Madrid (2002–2006), and Postdoctoral Researcher at Delft University of Technology (2006–2008). Her research interests focus on catalysis engineering applied to environmental protection with an emphasis on advanced oxidation technologies for water treatment and sustainable technologies for safer products. Ongoing research includes the preparation, characterization, and design of catalysts (nanoparticles, bidimensional materials, carbon-based materials, and MOFs), manufacturing and application of structured catalyst by 3D-printing technologies (Robocasting), kinetic modeling, computational fluid dynamics of catalytic reactors and, also, development of high-efficient catalytic technologies for wastewater treatment and green petrochemical industry. She has co-authored more than 45 peer-reviewed scientific paper in international journals, with one book contribution and over 70 contributions in conference proceedings. She has been involved in 13 funded research projects, in some cases as Principal Investigator.

Macarena Munoz (Dr.) is Senior Researcher (Ramón y Cajal fellow) of the Chemical Engineering Department at Autonoma University of Madrid (UAM). She graduated in Environmental Sciences in 2008 and was awarded her Ph.D. in Chemical Engineering from UAM in 2012. She has been fully dedicated to scientific research for over ten years. During this time, she has also been Guest Researcher in Clausthal Institute of Environmental Technology (Germany) and Aveiro University (Portugal), and took up a postdoctoral research position at Friedrich-Alexander Universität Erlangen-Nürnberg (Germany). Since January 2018, she has held the Ramón y Cajal position at UAM. Her current research line is focused on the intensification of advanced oxidation processes in order to make them more cost-efficient, sustainable, and technically feasible. The application of these processes is intended to eliminate pollutants of emerging concern and cyanotoxins from water. In 2019, she started a new line of research concerned with the elimination of microplastics from aqueous solutions. Her scientific researcher output includes 49 JCR publications, 2 journal covers, 2 book chapters, and 57 contributions to conferences, including a plenary conference. She has been involved in 11 competitive research projects. Regarding technology transfer, she is the first author of two national patents. She has served as Project Evaluator (ANEP (Spain); National Agency for Scientific and Technical Promotion of Argentina; Fundación de Ciencia de Israel) and she is a reviewer for 55 international journals.

Editorial

Editorial Catalysts: Special Issue on Trends in Catalytic Wet Peroxide Oxidation Processes

Asunción Quintanilla * and Macarena Munoz *

Chemical Engineering Department, Universidad Autónoma de Madrid, Ctra. Colmenar km 15, 28049 Madrid, Spain
* Correspondence: asun.quintanilla@uam.es (A.Q.); macarena.munnoz@uam.es (M.M.);
  Tel.: 34-91-497-3454 (A.Q. & M.M.); Fax: +34-91497-3516 (A.Q. & M.M.)

Received: 29 October 2019; Accepted: 31 October 2019; Published: 4 November 2019

The catalytic wet peroxide oxidation (CWPO) process is an advanced oxidation technology that has shown great potential for the decontamination of wastewater. CWPO allows the removal of recalcitrant organic compounds under mild conditions (temperatures and pressures in the range of 25–100 °C and 0.1–0.5 MPa, respectively) by using hydrogen peroxide (H_2O_2) as an oxidant, which is considered an environmentally friendly agent. This process requires a solid catalyst with redox properties to generate hydroxyl and hydroperoxyl radicals from the H_2O_2 decomposition. These radical species easily react with the pollutants, oxidizing them into biodegradable forms and finally into CO_2 and water.

This special issue gives an overview of the state-of-the-art CWPO research for the treatment of industrial and urban wastewaters and how this process can be integrated into the water treatment process [1]. It is illustrated that the high versatility of this low-cost technology, thanks to the CWPO operational flexibility, is easily adaptable to any kind of wastewater, either polluted by high-loaded recalcitrant organics in industrial wastewaters or by emerging pollutants at micro-concentration levels in urban waters., This versatility also stands on the application of different types of solid catalysts, which can be tailored according to the process requirements.

For this reason, intensive research effort has been focused on the development of catalysts capable of promoting the abatement of different pollutants in combination with an adequate stability for long-term use and high efficiency of H_2O_2 consumption. In this sense, supported gold nanoparticles have demonstrated to fit these requirements, and a rigorous revision of the main goals of CWPO in presence of gold catalyst can be found in the special issue [2]. However, deactivation cannot be completely avoided due progressive fouling of the catalyst by the condensation by-products formed upon reaction. An insight into the CWPO reaction mechanism in order to understand the formation, nature, and role of these species [3,4] as well as the hydroxyl radical production mechanism [5], has been also covered.

On the other hand, different innovative solutions show the current trends in the CWPO technology, mainly aimed at the development of an efficient process operated at ambient conditions, by assisting CWPO with UV light irradiation [6], solar light [7], air flow [8], or additional radical activators [9,10]; and also by operated under neutral pH with efficient production of hydroxyl radicals [11]. All these achievements, with significant impact on the operating cost of the CWPO units, were conditioned by the presence of a proper catalyst designed and tailored to provide the best performance.

Finally, we would like to acknowledge the work of excellence developed by the authors of all the contributions to this collection issue, the good aid provided by the involved editorial assistants, and the efforts and comments provided by the reviewers to improve the quality of the articles. Without them, this special issue would not have been possible.

Conflicts of Interest: The authors declare no conflict of interest.

References

1. Rueda Márquez, J.J.; Levchuk, I.; Sillanpää, M. Application of catalytic wet peroxide oxidation for industrial and urban wastewater treatment: A review. *Catalysts* **2018**, *8*, 673. [CrossRef]
2. Rodriguez, C.S.D.; Silva, R.M.; Carabineiro, S.A.C.; Maldonado-Hódar, F.J.; Madeira, L.M. Wastewater treatment by catalytic wet peroxidation using nano gold-based catalysts: A review. *Catalysts* **2019**, *9*, 478. [CrossRef]
3. Quintanilla, A.; Díaz de Tuesta, J.L.; Figueruelo, C.; Munoz, M.; Casas, J.A. Condensation by-products in wet peroxide oxidation: Fouling or catalytic promotion? Part I: Evidences of an autocatalytic process. *Catalysts* **2019**, *9*, 516. [CrossRef]
4. Quintanilla, A.; Díaz de Tuesta, J.L.; Figueruelo, C.; Munoz, M.; Casas, J.A. Condensation by-products in wet peroxide oxidation: Fouling or catalytic promotion? Part II: Activity, nature and stability. *Catalysts* **2019**, *9*, 518. [CrossRef]
5. Liu, Z.; Shen, Q.; Zhou, C.; Fang, L.; Yang, M.; Xia, T. Kinetic and mechanistic study on catalytic decomposition of hydrogen peroxide on carbon-nanodots/graphitic carbon nitride composite. *Catalysts* **2018**, *8*, 445. [CrossRef]
6. Sarto, S.; Paesal, P.; Tanyong, I.B.; Laksmana, W.T.; Prasetya, A.; Ariyanto, T. Catalytic degradation of textile wastewater effluent by peroxide oxidation assisted by UV light irradiation. *Catalysts* **2019**, *9*, 509. [CrossRef]
7. Xu, X.; Liu, S.; Cui, Y.; Wang, X.; Smith, K.; Wang, Y. Solar-driven removal of 1,4-dioxane using WO3/nγ-Al2O3 nano-catalyst in water. *Catalysts* **2019**, *9*, 389. [CrossRef]
8. Campos, A.M.; Riaño, P.F.; Lugo, D.L.; Barriega, J.A.; Celis, C.A.; Moreno, S.; Pérez, A. Degradation of crystal violet by Catalytic Wet Peroxide Oxidation (CWPO) with mixed Mn/Cu oxides. *Catalysts* **2019**, *9*, 530. [CrossRef]
9. Lorenzo, D.; Dominguez, C.M.; Romero, A.; Santos, A. Wet peroxide oxidation of chlorobenzenes catalyzed by goethite and promoted by hydroxylamine. *Catalysts* **2019**, *9*, 553. [CrossRef]
10. Magioglou, E.; Frontistis, Z.; Vakros, J.; Manariotis, I.D.; Mantzavinos, D. Activation of persulfate by biochars from valorized olive stones for the degradation of sulfamethoxazole. *Catalysts* **2019**, *9*, 419. [CrossRef]
11. Messele, S.A.; Bengoa, C.; Stüber, F.E.; Giralt, J.; Fortuny, A.; Fabregat, A.; Font, J. Enhanced degradation of phenol by a fenton-like system (Fe/EDTA/H2O2) at circumneutral pH. *Catalysts* **2019**, *9*, 474. [CrossRef]

© 2019 by the authors. Licensee MDPI, Basel, Switzerland. This article is an open access article distributed under the terms and conditions of the Creative Commons Attribution (CC BY) license (http://creativecommons.org/licenses/by/4.0/).

Review

Application of Catalytic Wet Peroxide Oxidation for Industrial and Urban Wastewater Treatment: A Review

Juan José Rueda Márquez [1,*], Irina Levchuk [2] and Mika Sillanpää [1]

[1] Laboratory of Green Chemistry, Lappeenranta University of Technology, Sammonkatu 12 (Innovation Centre for Safety and Material Technology, TUMA), 50130 Mikkeli, Finland; mika.sillanpaa@lut.fi
[2] Water and Wastewater Engineering Research Group, School of Engineering, Aalto University, PO Box 15200, FI-00076 Aalto, Finland; irina.levchuk@aalto.fi
[*] Correspondence: juan.rueda.marquez@lut.fi

Received: 12 November 2018; Accepted: 14 December 2018; Published: 19 December 2018

Abstract: Catalytic wet peroxide oxidation (CWPO) is emerging as an advanced oxidation process (AOP) of significant promise, which is mainly due to its efficiency for the decomposition of recalcitrant organic compounds in industrial and urban wastewaters and relatively low operating costs. In current study, we have systemised and critically discussed the feasibility of CWPO for industrial and urban wastewater treatment. More specifically, types of catalysts the effect of pH, temperature, and hydrogen peroxide concentrations on the efficiency of CWPO were taken into consideration. The operating and maintenance costs of CWPO applied to wastewater treatment and toxicity assessment were also discussed. Knowledge gaps were identified and summarised. The main conclusions of this work are: (i) catalyst leaching and deactivation is one of the main problematic issues; (ii) majority of studies were performed in semi-batch and batch reactors, while continuous fixed bed reactors were not extensively studied for treatment of real wastewaters; (iii) toxicity of wastewaters treated by CWPO is of key importance for possible application, however it was not studied thoroughly; and, (iv) CWPO can be regarded as economically viable for wastewater treatment, especially when conducted at ambient temperature and natural pH of wastewater.

Keywords: catalytic wet peroxide oxidation; heterogeneous Fenton; wastewater; cost; toxicity; iron leaching

---

1. Introduction

Water is a vital and limited resource, which is constantly under pressure from urbanisation, pollution, etc. The majority of these activities produce an over-exploitation of fresh water. For instance, at least 11% of the European population and 17% of its territory have been affected by water scarcity [1]. Even in highly developed countries, the majority of wastewater is discharged directly into the environment without adequate treatment, with detrimental impacts on human health, economic productivity, and the quality of freshwater resources and ecosystems [2]. In accordance with the Water Framework Directive [3], the good status of the water should have been achieved by 2015. However, only about half of European waters are able to meet the requirements of this directive [4].

Industrial and urban wastewater effluents have been recognised as one of the major sources of many environmental contaminants, such as polychlorinated biphenyls (PCBs) [5], polycyclic aromatic hydrocarbons (PAHs) [6], pharmaceutically active compounds (PhACs) [7], personal care products (PCPs) [7], pesticides [8], metals [9], antibiotics [10], and other pollutants of emerging concern. Neuroendocrine, mutagenic, and/or health effects on the aquatic environment when exposed to pollutants of emerging concern were reported [11]. Even at a low concentration (μg/L), some emerging contaminants (e.g. synthetic musks) are persistent and bio-accumulate due to their hydrophobicity [12], so an improvement of the existing wastewater treatment process is needed in order to prevent the spread of emerging pollutants into the environment.

Irrefutably, Advanced Oxidation Processes (AOPs) are promising methods for the degradation of resistant and recalcitrant compounds or their transformation into biodegradable form (partial mineralisation). It is generally accepted that during AOPs, the generation of highly reactive oxidising species, such as hydroxyl radicals, occurs. These oxidising species possess high electrochemical oxidation potential (standard oxidising potential for hydroxyl radicals varies between 2.8 V at pH 0 and 2.0 at pH 14 [13]) and a non-selective nature, leading to the degradation of organic contaminants, including those that are resistant to conventional oxidation processes, such as chlorination and ozonation [14]. When oxidising species react with organic pollutants in water, series of oxidation reactions are initiated causing, in an ideal case, complete mineralisation with the formation of CO_2, water and inorganic ions as final products. After achieving the complete mineralisation of contaminants and the generated by-products, the further treatment of water is not needed [15]. This way, the secondary loading of contaminants into the environment can be avoided and AOPs can be considered as "clean technology" [15]. It is highly possible that AOPs can be among the most used water treatment processes for the elimination of persistent organic compounds from wastewater in the near future [16,17]. However, not all AOPs are feasible for the treatment of real wastewater due to high electricity demand, a significant amount of oxidant, the necessity of pH adjustment for optimal operation, etc. Among several AOPs that are studied and used for the purification of wastewater, catalytic wet peroxide oxidation (CWPO) or the heterogeneous Fenton process is emerging as one with significant promise (Figure 1).

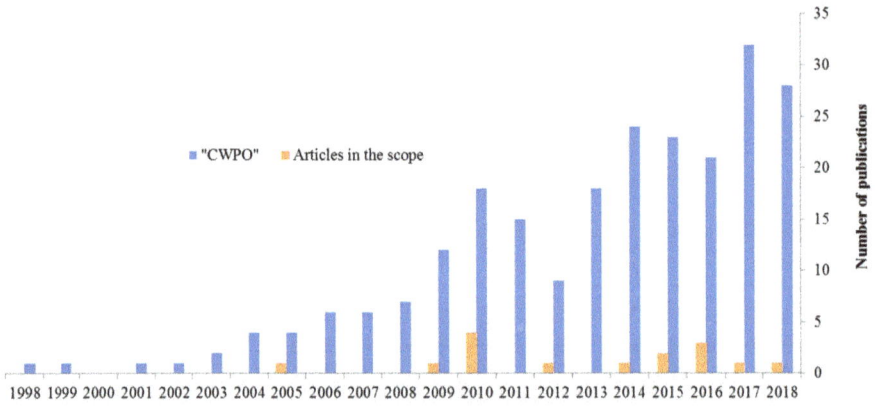

Figure 1. Number of scientific publications (Scopus) containing keywords: "CWPO" in the title and/or abstract and/or keywords of article (blue). Number of articles in the scope of the review (real wastewater matrix was used) in orange.

However, although the number of scientific publications is increasing, there are not many studies using real wastewaters (see Figure 1). The most common compounds that are used in CWPO tests are phenols and textile dyes [18–24] as model pollutants.

Taking into account that the composition of real industrial and municipal wastewaters is very complex, the matrix of wastewater may significantly affect the performance of CWPO in the removal of target pollutants. For instance, the removal efficiency of pharmaceuticals from industrial, urban, and hospital wastewater was reported to be lower than that from ultrapure water [25,26], due to the possible complexation of inorganic ions, such as chloride, carbonate, sulphate, etc. with iron or their role as scavengers [27,28]. The aim of this article is to provide systematisation and critical discussion on the feasibility of CWPO for the treatment of industrial (textile, petrochemical, olive oil mill, pharmaceutical, cosmetic, winery, and coffee processing industries) and urban wastewaters. Hence, only research papers that are devoted to treatment of real and/or synthetic wastewaters (prepared based on a matrix of real wastewater) by CWPO were chosen for this review. Special attention was also devoted to toxicity assays when real wastewater was used, because of the important possible impact on the receiving environment.

2. Main Principles and Mechanism of CWPO

CWPO is considered to be a low-cost technology [28] because it can be operated without lamps (leading to reduction of electrical consumption) and at atmospheric temperature and pressure. The organic pollutants that are present in wastewater are degraded by hydroxyl radicals (HO$^\cdot$) generated due to the partial decomposition of H_2O_2 promoted with an appropriate catalyst. Iron-based materials are the most commonly used catalysts for the CWPO process. Generally, catalysts are classified as supported and non-supported (Figure 2). Many studies focus on the development of new catalysts for CWPO in order to increase the stability of catalysts (avoiding iron leaching) and their efficiency in terms of organic compounds removal [29–32]. Some materials used in CWPO are synthesised using Cu^{2+}, Mn^{2+}, and Co^{2+}.

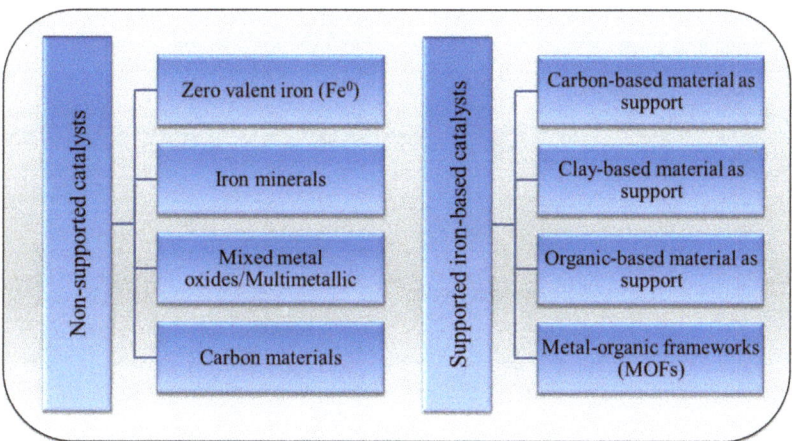

Figure 2. Classification of catalysts used for catalytic wet peroxide oxidation (CWPO).

In comparison with the widely studied homogeneous Fenton process, CWPO is especially attractive because it significantly reduces (e.g. zero valent iron) or does not generate sludge and it enables work in a wide pH range [33].

CWPO can be integrated into the water treatment process, as follows [34] (Figure 3):

(1) Increasing the quality of the industrial or urban wastewater effluent. In the final step of the wastewater treatment process, CWPO is able to remove residual contaminants, such as persistent toxic endocrine-disruption or refractory compounds, and to increase the quality of the treated effluent for water reuse or safe discharge.

(2) Increasing the biodegradability of industrial wastewater. In this case, CWPO can be applied before the biological process in order to increase the biodegradability of recalcitrant compounds and their suitability for biological treatment (conventional or not). It is important to mention that only non-biodegradable wastewaters are suitable for CWPO. The CWPO followed by biological processes can enhance the efficiency of the biological process and the viability of treatment from an economic point of view [35].

The concentration of organic contaminants (e.g. TOC or COD) in industrial wastewaters is significantly higher than that in urban effluents. Consequently, the operational conditions of CWPO applied for industrial wastewaters (prior biological treatment) and urban wastewater effluents (after biological treatment), such as temperature, catalyst load, and H_2O_2 consumption would significantly vary.

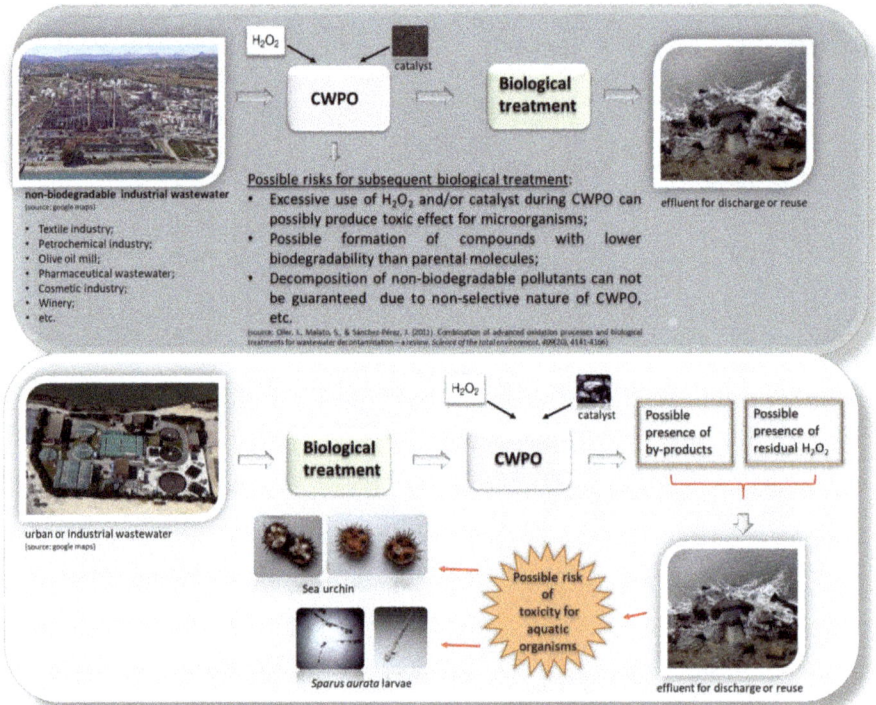

Figure 3. Schematic diagram showing possible integration of CWPO into the wastewater treatment process and potential risks.

It is generally accepted that the decomposition of organic contaminants during the CWPO process (heterogeneous Fenton) occurs mainly due to the presence of highly oxidative species, such as hydroxyl radicals [13,36,37], which are formed during the classical Fenton's reaction.

$$Fe^{2+} + H_2O_2 \rightarrow Fe^{3+} + HO\cdot + OH^- \qquad (1)$$

$$Fe^{3+} + H_2O_2 \rightarrow Fe^{2+} + HOO\cdot + H^+ \qquad (2)$$

Iron-based catalysts that are usually used for CWPO possess a relatively low adsorption capacity towards organic compounds. The oxidation potential of hydrogen peroxide towards organic pollutants in wastewater is also known to be relatively weak, so the highly oxidative species that are generated as a result of complex reactions between hydrogen peroxide and iron-based catalysts play a crucial role in CWPO efficiency. Taking into consideration that different iron-based catalysts can be used for CWPO (e.g. zero valent iron, iron minerals, supported iron-based materials), the efficiency of the process will strongly depend on type of iron specie on the surface of catalyst [38]. For instance, the presence of Fe^{2+} on the surface of catalyst plays an important role in the formation of hydroxyl radicals (reaction 1). Leaching of Fe^{2+}/Fe^{3+} into water during CWPO especially at low pH is another important factor, affecting the overall efficiency of the process. In a recent review [36], it was suggested that hydroxyl radicals, hydroperoxyl radicals, and high-valent iron species are among the main reactive oxygen species that are responsible for the decomposition of organic pollutants during CWPO. Generalised representation of the mechanism of CWPO catalysed by iron-based materials is shown in Figure 4A. However, the mechanism of CWPO catalysed by iron-based materials is not fully understood [36].

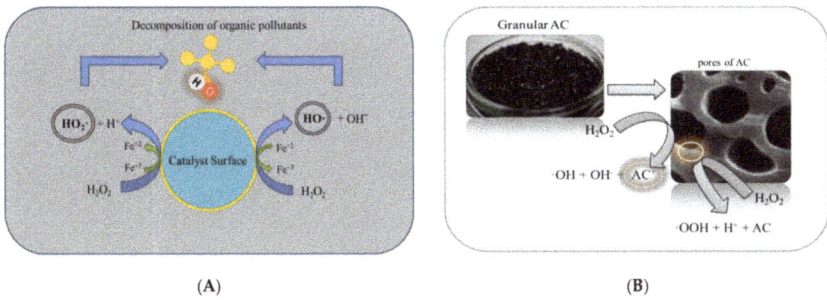

(A) (B)

Figure 4. Schematic representation of CWPO mechanism A—catalysed by iron-based materials; B—catalysed by carbon materials (in the absence of iron).

The CWPO process catalysed by carbon-based materials (without iron) was reported to be efficient in the elimination of organic compounds in water (Figure 4B) [39–41]. According to recent studies, the generation of hydroxyl radicals occurs during the decomposition of hydrogen peroxide by carbon materials (reactions 3 and 4), as follows [39,42,43]:

$$H_2O_2 + AC \rightarrow HO\cdot + OH^- + AC^+ \quad (3)$$

$$AC^+ + H_2O_2 \rightarrow AC + HOO\cdot + H^+ \quad (4)$$

It was suggested that a predominant pathway of organic contaminant decomposition during CWPO occurs due to the attack of organic pollutants that are freely dissolved in the pore volume of activated carbon (AC) by hydroxyl radicals [44]. The adsorbed fraction of organic compounds was found to be almost non-reactive [44]. According to Anfruns et al. [45], the H_2O_2 treatment for the regeneration of activated carbon is limited for non-polar and hydrophobic compounds. Recently, the mechanism of this process was suggested [13,46] to consist mainly of the following steps:

(1) reducing active sites on the surface of carbon materials promotes the decomposition of H_2O_2 and formation of HO· [13,46];

(2) H_2O_2 adsorbed on oxidized active sites leads to the formation of HOO· and H^+ [13,46];

(3) adsorbed HOO· and H^+ in contact with reducing active sites on the carbon surface can lead to the generation of atomic oxygen and water [13,46];

(4) the reaction of H_2O_2 with formed HOO·, HO·, and $O_2\cdot^-$ in the bulk can lead to the generation of HOO·, HO·, O_2, and H_2O [13,46]

(5) HOO·, HO·, and $O_2\cdot^-$ radicals react with each other, leading to the formation of H_2O, O_2, and low amounts of H_2O_2 [13,46].

3. CWPO for the Enhancement of Industrial Wastewater Biodegradability

About 50% of studies reviewed in this article were devoted to the application of CWPO to the enhancement of the biodegradability of industrial wastewater (textile, petrochemical, olive oil mill, pharmaceutical, cosmetic, winery, and coffee processing industries). The values of TOC and COD in the industrial wastewaters studied strongly varied. For instance, the COD of industrial wastewaters subjected to CWPO was in the range 0.3–58 g/L. In general, the biodegradability of the studied industrial wastewaters was poor, as indicated by a relatively low BOD/COD ratio (0.09–0.355).

In more than half of the studies on the CWPO of industrial wastewater, the initial pH of wastewater was adjusted to 3–4. The results of CWPO applied to the enhancement of industrial wastewater biodegradability are summarised in Table 1. It should be mentioned that wastewaters from textile dyeing, tannery, microelectronics, organic fertilizer production, dairy industries, etc. could be of particular interest for CWPO, and, to the best of our knowledge, remain missing.

3.1. Catalysts

In the majority of studies on the CWPO of industrial wastewaters, supported catalysts were used [47–52]. Mostly iron-based catalysts supported on silica [49,51], pillared clays (PILC) [47,48], and alumina [52] were applied, while copper-based catalyst supported on organic material [50] was also studied. To the best of our knowledge, non-supported catalyst (zero-valent iron) was only used for the treatment of industrial wastewater in one study [53]. This is not surprising, because supported catalysts are emerging as potential for CWPO, which is mainly due to the simplicity of catalyst separation after treatment and the fact that sludge is not generated.

The dose of catalyst that was used in studies on industrial wastewater treatment by CWPO varied from 0.5 to 5 g/L. Molina et al. [48] reported that iron loading (Fe/(Fe+Al) molar ratios 0.05–0.15) is more important than catalyst concentration (1.25–3.75 mg/L), indicating the key importance to the iron loading for the efficiency of the process. Iron concentration in the catalyst was also reported to be more important than the surface area of the catalysts [54].

For the practical application of CWPO to real wastewater, the stability of catalyst and its efficiency in the long term are crucial. Interestingly, the stability of catalysts may vary in a real wastewater matrix and model solution. Thus, the stability of Al-Fe PILC catalyst during CWPO was higher in industrial wastewater than in an aqueous solution of 4-Chlorophenol [48]. To the best of our knowledge, only two studies evaluated long-term catalyst efficiency for the enhancement of industrial wastewater biodegradability [49,52]. Melero et al. [49] studied the stability of Fe_2O_3/SBA-15 catalyst used in the treatment of industrial wastewater at a continuous up-flow fixed-bed reactor over a 55-hour period. A slight decrease in TOC removal and H_2O_2 consumption was observed after 20 hours of treatment. This observation was attributed to the possible modification of iron species during CWPO [55]. Despite this fact, the overall stability of the catalyst was high during the 55 hours of treatment, leading to 50–60% TOC elimination [49]. Interestingly, the leaching of iron was below 0.05 mg/L (detection limit of ICP-AES), suggesting the high stability of this catalyst [49]. Bautista et al. [52] demonstrated the high stability of Fe/γ-Al_2O_3 catalyst for the treatment of cosmetic wastewater over 100 hours. An increase in C and S on the surface of the catalyst was observed after 100 hours, which was attributed to possible the adsorption or deposition of organic compounds on the surface. Interestingly, no significant effect of C and S deposits on the efficiency of the catalyst was observed. Moreover, the leaching of iron over 100 hours was below 3% of the initial iron weight [52].

The leaching of iron from catalysts after CWPO of industrial wastewater was studied in the majority of the reviewed articles. Generally, the leaching of iron from catalysts increases as the pH decreases. For example, the concentration of dissolved iron from Fe^0 decreased from 13.8 to 0.39 mg/L with an increase of pH from 2 to 8 [53]. Moreover, with increase of iron concentration in the catalyst, the dissolution of iron (leaching) rises, but not proportionally [48]. The effect of the initial TOC concentration of wastewater on the leaching of iron from silica-supported iron oxide catalyst (Fe_2O_3/SBA-15) was studied by Pariente et al. [51]. It was demonstrated that, as the initial TOC of petrochemical wastewater increases, so does the leaching of iron from the catalyst.

A correlation between the percentage of eliminated TOC and amount of leached iron was reported [48]. This was attributed to the generation of by-products during CWPO, such as oxalic acid, which may significantly increase the leaching of iron from the catalyst due to possible iron complexation [56]. Pariente et al. [51] reported a decrease in iron leaching from the catalyst with an increase in temperature from 120 to 160 °C. This was explained by the fact that, at higher temperature, the decomposition of low molecular weight carboxylic acids (for instance, oxalic acid) is more efficient than that at a lower temperature.

3.2. Temperature

Temperature is an important factor to be taken into account during CWPO. In reviewed studies that are devoted to the enhancement of industrial wastewater biodegradability through the application of CWPO, the employed temperature of the process varied from 25 to 160 °C. Interestingly, CWPO of industrial wastewater was conducted at an ambient temperature only in two studies [50,53], while, in majority of the studies, the temperature was higher than 50°C [47–49,51,52]. An increase in reaction temperature might significantly enhance the decomposition of organic pollutants from wastewaters and the consumption of H_2O_2. COD removal from olive mill wastewater increased from 37 to 69% as the process temperature was raised from 25 to 70 °C [47]. The elimination of COD and TOC from cosmetic wastewater was significantly enhanced when the temperature of CWPO was elevated from 50 to 70 °C, while a further increase of temperature up to 85 °C did not result in a significant increase in organic pollutants removal [52]. Interestingly, the removal of TOC of petrochemical wastewater that was treated by CWPO at a temperature of 120–160 °C did not vary significantly with a change in temperature [51]. One should keep in mind that, as the temperature of the process increases, so does the cost of the treatment. Hence, optimization of operational conditions, such as the temperature of CWPO, is of high importance for practical application.

3.3. Effect of Initial Concentration of Organic Pollutants in Wastewater

When working in water treatment, one should keep in mind fluctuations in pollutants concentration, which can significantly affect the efficiency of the applied process. Domínguez et al. [54] studied the effect of initial organic loading (COD 3.5, 17 and 35 g/L) of winery wastewater on the efficiency of CWPO. Interestingly, it was demonstrated that the effect of the initial concentration of organic pollutants on the efficiency of CWPO is insignificant when a stoichiometric amount of H_2O_2 is added in accordance with the initial organic load [54].

The effect of the initial TOC (0.22–2.2 g/L) of petrochemical wastewater on the performance of intensified CWPO was studied [51]. A notable increase in TOC elimination was reported with a decrease in the initial TOC of the wastewater. However, it was suggested that the optimization of operating conditions for more concentrated wastewaters would allow the application of intensified CWPO.

3.4. Effect of pH

CWPO can be operated in a wide pH range, but the efficiency of CWPO can significantly vary at different pHs. For instance, the degradation of model compound (benzoic acid) by $Fe_3O_4@CeO_2$ was studied in a wide pH range (3.2–10.3) [57]. About 80% of model compound removal was achieved at acidic and neutral pH, while in alkaline conditions the performance of CWPO significantly decreased (below 50%). The wastewater's pH affects not only the performance of the process, but also the mechanism (homogeneous or heterogeneous Fenton) that is involved during CWPO catalysed by iron-based materials. Usually, a higher performance of CWPO catalysed by iron-based materials is obtained at pH 3–4. For instance, the elimination of COD from industrial wastewater (coal-chemical engineering wastewater effluent) during the CWPO (Fe^0/H_2O_2) process increased up to 98% with a decrease of pH from 8 to 3 [53]. Often, at pH below 3, the reaction slows down. It was demonstrated that, at acidic and neutral pH, the consumption of hydrogen peroxide during the CWPO of industrial wastewater is very similar, while the elimination of organic pollutants is higher in acidic conditions [54]. This can be explained by the fact that different a mechanism occurs at acidic and neutral/alkaline pH. At pH above 4, some hydrogen peroxide decomposes into water and oxygen [58]. In the pH range of 3–4, more iron dissolves from the catalyst (in the case of an iron-based catalyst), leading to the occurrence of the homogenous Fenton process in parallel with heterogeneous Fenton. The occurrence of a homogeneous Fenton reaction during CWPO is not always desirable, as it may decrease the operating time of the catalyst in the long-term perspective. In more than 70% of research papers on the CWPO of industrial wastewater reviewed in this article, the initial pH of the wastewater varied between 2.8 and 4. In some cases, the natural pH of wastewater was in this range, while, in majority of the studies, wastewater was acidified in order to improve the efficiency of CWPO. It should be mentioned that pH adjustment (decrease before and increase after treatment) of industrial wastewater prior to CWPO could significantly increase the cost of the treatment when applied on an industrial scale.

3.5. Effect of H_2O_2 Concentration

The initial concentration of H_2O_2 added to wastewater prior to the CWPO treatment of industrial wastewaters varied from 100 mg/L to 17.8 g/L (in reviewed articles). Such variation can be explained by the different initial loading of organic pollutants in wastewater. In the majority of reviewed articles, a stoichiometric ratio of H_2O_2 0.5–2 times the concentration of unknown contaminants (like TOC or COD) was used. Pliego et al. [59] reported that the stoichiometric amount of H_2O_2 required for the complete mineralization of COD in real wastewaters is 2.125 g per g of COD. Generally, the removal of organic pollutants from wastewaters by CWPO increases with a rise of added H_2O_2 concentration up to a certain level. However, when the concentration of added H_2O_2 is too high, the opposite effect is often reported [53]. This phenomenon can possibly be explained by the fact that an excessive amount of H_2O_2 plays the role of a hydroxyl radical scavenger, as shown in reactions 5 and 6 [60].

$$H_2O_2 + HO^{\cdot} \rightarrow H_2O + HO_2^{\cdot} \tag{5}$$

$$HO_2^{\cdot} + HO^{\cdot} \rightarrow H_2 + O_2 \tag{6}$$

The addition of hydrogen peroxide to the process can be conducted in two ways: (a) the single addition of H_2O_2 at the beginning of the process and (b) the gradual addition of H_2O_2. The gradual addition of H_2O_2 was reported to be more efficient than the addition of all the H_2O_2 at the beginning of CWPO, leading to the higher removal of organic pollutants from industrial wastewater and the almost full consumption of H_2O_2 during the treatment [54,61].

The effect of initial H_2O_2 concentration on the elimination of COD, TOC, and the toxicity (*Photobacterium phosphoreum*) of winery wastewater by CWPO (125 °C, graphite 5 g/L, pH 3.8) was studied [54]. The increase of initial H_2O_2 concentration in wastewater led to an increase in COD and TOC removal (up to H_2O_2/COD 1.6 stoichiometric), but when the dose of the initial H_2O_2 was between 0 and 0.5 times stoichiometric, the treated wastewater was relatively toxic. Interestingly, with a further increase of the initial H_2O_2 dose (1 and 1.6 times the stoichiometric amount), the toxicity of the wastewater after treatment significantly decreased, indicating that the toxic by-products were decomposed [54]. Thus, the optimum dose of H_2O_2 for elimination of toxicity was about stoichiometric.

3.6. Toxicity

To the best of our knowledge, a toxicity assessment of industrial wastewater that was treated by CWPO was carried out in only two studies [47,54]. Acute toxicity bioassays of *Vibrio fischeri* [47] and *Photobacterium phosphoreum* [54] were used. Interestingly, in some cases, the effluent of industrial wastewater after CWPO was significantly more toxic than raw wastewater, which can be attributed to the generation of by-products with higher toxicity than parent pollutants [54]. However, with an increase in reaction temperature, the toxicity of the final effluent decreased, indicating the decomposition of toxic by-products [54]. An increase in toxicity during CWPO was observed during the first two hours of the CWPO of industrial wastewater, followed by a decrease in toxicity with an increase in treatment time [47]. It should be mentioned that, in both studies, residual hydrogen peroxide was removed from the wastewater before toxicity assessment, so the possible synergetic effect of hydrogen peroxide and formed by-products was not evaluated.

3.7. Cost Estimation

Considering all of the advantages of CWPO for the pre-treatment (prior biological process) of non-biodegradable industrial wastewater, its practical application on an industrial scale might be beneficial. In order to evaluate the possibility of introducing CWPO on an industrial scale, cost estimation should be taken into account. Among the scientific articles devoted to the treatment of industrial wastewater by CWPO (reviewed in this study), a preliminary cost assessment was done only in one study [53]. The operating cost of CWPO (Fe^0/H_2O_2, 25 °C), followed by the aerobic biological treatment of coal-chemical engineering wastewater, was estimated to be 0.35 \$/m^3 [53]. It should be mentioned that a relatively low operating cost was achieved mostly by avoiding the adjustment of wastewater pH (initial pH 6.8) and conducting CWPO at an ambient temperature. When considering that the majority of the studies devoted to application of CWPO as pre-treatment step were conducted at elevated temperatures and with preliminary pH adjustment, the main operational costs in this case would be (1) energy cost for conducting CWPO at high temperatures, (2) acidification of wastewater, (3) H_2O_2, and (4) cost of catalyst. Sulfuric acid is usually used for adjustment of wastewater pH. The price of this reagent is about 0.25 €/kg [62]. Taking into account that the pH of wastewater treated by CWPO is usually close to neutral, the decrease of wastewater pH can be associated with a relatively high cost on the industrial scale. Moreover, the cost of possible water neutralization (usually conducted using NaOH, 0.55 €/kg) after CWPO is not taken into account.

The initial organic load in wastewaters prior to biological treatment is usually relatively high; therefore, the dose of hydrogen peroxide that is required for CWPO is also high. In some cases, CWPO as a pre-treatment step for wastewater cannot be economically feasible due to the high cost of hydrogen peroxide. We have estimated the cost of H_2O_2 as a function of the initial concentration of organic compounds in wastewater (Figure 5A), while considering that the price of H_2O_2 is 0.45 €/L [63,64]. Different catalysts can be used for CWPO; hence, the cost can vary significantly.

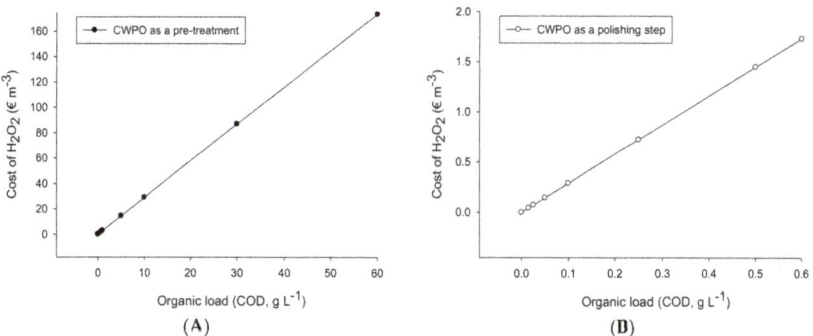

Figure 5. Cost of the H_2O_2 reagent as a function of the initial wastewater TOC (when stoichiometric amount of H_2O_2 is added) for CWPO as pre-treatment (**A**) and as a post-treatment (**B**).

4. CWPO as a Post-treatment Step for Urban and Industrial Wastewater Effluents

In the majority of studies that are devoted to the application of CWPO as a post-treatment step (after the biological process), synthetic and real industrial wastewater effluents were used. To the best of our knowledge, only in one study were urban wastewater effluent and hospital wastewater treated by CWPO [26]. The initial COD and TOC values (when measured) for synthetic and real industrial wastewater effluents that were treated by CWPO varied from 25 to 551 mg/L (COD) and from 15 to 27 mg/L (TOC), respectively. The TOC levels of urban wastewater effluent and hospital wastewater that were treated by CWPO were 2.6 mg/L and 110 mg/L, respectively [26]. The main outcomes of the studies where CWPO was applied as a post-treatment step are summarised in Table 2.

4.1. Catalysts

Iron-based catalysts that were supported on pillared clays [65] and alumina [26] were used as catalysts for the post-treatment of wastewater effluents by CWPO. The dose of iron-based supported catalysts varied from 2 to 5 g/L. However, granular activated carbon (GAC) without supported metals (such as iron, copper, etc.) was applied as a catalyst for the CWPO of real and synthetic industrial wastewater effluents in more than 60% of the studies focused on CWPO as the post-treatment method. For CWPO with carbon materials, a column of GAC was used (140.1 g of GAC). It is not surprising that carbon materials were used in the majority of the studies devoted to the post-treatment of wastewater effluents by CWPO, as they are highly beneficial for practical application due to the following properties [13]:

- High stability in wide temperature range;
- Stability under different pH conditions;
- High surface area;
- No leaching;
- Efficient for decomposition of H_2O_2; and,
- Low cost.

Moreover, depending on the source of carbon materials, they may contain metals, such as iron and copper, as impurities, which might affect the efficiency of CWPO. CWPO with carbon materials as catalyst is especially efficient as a post-treatment step when applied after Fenton/photo-Fenton and/or UV/H_2O_2 processes due to the high efficiency of H_2O_2 elimination and the removal of possibly toxic pollutants [42]. Therefore, CWPO with GAC as catalysts were efficient in the decrease of the initial TOC and COD of industrial wastewater effluent (previously treated by UV/H_2O_2 process) from 27 to 16.7 mg/L and from 59 to 26.6 mg/L, respectively, only after five minutes of contact time with GAC [66]. It should be noted that residual H_2O_2 after the UV/H_2O_2 process was sufficient for CWPO. When the initial load of organic pollutants in wastewater effluent is lower, even shorter contact time can be sufficient for the significant elimination of pollutants. For instance, already after 2.3 min of CWPO, the TOC of industrial wastewater effluent decreased from 15 to 3.75 mg/L, while COD decreased from 35 to 14.9 mg/L [43].

Interestingly, about 50% of the mineralisation (initial COD 551 mg/L) of biologically treated industrial wastewater was achieved after five hours of treatment with the Al-Ce-Fe-PILC catalyst [65]. The efficiency of CWPO with iron-based catalyst in the decomposition of widely used pharmaceuticals from hospital and urban wastewater effluents was recently demonstrated [26].

4.2. Temperature and pH

All studies that were devoted to the post-treatment of industrial wastewater effluents by CWPO reviewed in this article were conducted at an ambient temperature (20–25 °C). Only one study focusing on the elimination of pharmaceuticals from hospital and urban wastewater by CWPO was performed at a higher temperature (75 °C) [26].

When carbon materials were used as catalysts for the CWPO post-treatment of industrial wastewater effluents, the experiments were conducted without adjustment of wastewater pH. However, when iron-based catalysts were used, the pH of wastewaters was adjusted to 3–4.

4.3. H_2O_2 Concentration and Toxicity

The initial concentration of H_2O_2 used for CWPO applied as post-treatment step for industrial and urban wastewater effluents (in reviewed papers) varied between 79.3 mg/L and 3.4 g/L. It should be mentioned that the high efficiency of H_2O_2 consumption is one of the main advantages of CWPO [49], so, after the post-treatment of industrial wastewater effluents by CWPO using GAC, full consumption of H_2O_2 was reported [42,43,66,67]. Interestingly, very short contact time (between 2.3 and 6 min) was sufficient for the significant elimination of organic pollutants and complete decomposition of H_2O_2 [42,43,66,67]. Complete consumption of hydrogen peroxide during CWPO is extremely beneficial from a practical point of view due to the absence of residual hydrogen peroxide in wastewater effluent, so the possible toxicity of the final effluent (when CWPO is applied as post-treatment step) might be avoided. Interestingly, relatively high concentrations of residual H_2O_2 (about 70 mg/L) were reported after the treatment of urban and hospital wastewater effluents spiked with environmentally representative concentrations of pharmaceuticals (µg/L) [26]. It should be noted that the presence of hydrogen peroxide, even at low concentrations in discharged effluent, could be toxic for aquatic organisms in receiving water bodies [68]. It is therefore very important to ensure complete H_2O_2 consumption during CWPO when it is applied as a post-treatment step.

Assessment of the toxicity of industrial wastewater effluents after CWPO was performed using acute toxicity bioassays with *Vibrio fischeri* [67], *Sparus aurata* [43], and *Paracentrotus lividus* (fertilisation and embryo-larval development) [42]. It is worth noting that H_2O_2 was not removed from the wastewater prior to toxicity assessment in any of the reviewed studies that were devoted to CWPO post-treatment of wastewater, so the synergetic effect of pollutants and H_2O_2 in wastewaters was estimated. To the best of our knowledge, no toxicity assessment of urban wastewater effluents that were treated by CWPO has yet been reported. Generally, after CWPO, the toxicity of wastewater effluent decreases. For instance, Rueda-Márquez et al. [67] reported that the toxicity of refinery wastewater effluent after CWPO decreased enormously according to the most sensitive tested species (EC_{50}, *P. lividus* embryo-larval development). Based on a toxicity assessment of industrial wastewater effluents that were treated with CWPO, the effluents were recommended for safe discharge [42,43,67]. As far as authors are aware, there were no reports of toxicity assessments during the CWPO of real wastewater effluent when applied as a post-treatment step. Despite the fact that H_2O_2 is considered an environmental-friendly agent [7,69], the presence of H_2O_2 at high concentration in discharged wastewater effluents can be highly toxic for the aquatic environment [42,43,68].

4.4. Cost Estimation

Preliminary estimation of the operating and maintenance costs of CWPO that were applied to the post-treatment of wastewater effluents included only the cost of the catalyst and H_2O_2 reagent. Since the CWPO process was conducted at ambient temperature and at the natural pH of wastewater effluents, additional costs related to heating and/or pH adjustment were eliminated. The cost of the catalyst (GAC) was estimated to be 0.042 and 0.028 €/m^3 (including the cost of regeneration) [70,71]. Generally, the concentration of organic carbon in wastewater effluents was relatively low (15–27 mg/L of TOC). The occurrence of the oxidation process (CWPO) on the GAC surface prevented the saturation of the activated carbon, and so it caused a decrease in the regeneration cost (about 90% of the total cost). Therefore, the cost of the catalyst (GAC) was estimated to be 0.0035 €/m^3 (without regeneration cost). However, if the organic load of the wastewater is significantly higher (TOC > 3 g/L), the deposition of some reaction products on the surface of AC can occur [72], leading to the necessity of AC regeneration.

The concentration of H_2O_2 used during CWPO depends on organic load of the wastewater. Generally, when CWPO is applied as a post-treatment process, lower concentrations of H_2O_2 are used in comparison with CWPO that is applied as pre-treatment process. Taking into account that that the cost of H_2O_2 reagent is 0.45 €/L [63,64], the cost of H_2O_2 that is used during CWPO can be estimated to be 0.10–4.63 €/m^3 (see Figure 5D). From this estimation, it can be seen that when the required concentration of H_2O_2 is relatively high, CWPO is not economically viable. In the case of wastewater effluents, the concentration of H_2O_2 required for decomposition of organic pollutants that are present in water can be estimated according to [59], so an approximation of the H_2O_2 cost depending on the initial load of organic pollutants in wastewater can be made (Figure 5B). Thus, the total cost of CWPO catalysed by carbon materials applied as a post-treatment method for wastewater effluents (at ambient temperature and the natural pH of wastewater) can be estimated at 0.11–0.22 €/m^3 [42,43,67].

Table 1. CWPO for Increasing Biodegradability of Industrial Wastewater.

Reference	Type of Catalyst	Type of the Wastewater	Experimental Conditions	Main Outcomes
[52]	Fe/γ-Al$_2$O$_3$ (in form of powder)	Wastewater from cosmetic industry (TOC 691 mg/L and COD 2 376 mg/L)	Operating conditions: pH 3, 50–85 °C, concentration of catalyst 2,500–5,000 mg/L, concentration of H$_2$O$_2$ 2,272–9,088 mg/L	About 80% of COD was eliminated at 85 °C, H$_2$O$_2$ 2272 mg/L and space-time of 9.4 kg$_{cath}$/kg$_{COD}$. The H$_2$O$_2$ was fully consumed. Stability of catalyst during 100h was demonstrated. Leaching of Fe from catalyst was lower than 3%.
[51]	Fe$_2$O$_3$/SBA-15 (silica supported)	Diluted wastewater from petrochemical industry (TOC 0.22–2.2 g/L)	Operating conditions: 5 g of catalyst was used in fixed-bed reactor, 120–160 °C, 7, 14 and 21 g of H$_2$O$_2$/g of TOC (at 160 °C)	Removal of TOC was not affected by increase in temperature. As the temperature increased, the leaching of iron decreased. An increase of H$_2$O$_2$ concentration enhanced TOC removal (at 160 °C). Optimal conditions were 160 °C and 14 g of H$_2$O$_2$/g of TOC.
[53]	Fe0 (powder)	Coal-chemical engineering wastewater effluent (COD 341 ± 6 mg/L)	Operating conditions: 25, 50 and 70 °C, H$_2$O$_2$ 5–50 mmol/L, 25 °C	In optimal operational conditions (pH 6.8, Fe0 2g/L, H$_2$O$_2$ 25 mmol/L) 66% of COD removal was achieved.
[47]	Al-Fe-PILC	Olive mill wastewater (COD 12.5 g/L)	Operating conditions: 90 °C, Fe load (Fe/(Fe+Al) atmospheric pressure, Al-Fe-PILC 0.5 g/L, H$_2$O$_2$ 2.10–2 M, pH 5.2 (natural for WW)	In optimal operational conditions (50 °C, 8 h) about 50% of initial COD was eliminated. Moreover, toxicity of water (bioluminescent test with *Vibrio Vischeri*) decreased by 70%.
[48]	Al-Fe-PILC	Wastewater from cosmetic factory (COD 4200 mg/L, for the majority of experiments it was diluted 10 times)	Operating conditions: 90 °C, Fe load (Fe/(Fe+Al) molar ratio) 0.05–0.15, catalyst 1250–3750 mg/L, H$_2$O$_2$/COD ratios 0.5–2 stoichiometric doses (2.12 g H$_2$O$_2$/g COD)	Highest levels of COD removal (about 70%) from wastewater were achieved at highest Fe loading and catalyst dose. With increase of H$_2$O$_2$/COD ratio, the elimination of COD increased.
[49]	Fe$_2$O$_3$/SBA-15 nanocomposite (fixed bed)	Pharmaceutical wastewater (COD 1901 mg O2/L, TOC 860 mg/L)	Operating conditions: 60, 80 and 100 °C, pH 3 and 5.6, H$_2$O$_2$/C mass ratio 7 (5400 mg/L of H$_2$O$_2$) and 14 (10800 mg/L of H$_2$O$_2$), 2.9 g of catalyst	Optimal operating conditions at continuous up-flow fixed bed reactor were pH 3, initial H$_2$O$_2$ concentration 10,800 mg/L, feed flow rate 0.25 mL/min, 80 °C, amount of catalyst 2.9 g. Decrease of COD and TOC at optimal conditions was 81% and 59%, respectively.
[54]	graphite, activated carbon, carbon black	Winery wastewater (COD 35 ± 2.5 g/L, TOC 11.3 ± 0.9 g/L)	Operating conditions: 80, 100, 125 °C, pH 2.2–7, H$_2$O$_2$ doses 0–1.6 stoichiometric amount related to COD.	About 80% of COD elimination and a significant decrease in wastewater toxicity (*Photobacterium phosphoreum*) was obtained using 5g/L of graphite at natural pH of the wastewater (3.8), 125 °C and stoichiometric amount of H$_2$O$_2$ (added stepwise).
[50]	Cu$_3$(BTC)$_2$(H$_2$O)$_3$ BTC–benzene 1,3,5-tricarboxylic acid	Olive oil mill wastewater (COD 57.7 g/L)	Operating conditions: catalyst dose 0.97 g/L, H$_2$O$_2$ 113.2 mg/L, max temperature 32.85 °C	About 96% of polyphenol present in wastewater was removed after CWPO. Biodegradability of wastewater significantly increased after treatment.

Table 2. Catalytic wet peroxide oxidation (CWPO) as a post-treatment.

Reference	Type of catalyst	Type of the Wastewater	Experimental Conditions	Main Outcomes
[67]	GAC (supported in column)	Refinery wastewater effluent after H_2O_2/UVC. Two different influents: 1) TOC: 17 mg/L, COD 20 mg/L and 2) TOC: 19 mg/L, COD 15 mg/L.	Two experiments in ambient conditions (20 ± 2 °C) were passing through GAC (141.1 g/L). Initial concentrations of H_2O_2 during CWPO were 1) 160 mg/L and 2) 96 mg/L.	The H_2O_2 concentration after CWPO treatments was not detected (in either experiment). The concentrations of TOC and COD were 1) 1.75 and 9 mg/L and 2) 3.5 and 6.4 mg/L, respectively. The contact time for CWPO was 6 and 3.5 min for experiments 1) and 2). Toxicity evaluation of influent and effluent of CWPO was evaluated using *P. lividus* embryo larvae and fertilisation tests. The toxicity of the water after treatment decreased more than 220 times and reduced the Toxic Units from IV to 0.
[43]	GAC (supported in column)	Simulated industrial wastewater effluent in urban wastewater matrix after H_2O_2/UVC (TOC 15 mg/L and COD 35.4 mg/L)	The effluent (0.5 L) of photo-Fenton in ambient conditions (20 °C) was passing through AC column (141.1 g/L). Initial concentration of H_2O_2 was 79.3 mg/L.	TOC, COD and H_2O_2 were sufficiently removed by 57, 76.6 and 100%, respectively after 2.3 min of contact time. The final effluent was recommended for safe discharge in marine water bodies after toxicity evaluation using *Sparus aurata* larvae and *Vibrio fischeri*.
[66]	GAC (supported in column)	Plywood mill effluent (diluted 10 times) after H_2O_2/UVC treatment (TOC 27 mg/L and COD 59.6 mg/L).	The effluent of H_2O_2/UVC in ambient conditions (20 ± 2 °C) was passing through GAC (141.1 g/L). Initial concentration of H_2O_2 was 100 mg/L and pH 6.0.	TOC, COD and H_2O_2 were sufficiently removed by 56, 39 and 100%, respectively after 5 min of contact time. The pH of the water after treatment was 8.0.
[42]	GAC (supported in column)	Simulated industrial wastewater effluent in urban wastewater matrix after H_2O_2/UVC (TOC 21 mg/L and COD 39 mg/L)	The effluent of H_2O_2/UVC int ambient conditions was passing through GAC column (141.1 g/L). Initial concentration of H_2O_2 was 161 mg/L. The pH of the water was 7.4.	Concentration of TOC, COD and H_2O_2 after 3.5 min of contact time were 4.2, 16.4 and < D.L, respectively. The pH of the water after the experiment was 7.9. The toxicity of the final effluent was evaluated using *V. fischeri* and *P. lividus* (embryo larvae development and fertilisation test). The most sensitive test, embryo larvae development, demonstrated that the water decreased in toxicity after CWPO by around 350 times (based on EC50).
[65]	Al-Ce-Fe-PILC (pillared inter-layered clays)	Coffee wet processing wastewater after biological treatment (COD 551 mg O_2/L)	Operational conditions: 25 °C, Al-Ce-Fe-PILC 5 g/L, H_2O_2 0.1M, pH adjusted to 3.7	After CWPO of wastewater (5h) 50% of mineralisation, 70% of phenolic compound conversion.
[26]	Fe_3O_4/γ-Al_2O_3	Hospital wastewater (COD 365 mg/L, TOC 110 mg/L [73] and MWW effluent (TOC 2.6 mg/L) spiked with six pharmaceuticals	Operational conditions: 75 °C, catalyst dose 2 g/L, pH 3, H_2O_2 730 mg/L or 100 mg/L where concentration of spiked pharmaceuticals was 10 μg/L of each)	Complete elimination of spiked pharmaceuticals (at high concentrations) from hospital wastewater and urban wastewater effluent was achieved after 90 min (H_2O_2 730 mg/L). When pharmaceuticals were spiked at lower concentrations, complete degradation was reached after 30 min (H_2O_2 100 mg/L).

5. Conclusions, Knowledge Gaps, and Future Perspectives

This study discusses the feasibility of catalytic wet peroxide oxidation (CWPO) for pre-treatment (before biological treatment) and post-treatment (after the biological process) of industrial and urban wastewaters. Based on the reviewed literature, it can be suggested that CWPO is feasible as both a pre-treatment and post-treatment step for industrial and urban wastewaters. Interestingly, CWPO can be regarded as economically viable when performed at an ambient temperature and at the natural pH of the wastewater. For instance, the reported cost of CWPO as a pre-treatment step was 0.35 €/m^3 and 0.11–0.22 €/m^3 when CWPO is applied to post-treatment. The knowledge gaps and future perspectives that were identified in this study are presented below:

- Metal leaching and deactivation (e.g. due to mechanical and thermal degradation, poisoning, fouling, etc.) are among the main drawbacks of iron-based catalysts for practical application of CWPO. Based on revised literature it can be suggested that carbon materials are among the most promising catalysts for the practical application of CWPO for wastewater treatment. Properties of carbon materials, such as stability in a wide range of pH and temperature, high surface area, absence of leaching, possibility to control some surface properties, and relatively low cost of catalysts [74], makes them especially attractive for application.
- It can be expected that the elimination of emerging pollutants and the decrease of toxicity of municipal wastewater effluents by CWPO can be very efficient. However, there is a lack of studies that are devoted to the application of CWPO as post-treatment for municipal wastewater effluents.
- To the best of our knowledge, there is a lack of data on the toxicity assessment of wastewater during the CWPO process. Moreover, in all studies dealing with CWPO for the treatment of wastewater, only acute toxicity bioassays were used.
- Cost estimation is very important to the evaluation of CWPO feasibility for wastewater treatment. Cost assessment was reported in only a few studies (some reviewed in this work). Interestingly, cost evaluation was reported only when CWPO was conducted at an ambient temperature and the natural pH of wastewater.
- Despite the fact that CWPO was shown to be a promising treatment method, majority of studies with industrial or urban wastewaters were conducted at the laboratory scale. Moreover, among the revised studies, mostly batch or semi-batch reactors were used, while continuous catalytic systems, such as fixed bed reactors, were less studied. Taking into account that fixed bed reactors are promising from the practical point of view (especially for recovery and reuse of catalyst) and the reaction mechanism in batch and fixed bed reactors may vary due to different ratio between catalyst and water [75], it can be expected that in the future these will be more studied. Catalysts with magnetic properties can also be of high interest for the practical application of CWPO for wastewater treatment, which is mainly due to the simplicity of catalyst separation after treatment. However, investigations that are focused on industrial wastewater treatment by CWPO catalysed by magnetic catalysts are lacking.

Author Contributions: J.R., I.L. and M.S. did the literature search, elaborate discussion and wrote the paper.

Conflicts of Interest: The authors declare no conflict of interest.

References

1. European Commission. *Water Scarcity & Droughts in the European Union*; European Commission: Brussels, Belgium, 2016.
2. WWAP. *The United Nations World Water Development Report 2017. Wastewater: The Untapped Resource*; UNESCO: Paris, France, 2017.
3. European Commission. *Directive 2000/60/EC*; European Commission: Brussels, Belgium, 2000.

4. Ec-European Commission. *A Blueprint to Safeguard Europe's Water Resources, Communication from the Commission to the European Parliament, the Council, the European Economic and Social Committee and the Committee of the Regions*; Ec-European Commission: Brussels, Belgium, 2012.
5. Jing, R.; Fusi, S.; Chan, A.; Capozzi, S.; Kjellerup, B.V. Distribution of polychlorinated biphenyls in effluent from a large municipal wastewater treatment plant: Potential for bioremediation? *J. Environ. Sci.* 2018. [CrossRef]
6. Qiao, M.; Bai, Y.; Cao, W.; Huo, Y.; Zhao, X.; Liu, D.; Li, Z. Impact of secondary effluent from wastewater treatment plants on urban rivers: Polycyclic aromatic hydrocarbons and derivatives. *Chemosphere* **2018**, *211*, 185–191. [CrossRef] [PubMed]
7. Lara-Martín, P.A.; González-Mazo, E.; Petrovic, M.; Barceló, D.; Brownawell, B.J. Occurrence, distribution and partitioning of nonionic surfactants and pharmaceuticals in the urbanised Long Island Sound Estuary (NY). *Mar. Pollut. Bull.* **2014**, *85*, 710–719. [CrossRef] [PubMed]
8. Pintado-Herrera, M.G.; González-Mazo, E.; Lara-Martín, P.A. Atmospheric pressure gas chromatography–time-of-flight-mass spectrometry (APGC–ToF-MS) for the determination of regulated and emerging contaminants in aqueous samples after stir bar sorptive extraction (SBSE). *Anal. Chim. Acta* **2014**, *851*, 1–13. [CrossRef] [PubMed]
9. Fu, F.; Wang, Q. Removal of heavy metal ions from wastewaters: A review. *J. Environ. Manag.* **2011**, *92*, 407–418. [CrossRef] [PubMed]
10. Thai, P.K.; Ky, L.X.; Binh, V.N.; Nhung, P.H.; Nhan, P.T.; Hieu, N.Q.; Dang, N.T.T.; Tam, N.K.B.; Anh, N.T.K. Occurrence of antibiotic residues and antibiotic-resistant bacteria in effluents of pharmaceutical manufacturers and other sources around Hanoi. *Vietnam. Sci. Total Environ.* **2018**, *645*, 393–400. [CrossRef] [PubMed]
11. François, G.; Mélanie, D.; Marlène, F.; Michel, F. Effects of a municipal effluent on the freshwater mussel Elliptio complanata following challenge with Vibrio anguillarum. *J. Environ. Sci.* **2015**, *37*, 91–99. [CrossRef]
12. Deblonde, T.; Hartemann, P. Environmental impact of medical prescriptions: Assessing the risks and hazards of persistence, bioaccumulation and toxicity of pharmaceuticals. *Public Health* **2013**, *127*, 312–317. [CrossRef]
13. Ribeiro, R.S.; Silva, A.M.T.; Figueiredo, J.L.; Faria, J.L.; Gomes, H.T. Catalytic wet peroxide oxidation: A route towards the application of hybrid magnetic carbon nanocomposites for the degradation of organic pollutants. A review. *Appl. Catal. B Environ.* **2016**, *187*, 428–460. [CrossRef]
14. Tchobanoglous, G.; Burton, F.L.; Stensel, H.D. *Wastewater Engineering. Treatment and Reuse*; McGraw-Hill Higher Education: Singapore, 2004.
15. Ribeiro, A.R.; Nunes, O.C.; Pereira, M.F.R.; Silva, A.M.T. An overview on the advanced oxidation processes applied for the treatment of water pollutants defined in the recently launched Directive 2013/39/EU. *Environ. Int.* **2015**, *75*, 33–51. [CrossRef]
16. Gogate, P.R.; Pandit, A.B. A review of imperative technologies for wastewater treatment I: Oxidation technologies at ambient conditions. *Adv. Environ. Res.* **2004**, *8*, 501–551. [CrossRef]
17. Gogate, P.R.; Pandit, A.B. A review of imperative technologies for wastewater treatment II: Hybrid methods. *Adv. Environ. Res.* **2004**, *8*, 553–597. [CrossRef]
18. Diaz de Tuesta, J.L.; Quintanilla, A.; Casas, J.A.; Rodriguez, J.J. P-, B- and N-doped carbon black for the catalytic wet peroxide oxidation of phenol: Activity, stability and kinetic studies. *Catal. Commun.* **2017**, *102*, 131–135. [CrossRef]
19. Munoz, M.; de Pedro, Z.M.; Casas, J.A.; Rodriguez, J.J. Combining efficiently catalytic hydrodechlorination and wet peroxide oxidation (HDC–CWPO) for the abatement of organochlorinated water pollutants. *Appl. Catal. B Environ.* **2014**, *150*, 197–203. [CrossRef]
20. Munoz, M.; de Pedro, Z.M.; Casas, J.A.; Rodriguez, J.J. Improved wet peroxide oxidation strategies for the treatment of chlorophenols. *Chem. Eng. J.* **2013**, *228*, 646–654. [CrossRef]
21. Rashwan, W.E.; Fathy, N.A.; Elkhouly, S.M. A novel catalyst of ceria-nanorods loaded on carbon xerogel for catalytic wet oxidation of methyl green dye. *J. Taiwan Inst. Chem. Eng.* **2018**, *88*, 234–242. [CrossRef]
22. Rodrigues, C.S.D.; Carabineiro, S.A.C.; Maldonado-Hódar, F.J.; Madeira, L.M. Wet peroxide oxidation of dye-containing wastewaters using nanosised Au supported on Al2O3. *Catal. Today* **2017**, *280*, 165–175. [CrossRef]
23. Tehrani-Bagha, A.R.; Gharagozlou, M.; Emami, F. Catalytic wet peroxide oxidation of a reactive dye by magnetic copper ferrite nanoparticles. *J. Environ. Chem. Eng.* **2016**, *4*, 1530–1536. [CrossRef]

24. Ribeiro, R.S.; Silva, A.M.T.; Figueiredo, J.L.; Faria, J.L.; Gomes, H.T. The role of cobalt in bimetallic iron-cobalt magnetic carbon xerogels developed for catalytic wet peroxide oxidation. *Catal. Today* **2017**, *296*, 66–75. [CrossRef]
25. Cheng, M.; Zeng, G.; Huang, D.; Lai, C.; Liu, Y.; Zhang, C.; Wan, J.; Hu, L.; Zhou, C.; Xiong, W. Efficient degradation of sulfamethazine in simulated and real wastewater at slightly basic pH values using Co-SAM-SCS/H_2O_2 Fenton-like system. *Water Res.* **2018**, *138*, 7–18. [CrossRef]
26. Munoz, M.; Mora, F.J.; de Pedro, Z.M.; Alvarez-Torrellas, S.; Casas, J.A.; Rodriguez, J.J. Application of CWPO to the treatment of pharmaceutical emerging pollutants in different water matrices with a ferromagnetic catalyst. *J. Hazard. Mater.* **2017**, *331*, 45–54. [CrossRef] [PubMed]
27. Luo, Y.; Guo, W.; Ngo, H.H.; Nghiem, L.D.; Hai, F.I.; Zhang, J.; Liang, S.; Wang, X.C. A review on the occurrence of micropollutants in the aquatic environment and their fate and removal during wastewater treatment. *Sci. Total Environ.* **2014**, *473*, 619–641. [CrossRef] [PubMed]
28. Pignatello, J.J.; Oliveros, E.; MacKay, A. Advanced oxidation processes for organic contaminant destruction based on the Fenton reaction and related chemistry. *Crit. Rev. Environ. Sci. Technol.* **2006**, *36*, 1–84. [CrossRef]
29. Mena, I.F.; Diaz, E.; Moreno-Andrade, I.; Rodriguez, J.J.; Mohedano, A.F. Stability of carbon-supported iron catalysts for catalytic wet peroxide oxidation of ionic liquids. *J. Environ. Chem. Eng.* **2018**, *6*, 6444–6450. [CrossRef]
30. Bedia, J.; Monsalvo, V.M.; Rodriguez, J.J.; Mohedano, A.F. Iron catalysts by chemical activation of sewage sludge with $FeCl_3$ for CWPO. *Chem. Eng. J.* **2017**, *318*, 224–230. [CrossRef]
31. Gosu, V.; Sikarwar, P.; Subbaramaiah, V. Mineralisation of pyridine by CWPO process using nFe0/GAC catalyst. *J. Environ. Chem. Eng.* **2018**, *6*, 1000–1007. [CrossRef]
32. Mohedano, A.F.; Monsalvo, V.M.; Bedia, J.; Lopez, J.; Rodriguez, J.J. Highly stable iron catalysts from sewage sludge for CWPO. *J. Environ. Chem. Eng.* **2014**, *2*, 2359–2364. [CrossRef]
33. Wang, N.; Zheng, T.; Zhang, G.; Wang, P. A review on Fenton-like processes for organic wastewater treatment. *J. Environ. Chem. Eng.* **2016**, *4*, 762–787. [CrossRef]
34. Rizzo, L. Bioassays as a tool for evaluating advanced oxidation processes in water and wastewater treatment. *Water Res.* **2011**, *45*, 4311–4340. [CrossRef]
35. Blanco-Galvez, J.; Fernández-Ibáñez, P.; Malato-Rodríguez, S. Solar Photocatalytic Detoxification and Disinfection of Water: Recent Overview. *J. Sol. Energy Eng.* **2006**, *129*, 4–15. [CrossRef]
36. He, J.; Yang, X.; Men, B.; Wang, D. Interfacial mechanisms of heterogeneous Fenton reactions catalyzed by iron-based materials: A review. *J. Environ. Sci.* **2016**, *39*, 97–109. [CrossRef] [PubMed]
37. Rey, A.; Zazo, J.; Casas, J.; Bahamonde, A.; Rodriguez, J. Influence of the structural and surface characteristics of activated carbon on the catalytic decomposition of hydrogen peroxide. *Appl. Catal. A Gen.* **2011**, *402*, 146–155. [CrossRef]
38. Hou, L.; Zhang, Q.; Jérôme, F.; Duprez, D.; Zhang, H.; Royer, S. Shape-controlled nanostructured magnetite-type materials as highly efficient Fenton catalysts. *Appl. Catal. B Environ.* **2014**, *144*, 739–749. [CrossRef]
39. Santos, V.P.; Pereira, M.F.; Faria, P.; Órfão, J.J. Decolourisation of dye solutions by oxidation with H_2O_2 in the presence of modified activated carbons. *J. Hazard. Mater.* **2009**, *162*, 736–742. [CrossRef] [PubMed]
40. Huang, H.; Lu, M.; Chen, J.; Lee, C. Catalytic decomposition of hydrogen peroxide and 4-chlorophenol in the presence of modified activated carbons. *Chemosphere* **2003**, *51*, 935–943. [CrossRef]
41. Lücking, F.; Köser, H.; Jank, M.; Ritter, A. Iron powder, graphite and activated carbon as catalysts for the oxidation of 4-chlorophenol with hydrogen peroxide in aqueous solution. *Water Res.* **1998**, *32*, 2607–2614. [CrossRef]
42. Rueda-Márquez, J.J.; Pintado-Herrera, M.G.; Martín-Díaz, M.L.; Acevedo-Merino, A.; Manzano, M.A. Combined AOPs for potential wastewater reuse or safe discharge based on multi-barrier treatment (microfiltration-H_2O_2/UV-catalytic wet peroxide oxidation). *Chem. Eng. J.* **2015**, *270*, 80–90. [CrossRef]
43. Rueda-Márquez, J.J.; Sillanpää, M.; Pocostales, P.; Acevedo, A.; Manzano, M.A. Post-treatment of biologically treated wastewater containing organic contaminants using a sequence of H_2O_2 based advanced oxidation processes: Photolysis and catalytic wet oxidation. *Water Res.* **2015**, *71*, 85–96. [CrossRef]
44. Georgi, A.; Kopinke, F. Interaction of adsorption and catalytic reactions in water decontamination processes: Part Oxidation of organic contaminants with hydrogen peroxide catalyzed by activated carbon. *Appl. Catal. B Environ.* **2005**, *58*, 9–18. [CrossRef]

45. Anfruns, A.; Montes-Morán, M.A.; Gonzalez-Olmos, R.; Martin, M.J. H$_2$O$_2$-based oxidation processes for the regeneration of activated carbons saturated with volatile organic compounds of different polarity. *Chemosphere* **2013**, *91*, 48–54. [CrossRef]
46. Ribeiro, R.S.; Silva, A.M.T.; Figueiredo, J.L.; Faria, J.L.; Gomes, H.T. The influence of structure and surface chemistry of carbon materials on the decomposition of hydrogen peroxide. *Carbon* **2013**, *62*, 97–108. [CrossRef]
47. Azabou, S.; Najjar, W.; Bouaziz, M.; Ghorbel, A.; Sayadi, S. A compact process for the treatment of olive mill wastewater by combining wet hydrogen peroxide catalytic oxidation and biological techniques. *J. Hazard. Mater.* **2010**, *183*, 62–69. [CrossRef] [PubMed]
48. Molina, C.B.; Zazo, J.A.; Casas, J.A.; Rodriguez, J.J. CWPO of 4-CP and industrial wastewater with Al-Fe pillared clays. *Water Sci. Technol.* **2010**, *61*, 2161–2168. [CrossRef] [PubMed]
49. Melero, J.A.; Martínez, F.; Botas, J.A.; Molina, R.; Pariente, M.I. Heterogeneous catalytic wet peroxide oxidation systems for the treatment of an industrial pharmaceutical wastewater. *Water Res.* **2009**, *43*, 4010–4018. [CrossRef] [PubMed]
50. De Rosa, S.; Giordano, G.; Granato, T.; Katovic, A.; Siciliano, A.; Tripicchio, F. Chemical pretreatment of olive oil mill wastewater using a metal-organic framework catalyst. *J. Agric. Food Chem.* **2005**, *53*, 8306–8309. [CrossRef] [PubMed]
51. Pariente, M.; Melero, J.; Martínez, F.; Botas, J.; Gallego, A. Catalytic wet hydrogen peroxide oxidation of a petrochemical wastewater. *Water Sci. Technol.* **2010**, *61*, 1829–1836. [CrossRef] [PubMed]
52. Bautista, P.; Mohedano, A.; Casas, J.; Zazo, J.; Rodriguez, J. Oxidation of cosmetic wastewaters with H$_2$O$_2$ using a Fe/γ-Al2O3 catalyst. *Water Sci. Technol.* **2010**, *61*, 1631–1636. [CrossRef] [PubMed]
53. Fang, Y.; Yin, W.; Jiang, Y.; Ge, H.; Li, P.; Wu, J. Depth treatment of coal-chemical engineering wastewater by a cost-effective sequential heterogeneous Fenton and biodegradation process. *Environ. Sci. Pollut. Res.* **2018**, *25*, 13118–13126. [CrossRef] [PubMed]
54. Domínguez, C.M.; Quintanilla, A.; Casas, J.A.; Rodriguez, J.J. Treatment of real winery wastewater by wet oxidation at mild temperature. *Sep. Purif. Technol.* **2014**, *129*, 121–128. [CrossRef]
55. Martínez, F.; Melero, J.A.; Botas, J.Á.; Pariente, M.I.; Molina, R. Treatment of phenolic effluents by catalytic wet hydrogen peroxide oxidation over Fe2O3/SBA-15 extruded catalyst in a fixed-bed reactor. *Ind. Eng. Chem. Res.* **2007**, *46*, 4396–440557. [CrossRef]
56. Zazo, J.A.; Casas, J.A.; Mohedano, A.F.; Rodríguez, J.J. Catalytic wet peroxide oxidation of phenol with a Fe/active carbon catalyst. *Appl. Catal. B Environ.* **2006**, *65*, 261–268. [CrossRef]
57. Qin, H.; Xiao, R.; Shi, W.; Wang, Y.; Li, H.; Guo, L.; Cheng, H.; Chen, J. Magnetic core–shell-structured Fe3O4@ CeO2 as an efficient catalyst for catalytic wet peroxide oxidation of benzoic acid. *RSC Adv.* **2018**, *8*, 33972–33979. [CrossRef]
58. Du, W.; Xu, Y.; Wang, Y. Photoinduced degradation of orange II on different iron (hydr) oxides in aqueous suspension: Rate enhancement on addition of hydrogen peroxide, silver nitrate, and sodium fluoride. *Langmuir* **2008**, *24*, 175–181. [CrossRef] [PubMed]
59. Pliego, G.; Zazo, J.A.; Blasco, S.; Casas, J.A.; Rodriguez, J.J. Treatment of highly polluted hazardous industrial wastewaters by combined coagulation–adsorption and high-temperature Fenton oxidation. *Ind. Eng. Chem. Res.* **2012**, *51*, 2888–2896. [CrossRef]
60. Chou, S.; Huang, C. Application of a supported iron oxyhydroxide catalyst in oxidation of benzoic acid by hydrogen peroxide. *Chemosphere* **1999**, *38*, 2719–2731. [CrossRef]
61. Prasad, J.; Tardio, J.; Jani, H.; Bhargava, S.K.; Akolekar, D.B.; Grocott, S.C. Wet peroxide oxidation and catalytic wet oxidation of stripped sour water produced during oil shale refining. *J. Hazard. Mater.* **2007**, *146*, 589–594. [CrossRef] [PubMed]
62. Sánchez Pérez, J.A.; Román Sánchez, I.M.; Carra, I.; Cabrera Reina, A.; Casas López, J.L.; Malato, S. Economic evaluation of a combined photo-Fenton/MBR process using pesticides as model pollutant. Factors affecting costs. *J. Hazard. Mater.* **2013**, *244*, 195–203. [CrossRef]
63. Miralles-Cuevas, S.; Oller, I.; Agüera, A.; Sánchez Pérez, J.A.; Malato, S. Strategies for reducing cost by using solar photo-Fenton treatment combined with nanofiltration to remove microcontaminants in real municipal effluents: Toxicity and economic assessment. *Chem. Eng. J.* **2017**, *318*, 161–170. [CrossRef]
64. García, E.B.; Rivas, G.; Arzate, S.; Sánchez Pérez, J.A. Wild bacteria inactivation in WWTP secondary effluents by solar photo-fenton at neutral pH in raceway pond reactors. *Catal. Today* **2017**, *313*, 72–78. [CrossRef]

65. Sanabria, N.R.; Peralta, Y.M.; Montañez, M.K.; Rodríguez-Valencia, N.; Molina, R.; Moreno, S. Catalytic oxidation with Al–Ce–Fe–PILC as a post-treatment system for coffee wet processing wastewater. *Water Sci. Technol.* **2012**, *66*, 1663–1668. [CrossRef]
66. Rueda-Márquez, J.J.; Levchuk, I.; Uski, J.; Sillanpää, M.; Acevedo, A.; Manzano, M.A. Post-treatment of plywood mill effluent by Multi-Barrier Treatment: A pilot-scale study. *Chem. Eng. J.* **2016**, *283*, 21–28. [CrossRef]
67. Rueda-Márquez, J.J.; Levchuk, I.; Salcedo, I.; Acevedo-Merino, A.; Manzano, M.A. Post-treatment of refinery wastewater effluent using a combination of AOPs (H_2O_2 photolysis and catalytic wet peroxide oxidation) for possible water reuse. Comparison of low and medium pressure lamp performance. *Water Res.* **2016**, *91*, 86–96.
68. Drábková, M.; Matthijs, H.; Admiraal, W.; Maršálek, B. Selective effects of H_2O_2 on cyanobacterial photosynthesis. *Photosynthetica* **2007**, *45*, 363–369. [CrossRef]
69. St. Laurent, J.B.; de Buzzaccarini, F.; De Clerck, K.; Demeyere, H.; Labeque, R.; Lodewick, R.; van Langenhove, L. *Laundry Cleaning of Textiles*; Elsevier: Amsterdam, The Netherlands, 2007; pp. 57–102.
70. U.S. Department of the Interior Bureau of Reclamation. Reverse Osmosis Treatment of Central Arizona Project Water for the City of Tucson. Desalination Research and Development Program Report No. 36, Costs 20 Appendix C. 2004. Available online: https://www.usbr.gov/research/dwpr/reportpdfs/report036.pdf (accessed on 2 October 2018).
71. Nguyen, T.V.; Jeong, S.; Pham, T.T.N.; Kandasamy, J.; Vigneswaran, S. Effect of granular activated carbon filter on the subsequent flocculation in seawater treatment. *Desalination* **2014**, *354*, 9–16. [CrossRef]
72. Domínguez, C.M.; Ocón, P.; Quintanilla, A.; Casas, J.A.; Rodriguez, J.J. Highly efficient application of activated carbon as catalyst for wet peroxide oxidation. *Appl. Catal. B Environ.* **2013**, *140*, 663–670. [CrossRef]
73. Munoz, M.; Garcia-Muñoz, P.; Pliego, G.; Pedro, Z.M.D.; Zazo, J.A.; Casas, J.A.; Rodriguez, J.J. Application of intensified Fenton oxidation to the treatment of hospital wastewater: Kinetics, ecotoxicity and disinfection. *J. Environ. Chem. Eng.* **2016**, *4*, 4107–4112. [CrossRef]
74. Pliego, G.; Zazo, J.A.; Garcia-Muñoz, P.; Munoz, M.; Casas, J.A.; Rodriguez, J.J. Trends in the intensification of the Fenton process for wastewater treatment: An overview. *Crit. Rev. Environ. Sci. Technol.* **2015**, *45*, 2611–2692. [CrossRef]
75. Bergault, I.; Rajashekharam, M.; Chaudhari, R.; Schweich, D.; Delmas, H. Modeling and comparison of acetophenone hydrogenation in trickle-bed and slurry airlift reactors. *Chem. Eng. Sci.* **1997**, *52*, 4033–4043. [CrossRef]

© 2018 by the authors. Licensee MDPI, Basel, Switzerland. This article is an open access article distributed under the terms and conditions of the Creative Commons Attribution (CC BY) license (http://creativecommons.org/licenses/by/4.0/).

Review

Wastewater Treatment by Catalytic Wet Peroxidation Using Nano Gold-Based Catalysts: A Review

Carmen S.D. Rodrigues [1,*], Ricardo M. Silva [1], Sónia A.C. Carabineiro [2], F.J. Maldonado-Hódar [3] and Luís M. Madeira [1]

1. LEPABE - Laboratory for Process Engineering, Environment, Biotechnology and Energy, Faculty of Engineering, University of Porto, Rua Dr. Roberto Frias, 4200-465 Porto, Portugal; up201306722@fe.up.pt (R.M.S.); mmadeira@fe.up.pt (L.M.M.)
2. LCM – Laboratory of Catalysis and Materials, Associate Laboratory LSRE/LCM, Department of Chemical Engineering, Faculty of Engineering, University of Porto, R. Dr. Roberto Frias, 4200-465 Porto, Portugal; scarabin@fe.up.pt
3. Department of Inorganic Chemistry, Faculty of Sciences, University of Granada, Avenida de Fuente Nueva, s/n 18071 Granada, Spain; fjmaldon@ugr.es
* Correspondence: csdr@fe.up.pt; Tel.: +351-22-041-4851

Received: 30 April 2019; Accepted: 20 May 2019; Published: 23 May 2019

Abstract: Nowadays, there is an increasing interest in the development of promising, efficient, and environmentally friendly wastewater treatment technologies. Among them are the advanced oxidation processes (AOPs), in particular, catalytic wet peroxidation (CWPO), assisted or not by radiation. One of the challenges for the industrial application of this process is the development of stable and efficient catalysts, without leaching of the metal to the aqueous phase during the treatment. Gold catalysts, in particular, have attracted much attention from researchers because they show these characteristics. Recently, numerous studies have been reported in the literature regarding the preparation of gold catalysts supported on various supports and testing their catalytic performance in the treatment of real wastewaters or model pollutants by CWPO. This review summarizes this research; the properties of such catalysts and their expected effects on the overall efficiency of the CWPO process, together with a description of the effect of operational variables (such as pH, temperature, oxidant concentration, catalyst, and gold content). In addition, an overview is given of the main technical issues of this process aiming at its industrial application, namely the possibility of using the catalyst in continuous flow reactors. Such considerations will provide useful information for a faster and more effective analysis and optimization of the CWPO process.

Keywords: catalytic wet peroxidation; radiation; gold-based catalysts; wastewater treatment; advanced oxidation processes

---

1. Introduction

The world's population growth and increasing industrial development led to the intense usage of natural resources with the water bodies being used as a final destination for wastewater containing pollutants [1–3]. The discharge of untreated wastewater introduces persistent contaminants into the environment, some examples being metals, organic, and inorganic compounds [4–6], which have harmful effects on ecology and public health [7].

In an attempt to minimize the impacts of effluent discharges, the European Union Water Framework Directive (EU-WFD), in 2000, imposed maximum permissible values for ecotoxic or possibly ecotoxic substances [8]. Thus, it is mandatory to adopt practical, efficient, and low-cost effluent purification technologies [9,10], which will allow the complete elimination or, at least, reduction of the contaminants concentration up to the limit values imposed by legislation [10,11], before wastewaters are discharged into water bodies.

The wastewaters can be treated by physical-chemical processes, such as sedimentation, coagulation/flocculation, filtration, adsorption, ultrafiltration, reverse osmosis, ion exchange, or chemical precipitation [12,13], by biological degradation [13–15], and/or by conventional oxidative processes, which degrade the pollutant by the action of oxygen or other oxidants, such as hydrogen peroxide, ozone, and permanganate [16–18]. Physical-chemical processes are not very appealing because the pollutants are concentrated at another phase, which requires a subsequent treatment [10]. Biological degradation, although economically advantageous, is inefficient since the compounds present in effluents are very often toxic and/or non-biodegradable [19,20]. Moreover, conventional oxidative processes might not have enough capacity to completely oxidize refractory compounds with high chemical stability and, therefore, there is a high risk of intermediate products being formed during oxidation, which can be even more toxic than the initial ones [21–23].

Advanced oxidation processes (AOPs) are emergent and attractive treatment technologies to degrade compounds with high chemical stability, toxicity, and non-biodegradability [10,24]. AOPs generate the hydroxyl radical (HO•), responsible for oxidizing refractory organic compounds into non-toxic products, such as CO_2 and H_2O [10,25–27]. Given the high efficiency of the hydroxyl radical, the AOPs have been widely used, not only in wastewater treatment [9,19,28–30], but also in soil and sediment remediation [31,32], decontamination of gaseous effluents containing volatile organic compounds and elimination of odors [33–36], water and groundwater treatment [37–39], and conditioning of municipal sludge [40,41].

Several AOPs are available, as will be detailed in the next section, that use different oxidants, with or without catalysts, in the presence of absence of radiation. Herein, we will focus on the catalytic wet peroxidation (CWPO) process using nano gold-based catalysts for wastewater treatment. This process presents several advantages compared to other AOPs, namely: it uses environmentally friendly reagents, does not require sophisticated equipment, and is operated under mild conditions of pressure and temperature. Moreover, catalysis by gold presents additional advantages, such as non-leaching of the metal to the treated effluent and efficient and stable performance, which are important for industrial applications.

A survey of the catalyst properties, operating conditions, and their effect on the efficiency of the process will be discussed. To the best of the authors knowledge, such review has not yet been reported in the literature.

2. Advanced Oxidation Processes

As mentioned above, AOPs are based on the formation of the hydroxyl radical. This radical has a high oxidation potential (2.8 eV [42,43]), being immediately below the fluorine (see Table 1), and exhibits high oxidation reaction rates, compared to traditional oxidants, such as chlorine, hydrogen peroxide, or potassium permanganate [11].

Table 1. Standard oxidation potential of some chemicals species (adapted from [42,43]).

Chemical Species	Oxidation Potential (eV)
Fluorine	3.03
Hydroxyl radical	2.80
Atomic oxygen	2.42
Ozone	2.07
Hydrogen peroxide	1.77
Potassium permanganate	1.67
Hypobromous acid	1.59
Chlorine dioxide	1.50
Hypochlorous acid	1.49
Chlorine	1.36
Bromine	1.09
Iodine	0.54

The hydroxyl radicals are able to react with almost every type of organic compounds [27], leading, in some cases, to their complete oxidation into CO_2 and H_2O [10,25,26]. However, partial oxidation can be the main route, usually leading to more biodegradable products [25,26]. The oxidation of organic matter (RH) by hydroxyl radicals occurs by three mechanisms [19,42]: radical addition (Equation (1)), electron transfer to radicals (Equation (2)), and hydrogen abstraction (Equation (3)) that generates organic radicals, which yield peroxyl radicals by addition of molecular oxygen (Equation (4)).

$$HO^\bullet + RH \rightarrow RH_2O \tag{1}$$

$$HO^\bullet + RH \rightarrow RH^{\bullet+} + OH^- \tag{2}$$

$$HO^\bullet + RH \rightarrow R^\bullet + H_2O \tag{3}$$

$$R^\bullet + O_2 \rightarrow RO_2^\bullet \tag{4}$$

A large number of technologies are responsible for the generation of the hydroxyl radicals. Most of them use a combination of oxidants, such as ozone or/and hydrogen peroxide, semiconductors (like titanium dioxide or zinc oxide) or catalysts (e.g., transition metal ions), and irradiation (ultraviolet and/or visible, sunlight or ultrasounds), as shown in Figure 1 [11,19,24,44]. Processes in which the catalyst is dissolved in the effluent are called homogeneous, but when the catalyst is supported on a solid matrix they are designated as heterogeneous.

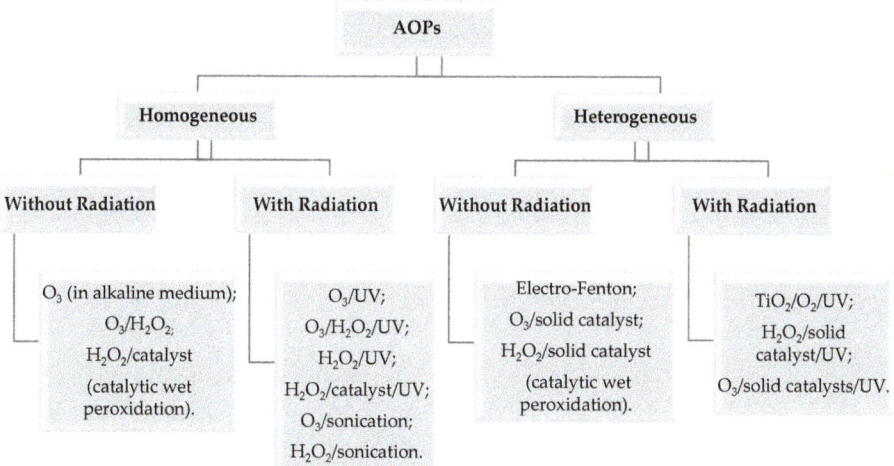

Figure 1. Some relevant advanced oxidation processes (AOPs) for wastewater treatment (adapted from Poyatos et al. [24]).

The benefits of AOPs are: the possibility of degrading pollutants in lowa wide range of concentrations, the easiness in combining with other processes, such as biological and adsorption, and also the fact that some of them are conducted at near ambient pressure and temperature [9,44]. Nevertheless, each AOP has characteristic drawbacks associated. As an example, when using ozone-based processes, sophisticated equipment is required, such as an ozone generator, a cooling system, pre-and post-treatment setups to dry the air fed to the ozonator and to reduce the residual ozone in the gas off, respectively [45], which increases the implementation and operation costs. The processes that use radiation have high expenses of energy consumption, in addition to the costs of the installation and equipment. However, when effective photocatalysts are used, this charge can be null in countries with high incidence of solar radiation, which can replace artificial sources. Another disadvantage of these processes is that the compounds present in the effluent can filtrate or absorb the radiation, limiting its reaction with the oxidant [46], which decreases the treatment efficiency.

The CWPO process is known for its simplicity, as it does not require any sophisticated equipment, involves safe and easy to handle reactants, has high efficiency, and low investment cost [47,48]. This review is focused on the wastewater treatment by CWPO using nano gold-based catalysts. The use of gold catalysts shown benefits as there is no loss of metal into solution and materials are stable and efficient, as will be further explained below.

3. Catalytic Wet Peroxidation

In the end of the 19th century, the CWPO process was firstly observed by H.J.H. Fenton, who described the highly oxidative properties of hydrogen peroxide in presence of iron ions during oxidation of tartaric acid [49]. Later, Haber and Weiss [27] discovered that the hydroxyl radical was the responsible for the degradation of the organic compounds. So, CWPO is based on the catalytic hydrogen peroxide decomposition by transition metallic cations (M) that generates hydroxyl radicals (see Equation (5)) in mild reaction conditions [27,50–53]. In this process, the catalyst is oxidized in the reaction with H_2O_2, generating $HO^{\bullet}$ (Equation (5)), being regenerated (reduced) with additional H_2O_2 molecules and even with the generated hydroperoxyl radicals ($HO_2^{\bullet}$), according to Equations (6) and (7) [52–55].

$$M^{n+} + H_2O_2 \rightarrow M^{(n+1)+} + HO^{\bullet} + HO^{-} \tag{5}$$

$$M^{(n+1)+} + H_2O_2 \rightarrow M^{n+} + HO_2^{\bullet} + H^{+} \tag{6}$$

$$M^{(n+1)+} + HO_2^{\bullet} \rightarrow M^{n+} + O_2 + H^{+} \tag{7}$$

The hydroxyl radical has an extremely short life-time but is very reactive as it can react with the excess of catalyst (Equation (8)) or even oxidant (Equation (9)) [50,52–55], being such reactions the undesired scavenging of the hydroxyl radicals.

$$M^{n+} + HO^{\bullet} \rightarrow M^{(n+1)+} + HO^{-} \tag{8}$$

$$H_2O_2 + HO^{\bullet} \rightarrow HO_2^{\bullet} + H_2O \tag{9}$$

The main limitations of homogeneous CWPO are the following: (i) the narrow pH range (2 to 4) in which the pollutants degradation efficiency is maximum [52,55,56], and (ii) the need to recover the catalyst after treatment, in order to comply with environmental regulations, as shown by some authors [57]. A subsequent unit is required afterwards, in which the generated sludge, containing organic compounds as well as metals, has to be further treated, becoming the overall process more complex and expensive [48,52,54–56]. In order to overcome this challenge, several studies have been reported in literature dealing with supporting metals on solid porous matrices. By doing so, the metal is deposited on the support, becoming a heterogeneous catalyst, which is present in solution in a solid form, forming a slurry (batch reactors), being easily recovered; alternatively, it can be packed in a fixed bed reactor.

The principles of the heterogeneous process are very similar to the homogeneous; however, complexity increases due to the diffusion/adsorption phenomenon. It is widely accepted that hydrogen peroxide is adsorbed on the matrix pores, but this is not completely proved [58].

The main reactions of heterogeneous CWPO (Equations (10)–(14)) are the same as the homogeneous analogue, but with the addition of the support (X):

$$X\text{-}M^{n+} + H_2O_2 \rightarrow X\text{-}M^{(n+1)+} + HO^{\bullet} + HO^{-} \tag{10}$$

$$X\text{-}M^{(n+1)+} + H_2O_2 \rightarrow X\text{-}M^{n+} + HO_2^{\bullet} + H^{+} \tag{11}$$

$$X\text{-}M^{(n+1)+} + HO_2^{\bullet} \rightarrow X\text{-}M^{n+} + O_2 + H^{+} \tag{12}$$

$$X\text{-}M^{n+} + HO^\bullet \rightarrow X\text{-}M^{(n+1)+} + HO^- \tag{13}$$

$$H_2O_2 + HO^\bullet \rightarrow HO_2^\bullet + H_2O \tag{14}$$

The wastewater treatment by CWPO has been extensively studied for decades, mostly using supported iron as a catalyst [29,52,58–61] (in this case, the process being called Fenton or Fenton-like). Recent studies report on catalysts where iron has been replaced by other metals, such as nickel, cobalt, copper, cerium, and manganese, as well as bimetallic particles [47,51,62–65]. However, quite often, such materials are not stable, leaching the metal to the effluent, making their reuse not possible, and their industrial application unfeasible.

In order to overcome the problem of the catalysts lack of stability, some authors report on effluents treated by CWPO using gold catalysts supported on porous matrices. Although this might seem less economically attractive, given the price of gold, these catalysts present high stability, with negligible metal leaching, and are efficient in hydrogen peroxide consumption and pollutants degradation [66–75]. In the next sections, we will discuss some methods of preparation of nano gold-based catalysts and the treatment of effluents by CWPO catalyzed by gold.

4. Nano Gold-based Catalysts

For gold to be an active catalyst, its synthesis must be carefully made in order to obtain well dispersed nanoparticles on the support. This preparation process of the gold catalysts starts by obtaining colloidal gold in suspension, by reducing Au^{3+} to Au^0 [76,77], using different reducing agents (such as alcohols, ascorbic and citric acid, amines, citrate, hydrazines, and toluene) [77,78]. As atomic gold is formed and its concentration increases, the solution becomes saturated and the metal gradually precipitates and forms nanoparticles. Nanoparticle formation is promoted by the addition of stabilizers [76,77], like amines, quaternary alkyl ammonium ions, phosphine, carboxyl acids, and thiols [77].

Usually, gold colloids are obtained by applying the Turkevich method [79], which consists on the reaction between $AuCl_4^-$ (using tetrachloroauric acid ($HAuCl_4$) or sodium tetrachloroauric ($NaAuCl_4$)) with sodium citrate as reducing agent and capping [80], resulting in gold nanoparticles with particle diameters of 10–12 nm [76,77]. However, for catalytic purposes, it is advantageous that gold particles have smaller sizes, between 2 and 10 nm [77]. This is achieved by reducing $AuCl_4^-$ with a strong reducing agent, such as $NaBH_4$ [77,78].

Another method for generating gold colloids was developed by Brust et al. [76]. It is based in the reaction of $HAuCl_4$ solution with $NaBH_4$ (reducing agent), in the presence of toluene and tetraoctylammonium bromide (TOAB) which acts as a transfer cation, stabilizing agent, and anti-coagulant. First, the migration of $AuCl_4^-$ from water to the organic phase (toluene) takes place, by ion metathesis of the counter anion on the phase transfer agent. Then, addition of sodium borohydride promotes the precursor reduction to metallic gold. This method produces gold nanoparticles with particle sizes between 2 to 6 nm [76,77].

The gold colloids can also be formed by dissolving $AuCl_4^-$ in a solvent (like benzyl alcohol or ethylene glycol [78]), other than water. Reduction occurs by thermal treatment or addition of reducing agents [77].

The catalysts preparation is finalized by deposition of gold on a support (normally a metal oxide or a carbon material). Both processes (colloid formation and deposition on the support) can occur simultaneously and can be achieved by using several methods, namely deposition/precipitation (DP), co-precipitation (CP), impregnation, vapor-phase deposition, grafting, sol-gel, ion-exchange, among others [80–83]. These most common techniques are described below.

4.1. Deposition/Precipitation

The DP method is one of the most widely used for gold catalysts preparation. This procedure was first described by Haruta [81], who adjusted the pH of $HAuCl_4$ solution in the range 6–10 with NaOH, then added the metal oxide used as support [80], and readjusted the pH. The suspension was stirred during 1 h for gold precipitation in the form of Au(OH)$_3$, that was deposited on the metal oxide surface. Finally, after the deposition step, the solid in suspension was recovered, washed, dried, and submitted to a thermal treatment at 250 °C in air atmosphere [81], or at 300 °C in hydrogen atmosphere [77], for gold reduction.

Haruta [81] referred the influence of pH on gold particle size. For pH about 6, the AuCl$_4^-$ is transformed into [Au(OH)$_n$Cl$_{4-n}$]$^-$ (n = 1 to 3) and the mean size of gold particles size is less than 4 nm. For pH in the range 7 to 8, the n value is close to 3, which is preferable for the preparation of the gold catalysts, depending on the support. At lower pH, the hydrolysis of the Au-Cl bond takes place in a smaller extent. Moreover, for values of pH below the oxide isoelectric point, its surface is positively charged and consequently adsorbs more negative charged gold species. This results in a higher concentration of chloride on the surface, which promotes high mobility of gold, leading to the formation of larger particles [80]. For pH values above the isoelectric point of the support, the adsorption of negatively charged gold species decreases drastically. Consequently, the gold loading is lower, and so is the chloride concentration, with smaller particles of Au being formed [80]. Figure 2a shows a HR-TEM image and histogram of gold nanoparticle size distribution of a catalyst prepared at pH 9 by deposition/precipitation. This procedure is reproducible, very reliable, and the obtained catalysts show high catalytic activity.

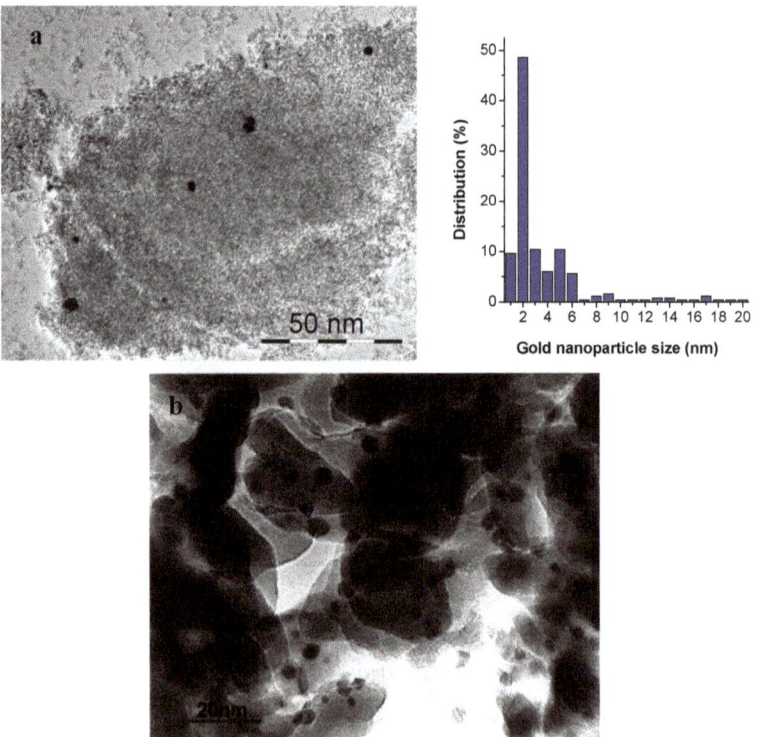

Figure 2. -HR-TEM image and histogram of gold particle size distribution of Au/Al$_2$O$_3$ catalyst (**a**) prepared at pH 9.0 by deposition/precipitation (DP) with NaOH and heating at 70 °C and -HR-TEM image of an Au/α-Fe$_2$O$_3$ catalyst (**b**) prepared by the co-precipitation method calcined at 400 °C. Nanogold particles are seen as dark spots. Adapted from Rodrigues et al. [67] and Hodge et al. [84], respectively.

4.2. Co-Precipitation

This method is based on simultaneous CP of hydroxide or carbonate and gold. For that, the gold precursor (HAuCl$_4$) and the soluble metal salt precursor (preferably a nitrate) are added to a Na$_2$CO$_3$ (and/or NH$_4$OH) solution and the suspension is mixed for a few minutes. After 1 h aging, the precipitates (Au and metal oxide) are washed and filtrated for five consecutive cycles, then dried overnight, and finally calcined in air atmosphere, to obtain a powder material [80–82,85]. This method differs from DP, in the sense that both oxide and gold are co-precipitated at the same time (in DP, Au is deposited on the already prepared support).

The preparation of catalysts by CP needs a concentration of metal salt around 0.1-0.4 M, pH range of 7–10 and temperature of precipitation and calcination between 47–87 and 227–397 °C. Within these conditions, a homogeneous dispersion of gold nanoparticles can be obtained [82].

CP is the most useful and simple method; however, its applicability is limited, as only metal hydroxides or carbonates can be co-precipitated with Au(OH)$_3$; moreover, reducible supports (α-Fe$_2$O$_3$, CO$_3$O$_4$, NiO, and ZnO) have to be employed in order to obtain a good dispersion of the gold nanoparticles [80,82]. Figure 2b shows a HR-TEM image of a Au catalyst prepared by this method.

4.3. Impregnation Method

This method consists in impregnating the support with a gold salt solution. This may be done by suspending the support on a large volume of metal salt, from which the solvent is removed, or by filling the pores of the support with the solution (this later procedure being called incipient wetness impregnation). Then, the precursor is dried and calcined at temperatures as high as 800 °C and reduced with hydrogen atmosphere at 120–250 °C, or aqueous oxalic acid at 40 °C, or aqueous magnesium citrate [80,83].

In the preparation of gold catalysts by this method, usually chloroauric acid (HAuCl$_4$·3H$_2$O) or auric chloride (AuCl$_3$ or Au$_2$Cl$_6$) are used as metal precursors. However, complex salts such as potassium aurocyanide (KAu(CN)$_2$) and the ethylenediamine complex [Au(en)$_2$]Cl$_3$ may also be employed. Regarding the supports, silica, alumina, and magnesia are often used, but titanium oxide, boehmite (AlO(OH)), magnesium hydroxide, or ferric oxide (α-Fe$_2$O$_3$) have also be employed [83]. Figure 3a shows a HR-TEM image of a Au/Al$_2$O$_3$ catalyst prepared by incipient wetness impregnation.

Although impregnation is a classical procedure in the preparation of platinum group metal catalysts, it is not often applicable to gold, since the obtained catalysts show larger gold particle sizes when compared to materials prepared by CP or DP techniques. Moreover, they show low catalytic activity and it is difficult to obtain a good dispersion of the gold on the metal oxides. On one hand, gold has less affinity for these supports and lower melting point (1063 °C) than those of Pd (1550 °C) or Pt (1769 °C). On the other hand, during calcination of the precursor at low temperature (below 600 °C) the crystals of HAuCl$_4$ are dispersed on the surface of the support and the chloride ion markedly enhances the coagulation of gold particles [80–82,86].

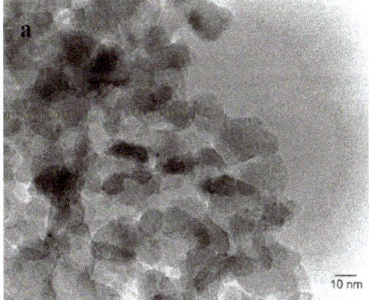

Figure 3. *Cont.*

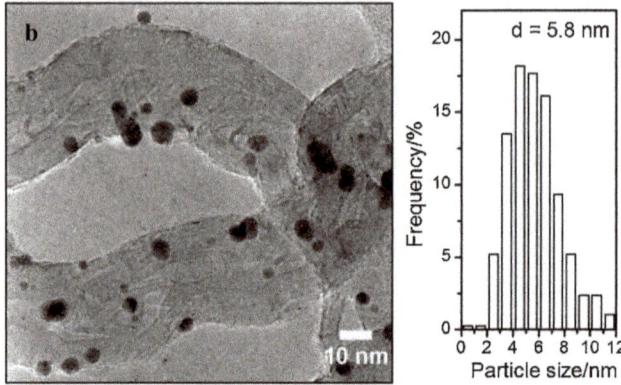

Figure 3. (a) HR-TEM image of an Au/Al$_2$O$_3$ catalyst prepared by incipient wetness impregnation, reduced at 250 °C using 5% (vol.) of hydrogen in nitrogen as gas phase. (b) HR-TEM image and respective histogram of gold particle size distribution of Au/CNT prepared by vapor-phase deposition (dried overnight at 90 °C). Gold nanoparticles are seen as dark spots. Adapted from Baatz et al. [87] and Lorençon et al. [88], respectively.

4.4. Vapor-phase Deposition and Grafting Methods

The procedures for preparation of gold-based catalysts by these two methods are similar, the only difference being in the use (or not) of solvent. In the vapor-phase deposition method (also called chemical vapor deposition), a vapor of an organic gold compound (such as dimethyl-gold(III)-acetyl acetone, dimethyl-gold(III) β-diketone, or gold acetylacetonate) is transported onto a high area support by an inert gas stream and chemically reacts with the support surface to form a precursor of gold. The organic gold compound adsorbed on the support is pyrolyzed in air atmosphere, to be decomposed into small gold particles [80,81,83]. This method can be applied to a variety of metal oxides, including acidic supports, like silica oxide [80,81,83]. Figure 3b presents a HR-TEM image and histogram of particle size distribution of a gold catalyst supported on carbon nanotubes prepared by the vapor-phase deposition method.

In the grafting method, a gold complex ([Au(PPh$_3$)]NO$_3$ and/or [Au$_9$(PPh$_3$)$_8$](NO$_3$)$_3$) in solution is grafted onto the surface of a number of precipitated wet hydroxides (manganese and cobaltous hydroxides being particularly effective), which have many OH groups at the surface, which react with gold [80,83]. Then, drying in vacuum at room temperature and temperature-programmed calcination in air atmosphere are carried out, in order to cause a simultaneous transformation of the precursor to gold particles and oxides [80]. The deposition of gold on activated carbon is only achieved with the grafting method, however, the gold particles have too large diameters, around 10 nm [82], which leads to an inferior catalytic activity.

4.5. Sol-Gel Method

According to several authors, in the sol-gel method a sol solution of the support is obtained by mixing the support precursor (like tetra-ethyl-ortho-silicate, aluminum tri-sec-butoxide, aluminum isopropoxide, or tetrabutoxy-titanium) with water, ethanol and methanol, and/or nitric acid. Then, the gold precursor (such as chloroauric acid, gold acetate, or hydrogen tetranitratoaurate) is added to the sol solution of the support, stirring vigorously for a variable time until the gel begins to be formed. The obtained gel is dried during 12–24 h at a temperature about 100–200 °C and a calcination step follows [80].

The catalysts prepared by this method show gold nanoparticles with sizes below 6 nm, involving materials resulting from soluble precursors which form three-dimensional networks with the addition of a base [80]. In Figure 4a it is possible to see a TEM image and gold particles size distribution of gold nanoparticles support on activated carbon prepared by the sol immobilization method.

Figure 4. (a) TEM image and particle size distribution of Au/C prepared by the sol immobilization method (dried at 60 °C until total evaporation of solution) and (b) TEM image of Au/Y prepared by ion-exchange at 25 °C and pH = 5.0. Gold nanoparticles are seen as dark spots. Adapted from Quintanilla et al. [68] and Lin et al. [89], respectively.

4.6. Ion-Exchange Method

The ion-exchange method consists in replacing the protons or other cations on surface, or within the structure of the support, by gold, and this leads first to atomically dispersed species and then, after calcination and reduction with hydrogen, to small gold particles [80,83].

This method is especially effective for depositing gold on zeolites, but the introduction of active species into the cavities of these supports, instead of placing gold on their external surface, presents several difficulties, for example, only limited cations or cationic complexes can be used ([Au(en)$_2$]$^{3+}$ – en = ethylenediamine - or [Au(NH$_3$)$_2$]$^+$) [80,83]. For this reason, the ion-exchange method is rarely used in the preparation of gold catalysts, although small metal particles are obtained.

Figure 4b shows a TEM image of gold supported on zeolite (Au/Y) catalyst prepared by the ion-exchange method.

In the next sections, we will present the treatment of wastewater by wet peroxidation using gold catalysts. The influence of catalyst properties and operating conditions, as well the catalyst stability, will be discussed.

5. Application of CWPO using Gold Catalysts in Wastewater Treatment

In the last century, there was an increasing interest in the use of gold catalysts by the scientific community. These materials have been used in chemical and environmental catalysis, in reactions such as CO oxidation [85,90,91], hydrogenation [92], water-gas-shift [93–95], combustion of volatile organic compounds [96–98], and organic compounds reduction [99] or oxidation [86,100–102]. Gold catalysts have also been used in wastewater treatment by catalytic wet peroxidation, which is the focus of this review, as said above. Gold has replaced catalysts that, although being efficient in the removal of pollutants, present the disadvantage of high metal leaching, like iron-based catalysts [69,70,103,104]. Gold does not leach, is stable and efficient, as mentioned above and will be further discussed ahead.

The preparation methods previously described influence the gold particles size and, consequently, the dispersion of the metal on the surface of the support. These two parameters are correlated as demonstrated by Equation (15) and have a strong effect on the catalytic activity, as well as other properties of gold-based catalysts like pore size, surface area (S_{BET}), mesoporosity of the support, and the oxidative state of gold.

Some authors correlate the effect of the gold amount and dispersion (and indirectly the particle size) with the catalytic performance, by evaluating the turnover frequency (TOF), which provides the number of molecules of target substrate degraded per gold atom and time unit (Equation (17)).

$$DM\ (\%) = \frac{6 \times n_s \times MM \times 1000}{\rho \times N \times dp} \times 100 \tag{15}$$

$$n\ (\text{moles of gold}) = \frac{Y_{Au} \times W_{cat}}{MM} \tag{16}$$

$$TOF\ (h^{-1}) = \frac{C}{\frac{DM}{100} \times n \times t} \tag{17}$$

where: C refers to the molecules of substrate degraded, DM is the gold dispersion, n is the number of moles of gold used, t is the time of reaction, n_s is the number of atoms at the surface per unit area (1.15×10^{19} m^{-2} for Au) [105], MM is the molar mass of gold (196.97 g/mol), ρ is the density of gold (19.5 g/cm^3), N is the Avogadro's number (6.022×10^{23} mol^{-1}), dp is the average gold particle size (nm), Y_{Au} is the amount of gold in the catalyst (wt.%), and W_{cat} is the mass of catalyst (g).

The efforts to achieve active and stable gold catalysts to be used in the treatment of effluents by CWPO, in view of industrial applications, have been reported in literature. Table 2 presents an outlook of the research made, showing which pollutants were degraded, the wastewater treatment conditions used, catalysts employed, and efficiency reached.

The efficiency of the catalytic wet peroxidation process for wastewater treatment is influenced by the catalyst properties and many operating conditions, such as gold loading, pH, temperature, hydrogen peroxide dose, catalyst concentration, and also the radiation intensity source (the latter in the case of the photo-assisted wet peroxidation). The influence of the catalyst properties and the effect of such operating conditions, as well as the stability of these catalysts, will be briefly described below.

Table 2. Gold catalyst used in CWPO of model compounds degradation or wastewater treatment, operational conditions, and performances reached.

Model Compound/Effluent	Catalyst	Operation Conditions	Efficiency of CWPO	Ref.
Orange II (OII) dye	Au/Al$_2$O$_3$ (0.7 wt.%)	pH = 3.0; T = 50 °C; [H$_2$O$_2$] = 6 mM; [catalyst] = 2.0 g/L; [OII] = 0.1 mM; t = 4 h	Dye removal = 98.9%; TOC removal = 49.8%; COD removal = 42.2%; H$_2$O$_2$ consumption = 95.0%; Specific Oxygen Uptake Rate = 27.8 mgO$_2$/(g$_{VSS}$·h); Inhibition of *Vibrio Fischeri* = 0.0%; Gold leaching < 0.04%	[66]
Acrylic Dyeing Wastewater	Au/Al$_2$O$_3$ (0.7 wt.%)	pH = 3.0; T = 50 °C; [H$_2$O$_2$] = 3.52 g/L; [catalyst] = 2.0 g/L; t = 4 h	Color removal = 34.4%; TOC removal = 42.9%; COD removal = 50.5%; H$_2$O$_2$ consumption = 98.8%; BOD$_5$:COD = 0.23; Gold leaching < 0.04%	
OII dye	Au/Al$_2$O$_3$ (0.7 wt.%)	pH = 3.0; T = 30 °C; [H$_2$O$_2$] = 6 mM; [catalyst] = 2.0 g/L; [OII] = 0.1 mM; t = 16 h	Dye removal = 99.4%; TOC removal = 48.2%; H$_2$O$_2$ consumption = 96.1%; Gold leaching < 0.04%	[67]
	Au/Fe$_2$O$_3$ (0.8 wt.%)		Dye removal = 51.4%; TOC removal = 36.9%; H$_2$O$_2$ consumption = 68.5%; Gold leaching < 0.04%	
	Au/Fe$_2$O$_3$ (4.0 wt.%) from WGC		Dye removal = 40.9%; TOC removal = 29.6%; Gold leaching < 0.04%	
	Au/TiO$_2$ (1.6 wt.%)		Dye removal = 68.9%; TOC removal = 32.4%; H$_2$O$_2$ consumption = 91.7%; Gold leaching < 0.04%	
	Au/ZnO (1.2 wt.%)		Dye removal = 62.6%; TOC removal = 31.9%; H$_2$O$_2$ consumption = 96.1%; Gold leaching < 0.04%	
Phenol	Au/TiO$_2$ (0.8 wt.%)	[phenol] = 5.0 g/L; [catalyst] = 2.7 g/L; V$_{H2O2}$ = 5 mL; t = 24 h; V$_{solution}$ = 45 mL.	TOF$_{phenol}$ = 0.07*10^6 (h^{-1}); TOF$_{TOC}$ = 0.07*10^6 (h^{-1}); TOF$_{H2O2}$ = 2.52*10^6 (h^{-1})	[68]
	Au(3)/C (0.5 wt.%)		TOF$_{phenol}$ = 1.19*10^6 (h^{-1}); TOF$_{TOC}$ = 1.08*10^6 (h^{-1}); TOF$_{H2O2}$ = 16.70*10^6 (h^{-1})	
	Au(5)/C (0.5 wt.%)		TOF$_{phenol}$ = 0.32*10^6 (h^{-1}); TOF$_{TOC}$ = 0.25*10^6 (h^{-1}); TOF$_{H2O2}$ = 4.07*10^6 (h^{-1})	
	Au(7)/C (0.5 wt.%)		TOF$_{phenol}$ = 0.25*10^6 (h^{-1}); TOF$_{TOC}$ = 0.25*10^6 (h^{-1}); TOF$_{H2O2}$ = 2.27*10^6 (h^{-1})	
	Au(10)/C (0.5 wt.%)		TOF$_{phenol}$ = 0.47*10^6 (h^{-1}); TOF$_{TOC}$ = 0.43*10^6 (h^{-1}); TOF$_{H2O2}$ = 1.87*10^6 (h^{-1})	
Phenol	Au/Hap (2.4 wt.% of Au)	pH = 2.0; T = 70 °C; V$_{H2O2}$ with 30 wt.% = 1 mL; [catalyst] = 0.1 g/L; [phenol] = 100 mg/L; t = 2 h	Phenol removal = ~92.5%	[70]

Table 2. Cont.

Model Compound/Effluent	Catalyst	Operation Conditions	Efficiency of CWPO	Ref.
Phenol	Au/TiO$_2$ – AD (2.8wt.%)	[phenol] = 200 mg/L; [H$_2$O$_2$] = 1520 mg/L; pH = 2.5; T = 80 °C; P = 1 atm; LHSV = 3.8 h^{-1}	Phenol removal $_{steady-state}$ = 100.0%; TOC removal $_{steady-state}$ = ~65.0%	[71]
	Au/TiO$_2$ – AD (3.2 wt.%)		Phenol removal $_{steady-state}$ = 100.0%; TOC removal $_{steady-state}$ = ~80.0%	
Phenol	Au/AC (0.8 wt.% of Au)	pH = 3.5; T = 80 °C; [H$_2$O$_2$] = 25 g/L; [catalyst] = 2.5 g/L; [phenol] = 5 g/L; t = 22h	Phenol removal = 100%; TOC removal = 70%	[72]
Phenol	Au/DNP (1 wt.% of Au)	pH = 4.0; T = 50 °C; [H$_2$O$_2$] = 1.44 g/L; [catalyst] = 320 mg/L; [phenol] = 1 g/L; t = 7 h	Phenol removal = 100%; H$_2$O$_2$ consumption = 100%; BOD$_5$:COD = 0.72	[73]
Phenol	Au/CeO$_2$ (1.0%)	pH = 4.0; Room temperature; [H$_2$O$_2$] = 200 mg/L; [Au] = 0.0025 mM; [phenol] = 100 mg/L; t = 24 h	Phenol removal = 7.0%; H$_2$O$_2$ consumption = 88.0%; Gold leaching = 0.8%	[74]
	Au/Fe$_2$O$_3$ (1.5%)		Phenol removal = 3.0%; H$_2$O$_2$ consumption = 8.0%; Gold leaching = 0.7%	
	Au/TiO$_2$ (1.5%)		Phenol removal = 3.0%; H$_2$O$_2$ consumption = 19.0%; Gold leaching = 0.5%	
	Au/C (0.8%)		Phenol removal = 7.0%; H$_2$O$_2$ consumption = 14.0%; Gold leaching = 5.8%	
	Au/npD (<1.0%)		Phenol removal < 1.0%; H$_2$O$_2$ consumption = 6.0%; Gold leaching = 0.5%	
	Au/HO-npD (1.0%)		Phenol removal = 93.0%; H$_2$O$_2$ consumption = 48.0%; Gold leaching = 0.7%	
Methyl Blue dye (MB)	Au/CNT (41.0 wt.%)	[MB dye] = 50 mg/L; [catalyst] = 0.5 g/L; [H$_2$O$_2$] = 500 mM; pH = 7.08; t = 120 min	MB removal = ~100%	[88]
1,1-diphenyl-2-picrylhydrazyl (DPPH)	Au/CNT placed in water/cyclohexane mixture (1/10 v/v) (41.0 wt.%)	[DPPH] = 0.2 mM; [catalyst] = 1 g/L; [H$_2$O$_2$] = 250 mM; t = 10 min; Room temperature; W/O = 1:10 v/v	DPPH removal = 100%	
Acid Orange 7 (AO7) dye	Au/CeO$_2$ (1 wt.% of Au)	[H$_2$O$_2$] = 20 mM; [catalyst] = 0.5 g/L; [dye] = 35 mg/L; t = 33 h	AO7 removal = 80%	[106]

Table 2. *Cont.*

Model Compound/Effluent	Catalyst	Operation Conditions	Efficiency of CWPO	Ref.
Methyl Orange dye (MO)	Au/TN (1.0 wt.%)	[MO] = 50 mg/L; [catalyst] = 2 g/L; [H_2O_2] = 0.15 M; pH = 3.0; T = 80 °C; t = 240 min	MO removal = 85%; TOC removal = 83%	[107]
Bisphenol A (BPA)	Au/SRAC (3.0 wt.%)	[BPA] = 114 mg/L; [catalyst] = 125 mg/L; [H_2O_2] = 530 mg/L; pH = 3.0; T = 30 °C	BPA removal = 89.0%; H_2O_2 consumption = 44.1%;	[108]
	Au/PSAC (3.0 wt.%)		BPA removal = 23.8%; H_2O_2 consumption = 8.3%	
	Au/CNF (3.0 wt.%)		BPA removal = 20.4%; H_2O_2 consumption = 14.5%	
	Au/FDU-15 (3.0 wt.%)		BPA removal = 32.4%; H_2O_2 consumption = 22.8%	
	Au/X40s (10.0 wt.%)		BPA removal = 14.5%; H_2O_2 consumption = 10.7%	
	Au/Fe_2O_3 (5.0 wt.%)		BPA removal = 10.1%; H_2O_2 consumption = 7.6%	
	Au/TiO_2 (1.5 wt.%)		BPA removal = 5.3%; H_2O_2 consumption = 10.8%	
	Au-Fe_2O_3/Al_2O_3 (0.5 wt.%)		BPA removal = 6.6%; H_2O_2 consumption = 15.3%	
	Au*SRAC (1.5 wt.%)	[BPA] = 89 mg/L; [catalyst] = 125 mg/L; [H_2O_2] = 530 mg/L; pH = 3.0; T = 30 °C	BPA removal = ~80.0%; H_2O_2 consumption = ~40.0%	

* TOF = turnover frequency

5.1. Influence of the Catalyst Properties

As already mentioned above, the catalytic properties directly affect the efficiency of the CPWO process. Ge et al. [106] concluded that the textural properties influenced AO7 dye removal, and achieved 80% for Au/CeO$_2$ with a lower S$_{BET}$ (55 m^2/g) and an intermediate gold content (1 wt.%).

Alvaro et al. [109] concluded that the morphological properties of gold supported on mesoporous titania had an influence in the decontamination of Soman wastewater. The authors reached the best decontamination degree (~100%) for a catalyst with medium surface area (90 m^2/g) and large pore diameter (7.1 nm) associated to a highest gold loading (0.70 wt.%), among the studied samples.

Navalon et al. [74] evaluated phenol oxidation by CWPO using gold supported on CeO$_2$, TiO$_2$, carbon, Fe$_2$O$_3$, npD, and HO-npD, and observed that catalysts with smaller gold particle size (< 1 nm) and intermediate gold loading (1.0%) led to the highest performance (total phenol disappearance and 48.0% of hydrogen peroxide consumption). Moreover, a small amount of gold (0.7%) was leached from the support to the solution.

However, in the three studies reported above, the authors do not indicate any explanation why the efficiency of CPWO was the best for the catalysts selected. The main characteristics influencing the catalysts performance in CPWO are morphology and porosity (adsorption capacity), gold loading, and particle size.

The optimization of the gold loading of a catalyst is essential for economic aspects, as mentioned above, and is determinant for catalysing the reaction that generates hydroxyl radicals, influencing the efficiency of CWPO. In the work of Rodrigues et al. [67], the efficiency of CWPO decayed dramatically when the loading of gold on iron oxide increased from 0.8 to 4.0 wt.%, reducing from 99.7 to 36.6% and from 75.8 to 24.0% for OII dye and TOC removals, respectively. Moreover, a significant reduction was observed in the production of hydroxyl radicals with increasing gold content.

However, in a work using CWPO assisted with radiation to treat an OII dye solution with gold on iron oxide, the gold content had no effect on the colour removal [110]. This is due to the fact that the dye is degraded in the presence of UV/vis radiation alone. However, the authors observed a slight reduction in mineralization from 68.2 to 58.4% when the gold content increased, as well as in the production of hydroxyl radicals [110].

In both studies mentioned above, the authors pointed out an explanation for the decay of the process performance with the increase of the gold content; if in excess, gold reacts with the hydroxyl radical (HO$^\bullet$ + Fe$_2$O$_3$-Au^0 → Fe$_2$O$_3$-Au$^+$ + HO$^-$), being less available to oxidize the dye, and the reaction by-products [66,110].

The degradation of methyl orange decreased from ~40 to ~10% when the gold loading increased from 1.0 to 4.0 wt.%, in a study dealing with the removal of this dye by CWPO, at 25 °C, using gold supported on modified titanium nanotubes [107]. The authors attributed this decreased of process efficiency to the fact that the catalyst with lower gold content had a smaller particle size and gold was uniformly distributed on the surface of the support [107].

On the other hand, and as shown in Figure 5a, the degradation of AO7 dye increased with gold loading until 1 wt.%; however, a further increase of catalyst content to 2 wt.% impaired the oxidation of the dye by CWPO, using gold on cerium oxide as catalyst [106], increasing the ratio between the AO7 concentration after 30 h of reaction (C) and the initial AO7 concentration (C$_0$), C/C$_0$, from 0.2 to 0.4. Furthermore, the combination of CWPO with visible radiation showed an optimum for 1 wt.% of gold loading for shorter reaction times (less than 5 h), more notorious in the period from 2 to 5 h (see Figure 5b) [106].

In contrast with the above-mentioned studies that showed the existence of an optimal gold loading, Yang et al. [108] observed that the removal of bisphenol A and consumption of hydrogen peroxide increased from 21.1 and 9% to 89.0 and 45%, respectively, after 12 h of reaction, when the gold content in an Au/AC catalyst increased from 0 to 3 wt.%. Moreover, Sempere et al. [75] observed a decrease in the turnover frequency (calculated according to Equations (15)–(17)) of phenol and hydrogen peroxide from ~100 and 320 h^{-1} to ~70 and 180 h^{-1}, respectively, when the gold loading increased from 0.1

to 0.5 wt.% in Au/FH$_2$, and subsequent annealing treatment with hydrogen for sunlight assisted CWPO. The authors pointed out an explanation for the decay of the catalytic activity with increasing particle size, i.e., the catalyst having a lower gold content had a smaller particle diameter and showed more efficiency.

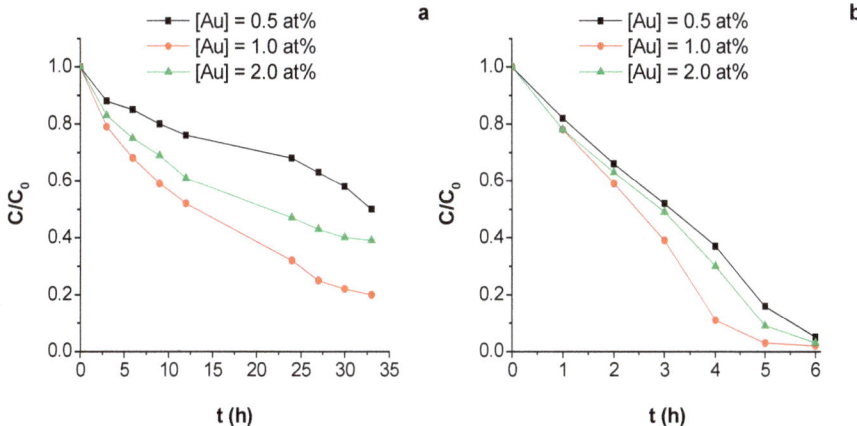

Figure 5. C/C$_o$ ratio during AO7 degradation by CWPO (**a**) and assisted with visible radiation (**b**) processes at different gold loadings ([AO7] = 35 mg/L, [H$_2$O$_2$] = 20 mM and [Au/CeO$_2$] = 0.5 g/L). Adapted from Ge et al. [106].

The studies described above allow us to conclude that the effect of the gold loading in the efficiency of CPWO depends on the type of catalyst used and the compound/wastewater to be treated.

Quintanilla et al. [68] evaluated the degradation of phenol, the mineralization and H$_2$O$_2$ consumption by wet peroxidation using Au/C, Au/Fe$_2$O$_3$, and Au/TiO$_2$. They verified that activated carbon is the preferable support because it has a higher adsorption capacity. Also, Au/C with less amount of gold (0.13 wt.%), lower gold size (5.1 nm), and higher fraction of Au$^{\delta+}$ in the catalyst surface (31%), presented the highest TOF for phenol oxidation (1.19 × 10^4 h^{-1}), TOC reduction (1.08 × 10^4 h^{-1}), and H$_2$O$_2$ consumption (16.70 × 10^4 h^{-1}), see Table 3. This catalyst (Au(3)/C) had gold particles with about 5 nm size, which is beneficial for the catalytic performance.

Table 3. Gold loading, particle size and percentage of exposed surface gold species of the catalysts, and turnover frequency (TOF) values of phenol, TOC oxidation, and hydrogen peroxide decomposition (experimental conditions: [Phenol] = 5 g/L, [Catalyst] = 2.7 g/L, V$_{H2O2}$ = 5 mL, t = 24 h and V$_{solution}$ = 45 mL). Adapted from Quintanilla et al. [68].

Catalyst	[Au]$_{total}$ (wt.%)	Au^0 Fraction (%)	Au$^{\delta+}$ Fraction (%)	Au Size (nm)	TOF × 10^{-4} (h^{-1})		
					Phenol	TOC	H$_2$O$_2$
Au/TiO$_2$	0.80	79	21	3.1 ± 1.8	0.07	0.07	2.52
Au(3)/AC *	0.13	69	31	5.1 ± 2.0	1.19	1.08	16.70
Au(5)/AC *	0.47	72	28	4.9 ± 1.0	0.32	0.25	4.07
Au(7)/AC *	0.48	71	29	6.8 ± 1.7	0.25	0.25	2.27
Au(10)/AC *	0.50	69	31	9.1 ± 1.1	0.47	0.43	1.87

* The numbers correspond to the initial average size (nm) of gold in the colloidal solution used for the catalyst preparation.

Rodrigues et al. [67] tested gold supported on titanium, zinc, aluminum, and iron oxides to treat a dye solution by CWPO. They concluded that the Au/Al$_2$O$_3$ catalyst with higher surface area (S$_{BET}$ = 210 m^2/g), lower amount of gold (0.7 wt.%), and an intermediate gold particle size (3.6 nm), had the best performance (removal of dye and TOC of ca. 98% and 47%, respectively, consumption of

H_2O_2 ~96%, and higher hydroxyl radicals generation) and the highest TOF value for OII dye removal (75.5 × 10^{-6} s^{-1}). The best efficiency of CPWO for Au/Al_2O_3 was associated to its higher adsorption capacity. The same was concluded for these same catalysts when light-assisted CWPO was used for OII dye degradation [110]. Au/Al_2O_3 permitted total discoloration and TOC removal of about 80%, combining further formation of hydroxyl radicals [110]. In the same way, Drašinac et al. [107] observed that the morphological properties of the catalysts and gold properties played an important role in methyl orange (MO) dye degradation. These authors reached the best CPWO performance (removals of 83 and 85% for TOC and MO dye, respectively) with gold supported on modified titanium nanotubes with a gold nanoparticle size of 7 nm and 1.1 wt.% loading, which had the highest total pore volume (1.31 cm^3/g), pore diameter (14.8 nm), and surface area (335 m^2/g). The authors reported that, in addition to the low gold content in the catalyst, the smaller particle diameter and the uniform distribution on the surface of the support benefit the catalytic process.

Yang et al. [108] reached the best performance of CWPO (89.0 and 44.1% for bisphenol A (BPA) removal and oxidant conversion, respectively) using a gold supported on styrene-based activated carbon (Au/SRAC) catalyst, wcih had an intermediate gold nanoparticle size (4.4 nm) and loading (3.0 wt.%). The authors pointed out the small gold size of the material as being beneficial in the degradation of the compound and oxidant conversion. On the other hand, Han et al. [70] reached the maximum removal of phenol (82%) and space-time conversion (0.53 mmol h^{-1} L^{-1}) when using a gold supported on hydroxyapatite (Au/Hap) catalyst with higher gold particle size (4.9 nm) and intermediate loading (2.4 wt.%). The authors stated that the best catalytic activity of this sample was due to the gold particle size close to 5 nm.

These studies showed that the textural properties of the catalysts, as well as the gold particle size, play an important role in the efficiency of the catalytic process.

5.2. Effect of the Operating Conditions

The efficiency of the catalytic wet peroxidation process for wastewater treatment is also influenced by many operating conditions, such as catalyst dose, pH, temperature, hydrogen peroxide concentration, and also the radiation intensity source (the latter in the case of the photo-assisted wet peroxidation). The effect of such operating conditions will be briefly described below.

5.2.1. Catalyst Dose

The efficiency of the process increases with the catalyst concentration, since more gold will be available in the reaction medium to catalyze CWPO, generating more hydroxyl radicals. However, above a certain concentration of catalyst there is very often a negative effect, once scavenging of hydroxyl radicals by the excess of gold occurs (Equation (13)). The ideal concentration of catalyst depends on the type of effluent to treat, being necessary to optimize it. In the work developed by Domínguez et al. [72], a linear increasing dependence was obtained for initial reaction rates of phenol and oxidant disappearance for a Au/AC concentration in the range of 0–6 g/L.

As reported by Domínguez et al. [72], the work of Martín et al. [73] showed that an increase in the Au/DNP catalyst concentration from 50 to 320 mg/L proportionately increased the initial reaction rate of phenol degradation and oxidant consumption (see Figure 6). The same tendency was observed by Navalon et al. [111], as the initial phenol degradation and H_2O_2 decomposition rates increased for Au/HO-npD concentrations, in the range of 0 to 400 mg/L, for solar light assisted CWPO.

On the other hand, Rodrigues et al. [66] reached an optimum dose of Au/Al_2O_3 catalyst of 2.0 g/L that maximized the OII degradation and mineralization, as well as the formation of hydroxyl radicals (see Figure 7) when CWPO was applied to the dye solution. Moreover, the oxidant decomposition increased with catalyst dose until 2.0 g/L, but remained equal for the highest concentration tested (see Figure 7a). No gold leaching was found in any of the tests.

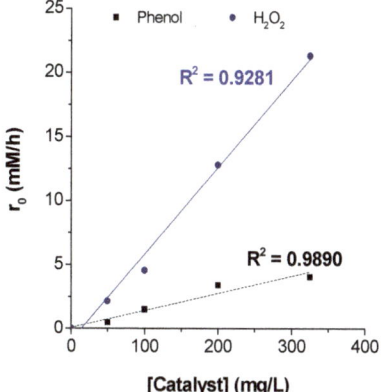

Figure 6. Effect of catalyst dose on the initial reaction rate of phenol oxidation and oxidant decomposition ([phenol]$_{initial}$ = 1.0 g/L, [H$_2$O$_2$]$_{initial}$ = 1.44 g/L, T = 50 °C and pH$_{initial}$ = 4.0). Adapted from Martín et al. [73].

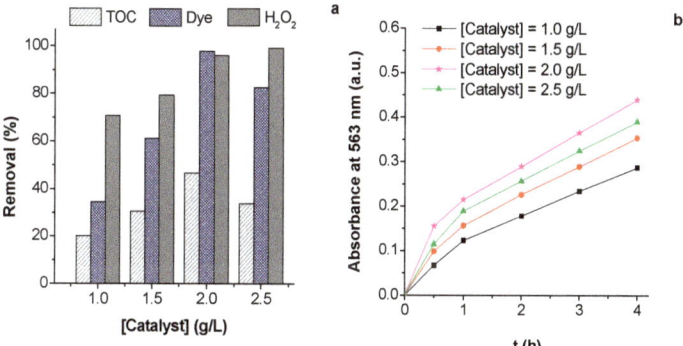

Figure 7. Influence of catalyst concentration in orange II dye and TOC removals and H$_2$O$_2$ consumption after 4 h (**a**) and evolution of hydroxyl radical formation (**b**) ([OII]$_a$) = 0.1 mM or [OII]$_b$) = 0.0 mM, [H$_2$O$_2$] = 6.0 mM, T = 30 °C and pH$_{initial}$ = 3.0). Adapted from Rodrigues et al. [66].

Similarly to what was reported earlier by Rodrigues et al. [66], in a subsequent work of the same authors [110], the effect of Au/Al$_2$O$_3$ concentration in OII oxidation by CWPO assisted by UV/visible radiation was evaluated. The maximum dye and TOC removals were achieved for a dose of 2.0 g/L, being 96.8 and 85.9%, respectively, after 2 h of reaction. The oxidant consumption increased with catalyst dose in the range of 1.0 to 2.5 g/L. Furthermore, for all catalyst doses, no gold leaching was found.

For an industrial application of CWPO in the treatment of effluents, the optimization of the catalyst concentration is crucial, not only for the efficiency of the process, but also in economic aspects. So, it is necessary to use the lowest catalyst quantity in order to reduce the costs of the treatment process, since gold catalysts are expensive, compared to other catalysts containing iron, copper, and others.

5.2.2. Hydrogen Peroxide Concentration

The initial concentration of H$_2$O$_2$ also plays a very important role in the oxidation of organic compounds in CWPO processes and in the operating costs of such treatment procedures; thus, it is necessary to determine the optimum dose of this reagent.

The improvement of the process by the addition of H₂O₂ is mostly due to the increased production of hydroxyl radicals, as described in Equations (10), (11), and (20). However, at high peroxide concentrations, the reaction between excess H₂O₂ and the strong oxidant HO• species becomes more relevant and, as a consequence, no subsequent improvement on the heterogeneous CWPO rate can be noticed, as the produced HO₂• radicals are less reactive than the HO• radicals (Equation (14)) [112]. Contrarily, if the concentration is low, the oxidation degree is small and there is the possibility of formation of unwanted intermediate products, which, in most cases, are more toxic and less biodegradable than the original compounds. Inherently, it is common to observe the existence of an optimum oxidant (hydrogen peroxide) dose in either wet peroxidation or radiation-assisted wet peroxidation processes.

The existence of an optimum oxidant dose was reported by several authors for CWPO catalyzed by gold on different supports [66,88,108,110,111]. In the work developed by Rodrigues et al. [66], the effect of this parameter was evaluated in the range of 3.0 to 12.0 mM. An increase in the removal of dye and TOC was observed, as well as in the formation of hydroxyl radicals with hydrogen peroxide concentration until 6.0 mM, but the efficiency of the process was reduced for higher oxidant doses (see Figure 8). The consumption of H₂O₂ also increased until 6 mM and remained constant for the higher doses. The authors reached 46.6 and 97.8% of TOC and dye removals, respectively, and ~100% for hydrogen peroxide consumption after 4 h of reaction with the optimized oxidant dose.

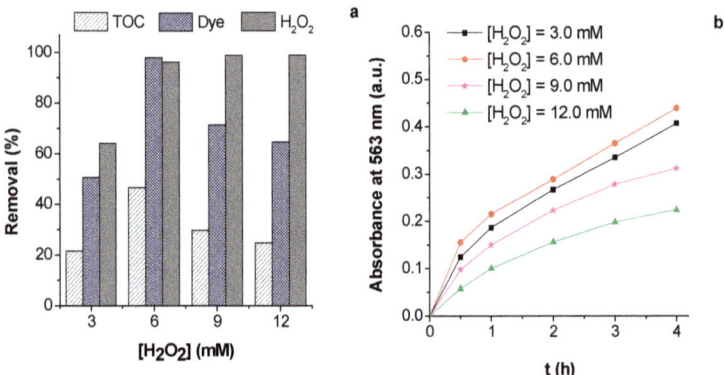

Figure 8. Effect of hydrogen peroxide concentration in dye and TOC removals and hydrogen peroxide consumption after 4 h of reaction (**a**) and hydroxyl radicals formation during CWPO (**b**) for a Au/Al₂O₃ catalyst ([OII]$_a$) = 0.1 mM or [OII]$_b$) = 0.0 mM, [catalyst] = 2.0 g/L, T = 30 °C and pH$_{initial}$ = 3.0). Adapted from Rodrigues et al. [66].

Another study evaluated the degradation of OII dye by photo-assisted CWPO, and the reduction of color, mineralization, and oxidant consumption increased when the oxidant dose increased from 1.5 to 3.0 mM [110]; for concentrations of 6.0 and 12.0 mM, a negative effect was observed for dye and TOC removals. For the optimal oxidant concentrations, efficiencies of 85.9, 96.8, and 94.5% were reached for dye and TOC removals and H₂O₂ consumption, respectively. Moreover, authors reported that the catalyst used (gold on alumina) did not show any leaching for any of the concentrations of hydrogen peroxide evaluated.

Yang et al. [108] tested gold supported on carbon as a CWPO catalyst for the oxidation of BPA. A smaller oxidant dose (275 mg/L) allowed to remove ~50% of the model compound and ca. 35% of oxidant consumption was found after 12 h of reaction. An improvement of catalytic activity was observed when the hydrogen peroxide concentration was increased to 530 mg/L, allowing, after 12 h, to reach a bisphenol A reduction of ca. 70% and a consumption of oxidant of about 40%. However, a further increase in the oxidant dosage (835 mg/L) did not influence the efficiency of CWPO. A similar tendency was reported by Lorençon et al. [88] for the CWPO catalyzed with Au/CNT for the degradation of a lipophilic compound (DPPH). The removal of DPPH increased with H_2O_2 concentration until 250 mM and remained practically constant for higher oxidant doses, as shown in Figure 9. The authors attained total degradation of DPPH for an optimal oxidant concentration of 250 mM. These two works highlighted once again the existence of an optimum oxidant dose for the gold catalysed CWPO process.

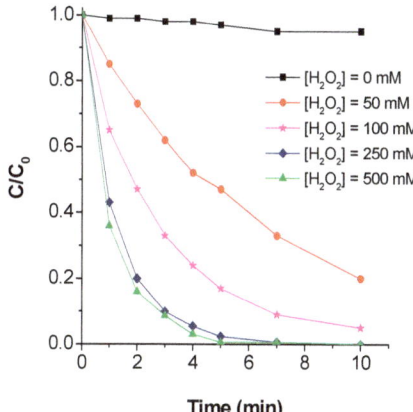

Figure 9. Influence of initial hydrogen peroxide concentration in the degradation of a lipophilic compound (DPPH) by CWPO using Au/CNT as catalyst ([DPPH]$_0$ = 0.2 mM, [catalyst] = 1.0 mg/mL, T = room temperature, stirring = 600 rpm). Adapted from Lorençon et al. [88].

Contrarily to what was reported in the studies mentioned above, in which an optimum oxidant dose was found, in the work of Martín et al. [73] a reduction in the concentration (or even up to total absence) of phenol and intermediate compounds (catechol, quinone, and hydroquinone) was found, resulting from the oxidation of the pollutant when the H_2O_2 concentration was increased in the tested range (362–1447 mg/L). On the other hand, the biodegradability of the effluent improved, increasing the BOD$_5$:COD ratio from ~ 0.05 to 0.4, when the oxidant dose increased from 362 to 1447 mg/L. However, higher doses were not tested.

The optimization of hydrogen peroxide amount is not only important for economic reasons, as the reagent is relatively expensive, but also to guarantee that H_2O_2 solution is not in excess. On one hand, too much H_2O_2 is detrimental to the subsequent biological treatment, if required, having deleterious effects on the microorganisms and leading to a decrease in the efficiency of the biological process. On the other hand, an excess of H_2O_2 contributes to the COD of the treated effluent, a commonly legislated parameter, and can give an erroneous indication of the possibility of the effluent discharge into water bodies.

5.2.3. Initial pH

The efficiency of wet peroxidation is also strongly dependent on the pH of the medium. A pH < 2.5 allows the scavenging reaction between the hydroxyl radical and H^+ to take place (Equation (18)) [113]. Furthermore, at neutral or alkaline conditions, hydrogen peroxide self-decomposition into water and oxygen (Equation (19)) is promoted, decreasing the amount of available oxidant to yield hydroxyl radicals to promote organics degradation.

$$H^+ + HO^\bullet + e^- \rightarrow H_2O \qquad (18)$$

$$2\,H_2O_2 \rightarrow 2\,H_2O + O_2 \qquad (19)$$

In the homogeneous process, for very acidic pH values, Au$^+$ is present in lesser amounts, while higher pHs lead to precipitation of gold in insoluble form (AuHO), resulting in reduction of the amount of Au available, which leads to small formation of radicals (Equation (5)). These drawbacks are overcome when gold is supported on a porous support (heterogeneous process), once the metal is inside the pores and confined within the structure of the solid matrix [52,114]. This reduces the precipitation of gold that occurs in the homogeneous process; thus, the catalyst is available to decompose the hydrogen peroxide and generate the hydroxyl radical.

However, Domínguez et al. [72] observed that the CWPO of phenol using gold supported on activated carbon was efficient in acid and neutral pH range (3.5–7.5) and decreased significantly for pH = 10.5. The authors achieved, after 24 h of reaction, removals of phenol and TOC of ~100 and ~60%, respectively, with an efficiency of hydrogen peroxide use (η, evaluated by the ratio between the amount of TOC removal and oxidant consumption) of 0.8 for pH values between 3.5 to 7.5.

A similar behavior was found by Martín et al. [73]. These authors showed that gold supported on diamond nanoparticles was catalytically active in the pH range between 4 and 7, with almost all phenol being degraded. They also observed a strong decay in its removal for the pH of 8 and 9 and total consumption of hydrogen peroxide for all values of pH tested (Figure 10). These authors reported that the biodegradability (evaluated by the BOD$_5$:COD ratio) was higher (~0.7) for the lower pH value tested (4.0), which was associated with no phenol detection and the lowest concentration of catechol and hydroquinone (intermediate compounds resulting from the oxidation of phenol) after CWPO treatment at this pH value, and decreased to values near 0.4 for higher pHs (5.0–7.0) that present phenol and higher concentrations of catechol and hydroquinone in solution after treatment.

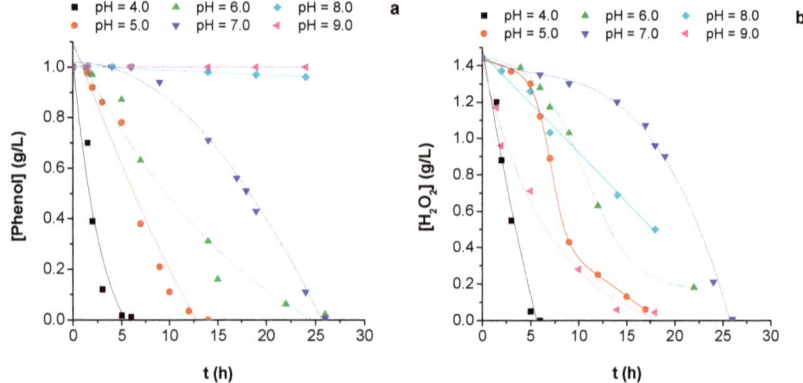

Figure 10. Effect of pH in phenol (**a**) and hydrogen peroxide (**b**) concentration during CWPO reaction with gold supported on diamond as catalyst at different pH values. Lines were added merely to better illustrate the data trends ([phenol]$_{initial}$ = 1.0 g/L, [H$_2$O$_2$]$_{initial}$ = 1.44 g/L, [catalyst] = 320 mg/L and T = 50 °C). Adapted from Martín et al. [73].

Studies of OII degradation by CWPO, without [66] and with radiation [110], using gold supported on alumina showed that the initial pH has an effect on the efficiency of the process. The authors observed optimum activity for initial pH = 3, which maximized the removal of OII (>97%) and TOC (85.9%) (see Figure 11), as well as the generation of hydroxyl radicals. The formation of radicals was evaluated in runs carried out without OII dye and when hydroxyl radicals were formed, and thus became in contact with 1,5-diphenyl carbazide, 1,5-diphenyl carbazone was formed, which presented a brown color that was measured at 563 nm. However, the oxidant consumption increased with pH (see Figure 11). As the support used in this study was alkaline, it increased the pH of the medium and, for the optimum initial pH found, the pH after 2 h of reaction was 4, which is in agreement with the best pH value reported by Domínguez et al. [72] and Martín et al. [73]. In these studies, all oxidant was consumed and the gold did not leach into solution during the oxidation for the initial pH range tested.

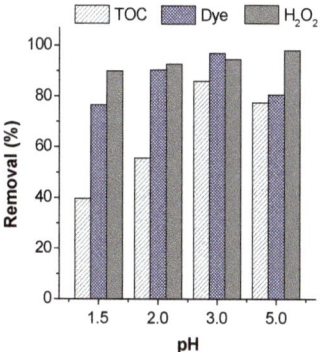

Figure 11. Removals of TOC and dye and hydrogen peroxide consumption, after 4 h of CWPO, with radiation, using Au/Al$_2$O$_3$ as catalyst ([OII] = 0.1 mM, [H$_2$O$_2$]$_{initial}$ = 6 mM, [catalyst] = 2.0 g/L, T = 30 °C and I = 500 W/m^2). Adapted from Rodrigues et al. [110].

In the work developed by Ferentz et al. [71], it was evaluated the degradation of phenol by CWPO using gold supported on titanium oxide. The TOC conversion increased with the pH of the phenol solution until 3.5, reduced about 20% for pH in the range of 4–8 and had a significant decrease for pH higher than 9. The authors explanation for the high pH effect on the efficiency of the process was associated with the adsorption of H$_2$O$_2$ on the titanium surface, which increases with the pH decrease. This is due to the dissociation of hydroxyl groups from the titanium surface, leading to the creation of Lewis acid sites (Ti^{4+}) that attach the hydrogen peroxide, leaving the surface of Ti(H$_2$O$_2$)$^{4+}$ prone to O-O bond cleavage forming the hydroxyl radical.

Navalon et al. [74] observed that the performance of CPWO, when using Au/HO-npD as a catalyst, was very affected by the pH of the phenol solution and the catalyst was abruptly inefficient at pH above 5. This fact is explained by the change of the catalyst charge from positive (pH < 5) to negative (pH > 5). On the other hand, in this study, high gold leaching (47%) was found, at pH less than 3, that was much lower for pH higher than 3 (0.7% at pH 4).

The effect of pH was evaluated in others studies that reached the best performances when using neutral or alkaline conditions (7.0-11.0) [70,75,108,111,115].

As mentioned above, the pH influences the efficiency of CPWO, with the use of gold catalysts. In addition to the decomposition of H$_2$O$_2$ in water and oxygen (Equation (19)) in alkaline conditions, the pH affects the surface chemical properties of the support that influence the adsorption of the oxidant and, consequently, the generation of hydroxyl radicals, as reported in the study developed by Ferentz et al. [71]. Changes in the colloids charge can also occur, which also affect the adsorption of pollutants and oxidant as mentioned by Navalon et al. [74].

Concerning the industrial application perspective, the gold catalysts that allow the use of oxidative processes at neutral pH are more advantageous, because they reduce the costs associated with the acid consumption needed to decrease the pH to the acid range, and the base necessary to neutralize the effluent after the treatment, before it is discharged into the water bodies or subsequent treatment processes, as biological degradation.

5.2.4. Temperature

The temperature has a large influence on the efficiency of the CWPO process. The possibility to increase the operating temperature, as a way of improving the efficiency of the process, has been scarcely investigated, because the idea of thermal decomposition of H$_2$O$_2$ into O$_2$ and H$_2$O seems to be widely accepted as a serious drawback [11]. However, according to the Arrhenius law, higher temperatures (often up to ca. 50–70 °C) can lead to a more efficient use of H_2O_2 upon enhanced generation of HO• radicals, at low metal concentrations. A decrease of the metal dose is important, since it improves the efficiency of H_2O_2 use, by minimizing competitive scavenging reactions [116]. Moreover, increasing the temperature accelerates oxidation of the organic compounds by the radicals.

Therefore, an increase in the temperature can be considered as a way to intensify the treatment process. Domínguez et al. [72] observed a positive effect of temperature, in the range of 50–80 °C, in the removal of phenol by CWPO using gold supported on activated carbon. The same tendency was reported by Drašinac et al. [107], as the methyl orange dye removal increased from ~30% at 25 °C to 85% at 80 °C, after 250 min of reaction, when the process was catalyzed by gold supported on modified titanium nanotubes.

Martín et al. [73] found that the reaction rate of phenol degradation and hydrogen peroxide consumption, when using an Au/DNP catalyst, increased with temperature in the range of 40–100 °C, reducing the reaction time from ~30 h to ~2.5 h, respectively, always reaching 100% compound removal and total consumption of oxidant. The same tendency was observed in a study that evaluated the methyl orange dye degradation, as its removal and TOC reduction increased with temperature in the range of 25 to 80 °C (see Figure 12), reaching 85 and 83% for MO and TOC reduction, respectively, for the optimal temperature (80 °C) [107].

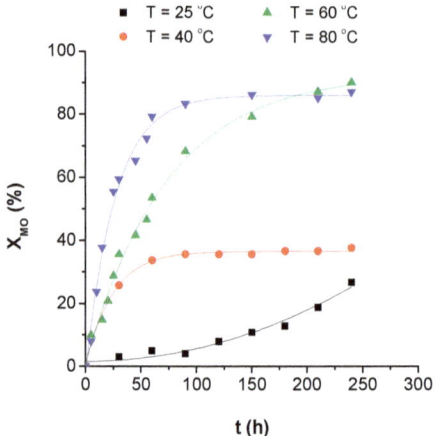

Figure 12. Methyl Orange (MO) decolorization during CWPO using gold supported on titanium nanotubes as catalyst, at different temperatures. Lines were added merely to better illustrate the data trends ([MO]$_{initial}$ = 50 mg/L, [H_2O_2]$_{initial}$ = 0.15 M, [catalyst] = 2.0 g/L and pH$_{initial}$ = 3.0). Adapted from Drašinac et al. [107].

In contrast, the effect of decomposition of hydrogen peroxide in water and oxygen was observed in the works of Yang et al. [108] and Rodrigues et al. [66,110]. Yang et al. [108] observed a significant increase in bisphenol A degradation and consumption of the hydrogen peroxide during CWPO, using Au/SRAC as catalyst, when the temperature increased from 30 to 40 °C (see Figure 13). Also when the temperature was raised to 50 °C, a little increase in the process efficiency was obtained. However, increasing the temperature to 60 °C, showed no improvement in the degradation of bisphenol A.

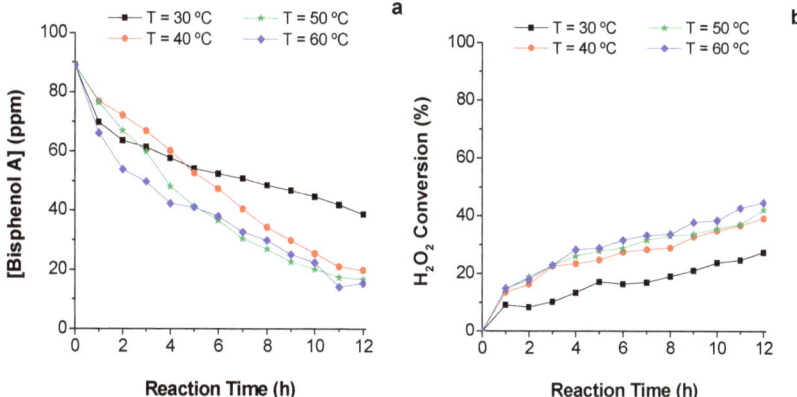

Figure 13. Effect of temperature in bisphenol A (BPA) concentration (**a**) and hydrogen peroxide conversion (**b**) during CWPO using Au/SRAC as catalyst ([BPA]$_{initial}$ = 89 ppm, [H$_2$O$_2$]$_{initial}$ = 530 ppm, [catalyst] = 125 ppm and pH$_{initial}$ = 3.0). Adapted from Yan et al. [108].

The application of wet peroxidation at high temperature can be beneficial in effluents that are generated at high temperature, as in the case of textile dye effluents [73], minimizing the energy costs for heating. However, for wastewater discharged from industrial processes at lower temperature, it may be more advantageous to apply a treatment process at moderate temperatures. In any case, a careful cost/benefit analysis should be made for each particular situation.

5.2.5. Effect of radiation use

In the literature, there are studies reporting the application of CWPO assisted with radiation. Table 4 shows an outline of these studies, reporting the operating conditions used and the efficiencies achieved. The use of radiation increases the rate of oxidation since there are additional mechanisms for the formation of free radicals, according to the following three processes: i) the catalytic decomposition H$_2$O$_2$ in the presence of radiation (Equation (20)) [42,117], ii) the decomposition of hydrogen peroxide by incidence of radiation (Equation (21)), and iii) photolysis of gold hydroxide (Equation (22)).

$$X\text{-}Au^{n+} + H_2O_2 + h\nu \rightarrow X\text{-}Au^{(n+1)+} + HO^\bullet + HO^- \qquad (20)$$

$$H_2O_2 + h\nu \rightarrow 2HO^\bullet \qquad (21)$$

$$X\text{-}Au(OH)^{n+} + h\nu \rightarrow X\text{-}Au^{n+} + HO^\bullet \qquad (22)$$

On the other hand, an improvement of the performance in radiation-assisted CWPO can also occur due to direct photolysis of the organic compounds to degrade.

Some authors report that an increased radiation intensity has a positive effect on the performance of the treatment process. Navalon et al. [74] showed that an increase of laser power from 0 to 70 mJ/pulse improved the phenol degradation when Au/OH-npD catalyst was used. Rodrigues et al. [110] observed an increase of color and TOC removal, as well as of hydrogen consumption, with the radiation intensity of a TQ150 mercury lamp when Au/Al$_2$O$_3$ was used as photo-catalyst.

Table 4. Gold catalysts used in CWPO assisted with radiation for model compounds degradation or wastewater treatment, operational conditions, and performances reached.

Model Compound/Effluent	Catalyst	Operation Conditions	Efficiency of CWPO assisted with Radiation	Ref.
Phenol	Au/FH$_2$ (0.1%)	pH = 4.0; Room temperature; [phenol] = 100 mg/L; t = 3.5 h; Radiation: Sunlight	Phenol removal = 100%; H$_2$O$_2$ consumption = ~60%	[75]
	Au/FN$_2$ (0.5%)	pH = 4.0; Room temperature; [H$_2$O$_2$] = 200 mg/L; [phenol] = 100 mg/L; t = 8 h; Radiation: Sunlight	Phenol removal = ~20%; H$_2$O$_2$ consumption = ~20%	
	Au/F (0.5%)		Phenol removal = ~10%; H$_2$O$_2$ consumption = ~15%	
Acid Orange 7 dye (AO7)	Au/CeO$_2$ (1.0 at.%)	pH = 3.0; T – 30 °C; [H$_2$O$_2$] = 20 mM; [catalyst] = 0.5 g/L; [AO7] = 35 mg/L; t = 6 h; Radiation: Visible light	Dye removal = 100%	[106]
Orange II (OII) dye	Au/Al$_2$O$_3$ (0.7 wt.%)	pH = 3.0; T = 30 °C; [H$_2$O$_2$] = 6 mM; [catalyst] = 2.0 g/L; [OII] = 0.1 mM; t = 2 h; Radiation: UV/visible light (500 W/m^2)	Dye removal = 96.8%; TOC removal = 80.5%	[110]
	Au/Fe$_2$O$_3$ (0.8 wt.%)		Dye removal = 97.8%; TOC removal = 68.2%	
	Au/Fe$_2$O$_3$ (4.0 wt.%) from WGC		Dye removal = 96.9%; TOC removal = 58.4%	
	Au/TiO$_2$ (1.6 wt.%)		Dye removal = 98.5%; TOC removal = 73.5%	
	Au/ZnO (1.2 wt.%)		Dye removal = 99.8%; TOC removal = 73.4%	
	Au/Al$_2$O$_3$ (0.7 wt.%)	pH = 3.0; T = 50 °C; [H$_2$O$_2$] = 3 mM; [catalyst] = 2.0 g/L; [OII] = 0.1 mM; t = 2 h; Radiation: UV/visible light (500 W/m^2)	Dye removal = 99.3%; TOC removal = 90.9%; H$_2$O$_2$ consumption = 98.6%; Gold leaching < 0.5 mg/L	
Acrylic dyeing wastewater	Au/Al$_2$O$_3$ (0.7 wt.%)	pH = 3.0; T = 50 °C; [H$_2$O$_2$] = 104 mM; [catalyst] = 2.0 g/L; t = 2 h; Radiation: UV/visible light (500 W/m^2)	Color removal = 100%; TOC removal = 72.4%; COD removal = 70.0%; BOD$_5$:COD = 0.5; Specific Oxygen Uptake Rate = 17.9 mgO$_2$/(g$_{VSS}$.h); Inhibition of Vibrio Fischeri = 0.0%	
Phenol	Au/HO-npD (1.0 wt%)	pH = 4.0; T = 30 °C; [H$_2$O$_2$] = 2.5 g/L; [catalyst] = 400 mg/L; [phenol] = 100 mg/L; t = 2 h; Radiation: Sunlight	Phenol removal = 100%; H$_2$O$_2$ consumption = 100%; COD removal = 69.7%; BOD$_5$:COD = 0.4	[111]
Phenol	Au/HO-npD (1.0 wt%)	pH = 4.0; [H$_2$O$_2$] = 200 mg/L; [catalyst] = 160 mg/L; [phenol] = 100 mg/L; t = 2 h; Radiation: Laser Flash (70 mJ/pulse)	Phenol removal = 100%; H$_2$O$_2$ consumption = ~90%	[115]
	Au/CeO$_2$ (1.0 wt%)	pH = 4.0; [H$_2$O$_2$] = 200 mg/L; [catalyst] = 160 mg/L; [phenol] = 100 mg/L; t = 3 h; Radiation: Laser Flash (70 mJ/pulse)	Phenol removal = ~15%; H$_2$O$_2$ consumption = ~100%	
	Au/TiO$_2$ (1.0 wt%)		Phenol removal = ~10%; H$_2$O$_2$ consumption = ~80%	

5.3. Catalyst Stability

The deactivation of catalysts in CWPO is mostly associated with the loss of metal by leaching from the solid support to the effluent during the treatment process, but also with possible gold nanoparticle sintering and/or pore blockage. For application of catalytic wet peroxidation in the treatment of real effluents, a crucial aspect to be taken into account is the reutilization of the catalysts, without reduction of their efficiency and stability in consecutive cycles of use, making it imperative to evaluate how stable and durable they are.

Ferentz et al. [71] evaluated the long-term stability of Au/TiO$_2$, with 2.8 and 3.2 Au wt.%, in a fixed bed reactor. The first catalyst was stable during 50 h, achieving constant removals of TOC and phenol >90 and >99%, respectively. In the period of 50 to 75 h, the efficiency of CWPO dropped to ~65% of

TOC removal and remained constant for higher reaction times, corresponding to the performance reached with pristine TiO$_2$. The authors associated the deactivation of the catalyst to the gold particles growth (as the size increased from 3–4 nm in the fresh catalyst to 20–30 nm after CWPO). For the second catalyst (3.2 wt.% Au/TiO$_2$), the efficiency of the process, in terms of TOC removal, decreased from ~95% to ~80% during the first 50 h of reaction and remained practically constant for higher reaction times. This loss of catalytic activity was attributed to: i) an increase of the gold size from 7–8 nm to 13–15 nm for fresh and used catalysts after 300 h, respectively, and ii) adsorption of dicarboxylates in gold, with the equilibrium established at 50 h.

In the work developed by Domínguez et al. [72], the activated carbon supported gold deactivated in the first cycle of CWPO of phenol. The removal of the model compound observed by the authors in next three cycles was attributed to the activity of activated carbon alone (see Figure 14). Since the authors did not observe any gold leaching for the solution, the catalyst deactivation was attributed to the presence of dicarboxylic acids (by-products formed in the oxidation of phenol), which adsorb on gold nanoparticles. In order to recover the catalytic activity, the catalyst was regenerated after the first cycle by: i) alkaline washing to pH 14 with Na$_2$CO$_3$, in order to dissolve the absorbed species, and subsequent washing with distilled water until neutralization, and ii) oxidative thermal treatment at 200 °C, during 14 h, in air atmosphere, which allowed to burn-off the carboxylic acids adsorbed on the gold nanoparticles. The first regeneration process led to an activity recovery of about 60% and the catalyst loss was 2 wt.% of gold by leaching. For the second process, a more significant loss of gold (10 wt.%) was observed, however, the catalyst activity was fully restored (see Figure 14).

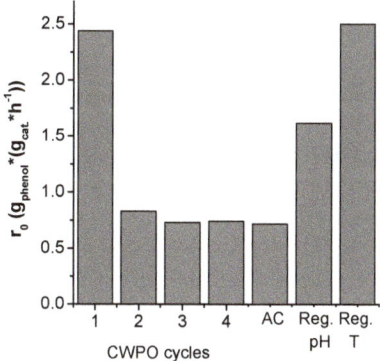

Figure 14. Initial rate of phenol degradation during four consecutives cycles of CWPO using gold supported on activated carbon and after regeneration ([phenol]$_{initial}$ = 5 g/L, [H$_2$O$_2$] = 25 g/L, pH = 3.5 and T = 80 °C). Adapted from Domínguez et al. [72].

In contrast to Ferentz et al. [71] and Domínguez et al. [72] that reported a loss of the catalytic activity of gold catalysts during CWPO, several studies in the literature refer to the stability of gold on different solid supports [66,67,70,73,108,110] when used in subsequent reutilization cycles. In the investigations performed by Rodrigues et al. [66,67,110], the gold supported on alumina, zinc oxide, titanium oxide, and iron oxide was stable during 3–5 consecutive cycles in acid medium (pH = 3.0), with OII and TOC removals and H$_2$O$_2$ consumption remaining unchanged during the cycles. The authors observed no catalyst loss of gold by leaching during the reactions.

Gold supported on activated carbon also did not deactivate in acidic pH (3.0) for degradation of bisphenol A by CWPO, with the removal of BPA and the consumption of oxidant being more or less constant, in ~80 and ~40%, respectively, during four consecutive cycles. This demonstrates that the catalyst can be reused several times [108]. The same tendency was observed by Han et al. [70], who evaluated the stability of gold supported on hydroxyapatite, showing that the conversion of phenol was constant after five cycles, either at pH 2.0 (>90%) or 5.0 (~80%) (see Figure 15). Similar results were obtained by Sempere et al. [75] that reused diamond supported gold (submitted to a thermal treatment at 420 °C and subsequent annealing treatment with hydrogen) three times in sunlight assisted CWPO in the oxidation of phenol. This catalyst did not lose the catalytic activity and the leaching of gold was negligible (< 1% of the initial gold in the first cycle of utilization) or was not observed (in the second and third cycles).

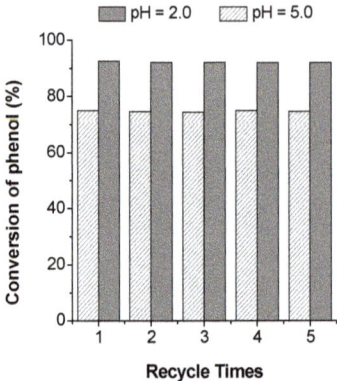

Figure 15. Conversion of phenol by CWPO using gold supported on hydroxyapatite as catalyst in consecutive five cycles of reutilization at pH = 2.0 and 5.0 ([phenol] = 100 mg/L, T = 70 °C, [catalyst] = 0.1 g/L and $V_{H2O2 \text{ with } 30 \text{ wt.\%}}$ = 1 mL). Adapted from Han et al. [70].

Martín et al. [73] evaluated the reuse of a diamond supported gold (Au/DNP) catalyst in phenol oxidation by CWPO during four cycles. The authors exhaustively washed the material with water at pH = 10 and, finally, with distilled water, in order to eliminate the deactivation of the catalyst by adsorption of carboxylic acids (intermediate products generated by phenol oxidation) on gold, as also pointed out by Domínguez et al. [72] as the main reason for deactivation. The Au/DNP, after a simple treatment by washing, can be reused during four times, reaching, in all cycles, not only total conversion of phenol and consumption of H_2O_2 in the end of reaction, but the same temporal profiles [73] (see Figure 16). However, the washing of the catalyst led to a loss of gold to the solution, but the leaching decreased with an increase of the cycles (from 3 to <0.1 wt.% after 1st use and 4th use, respectively), so that the performance of the catalyst was not affected by the small leaching and no deactivation was found [73].

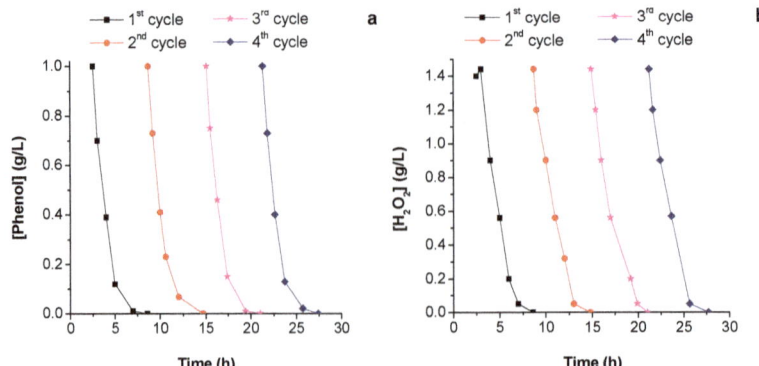

Figure 16. Evolution of phenol (**a**) and hydrogen peroxide (**b**) concentration during consecutive reuse cycles ([phenol] = 1 g/L, T = 50 °C, pH = 4.0, [Au/DNP] = 320 mg/L and [H_2O_2] = 1.44 g/L). Adapted from Martín et al. [73].

Navalon et al. [74] reused Au/HO-npD catalysts during three consecutive cycles, washing with water at pH 10 after each cycle. The authors did not observe any significant change in phenol and H_2O_2 disappearance profiles. Additionally, a run was carried out with a large excess of pollutant (40 g/L) and 0.5 mg/L of catalyst, in order to have an equivalent of 400 consecutive reuse cycles and use 5.5 equivalent of oxidant with respect to phenol. This run allowed to conclude that Au/HO-nDP had the ability to remove 36% of the pollutant before being deactivated, however, an alkaline washing before the second and third cycles allowed the catalyst to recover the catalytic activity.

6. Conclusions

This review showed that gold catalysts can be efficiently used in CWPO processes and that their catalytic activity depends on several operational variables of the process (such as pH, temperature, oxidant and catalyst concentrations, and gold content), as well as on the properties of the catalysts. We believe that such gathered information will provide useful insights that might lead to a more rapid and effective optimization of catalytic wet peroxidation processes using these materials.

Despite the recent progresses in this area, it is still crucial that work continues to be done to better apply these catalysts in the treatment of real effluents by CWPO and to deepen the knowledge coming from the laboratory studies for the scale-up of the process. The main limiting step in the application of this technology might be the costs of the treatment, since gold has a high cost (compared to other more common materials); however, gold also shows advantages, namely, high stability, high efficiency, and absence of leaching into the solution, which might justify the investment.

Acknowledgments: This work was financially supported by projects UID/EQU/00511/2019 - Laboratory for Process Engineering, Environment, Biotechnology and Energy - LEPABE – and by Associate Laboratory LSRE-LCM – UID/EQU/50020/2019 – funded by national (Portuguese) funds through FCT/MCTES (PIDDAC). SACC is thankful to FCT for Investigador FCT program (IF/01381/2013/CP1160/CT0007), with financing from the European Social Fund and the Human Potential Operational Program (POCH).

Conflicts of Interest: The authors declare no conflict of interest.

Abbreviations

AO7	Acid Orange dye
AOPs	Advanced Oxidation Processes
Au/AC	Gold on activated carbon
Au/C	Gold on carbon
Au/CNF	Gold on carbon nanofibers
Au/CNT	Gold on carbon nanotubes
Au/X40s	Gold on coconut shell carbon
Au/DNP	Gold on diamond nanoparticles
Au/F	Gold on diamond after thermal treatment at 420 °C in air atmosphere
Au/FH$_2$	Gold on diamond after thermal treatment at 420 °C in air atmosphere and at 500 °C in hydrogen atmosphere
Au/FN$_2$	Gold on diamond after thermal treatment at 420 °C in air atmosphere and at 500 °C in nitrogen atmosphere
Au/Hap	Gold on hydroxyapatite
Au/npD	Gold on nano power diamond
Au/HO-npD	Gold on nano power diamond previously treated with Fenton reagent
Au/FDU-15	Gold on ordered mesoporous carbon
Au/PSAC	Gold on pitch-based spherical activated carbon
Au/SRAC	Gold on styrene-based activated carbon
Au/TN	Gold on titanium nanotubes functionalization with hydrogen peroxide
Au/TiO2-AD	Gold on titanium oxide prepared by adsorption method
BOD$_5$	Biological oxygen demand after 5 days
BPA	Bisphenol A
CWPO	Catalytic Wet Peroxidation

COD	Chemical oxygen demand
DM	Gold metal dispersion
DPPH	1,1-diphenyl-2-picrylhydzazyl
EU-WFD	European Union Water Framework Directive
HR-TEM	High-resolution transmission electron microscopy
$h\nu$	Ultraviolet radiation
M	Transmission of metallic cations
MB	Methyl Blue dye
MO	Methyl Orange dye
NHE	Normal hydrogen electrode
OII	Orange II dye
RH	Organic matter
TOC	Total organic carbon
TOF	Turn off frequency
WGC	World Gold Council
X	Support

References

1. Zeng, G.-M.; Li, X.; Huang, J.-H.; Zhang, C.; Zhou, C.-F.; Niu, J.; Shi, L.-J.; He, S.-B.; Li, F. Micellar-enhanced ultrafiltration of cadmium and methylene blue in synthetic wastewater using sds. *J. Hazard. Mater.* **2011**, *185*, 1304–1310. [CrossRef]
2. Cundy, A.B.; Hopkinson, L.; Whitby, R.L.D. Use of iron-based technologies in contaminated land and groundwater remediation: A review. *Sci. Total Environ.* **2008**, *400*, 42–51. [CrossRef] [PubMed]
3. Chong, M.N.; Jin, B.; Chow, C.W.K.; Saint, C. Recent developments in photocatalytic water treatment technology: A review. *Water Res.* **2010**, *44*, 2997–3027. [CrossRef] [PubMed]
4. O'Connor, G.A. Organic compounds in sludge-amended soils and their potential for uptake by crop plants. *Sci. Total Environ.* **1996**, *185*, 71–81. [CrossRef]
5. Li, X.; Zeng, G.-M.; Huang, J.-H.; Zhang, D.-M.; Shi, L.-J.; He, S.-B.; Ruan, M. Simultaneous removal of cadmium ions and phenol with meuf using sds and mixed surfactants. *Desalination* **2011**, *276*, 136–141. [CrossRef]
6. Fatta-Kassinos, D.; Kalavrouziotis, I.K.; Koukoulakis, P.H.; Vasquez, M.I. The risks associated with wastewater reuse and xenobiotics in the agroecological environment. *Sci. Total Environ.* **2011**, *409*, 3555–3563. [CrossRef] [PubMed]
7. Xu, P.; Zeng, G.M.; Huang, D.L.; Feng, C.L.; Hu, S.; Zhao, M.H.; Lai, C.; Wei, Z.; Huang, C.; Xie, G.X.; et al. Use of iron oxide nanomaterials in wastewater treatment: A review. *Sci. Total Environ.* **2012**, *424*, 1–10. [CrossRef]
8. European Parliament & Council. *Water Framework Directive 2000/60/ce*; European Parliament & Counci: Brussels, Belgium, 2000; pp. 1–73.
9. Oturan, M.A.; Aaron, J.-J. Advanced oxidation processes in water/wastewater treatment: Principles and applications. A review. *Crit. Rev. Environ. Sci. Technol.* **2014**, *44*, 2577–2641. [CrossRef]
10. Bokare, A.D.; Choi, W. Review of iron-free Fenton-like systems for activating H_2O_2 in advanced oxidation processes. *J. Hazard. Mater.* **2014**, *275*, 121–135. [CrossRef]
11. Gogate, P.R.; Pandit, A.B. A review of imperative technologies for wastewater treatment I: Oxidation technologies at ambient conditions. *Adv. Environ. Res.* **2004**, *8*, 501–551. [CrossRef]
12. Seow, T.W.; Lim, C.K.; Norb, M.H.M.; Mubarak, M.F.M.; Lam, C.Y.L.; Yahya, A.; Ibrahim, Z. Review on wastewater treatment technologies. *Int. J. Appl. Environ. Sci.* **2016**, *11*, 111–126.
13. Ramalho, R.S. *Introduction to Wastewater Treatment Processes*; Academic Press: New York, NY, USA, 1977.
14. Pant, D.; Adholeya, A. Biological approaches for treatment of distillery wastewater: A review. *Bioresour. Technol.* **2007**, *98*, 2321–2334. [CrossRef]
15. Demirel, B.; Yenigun, O.; Onay, T.T. Anaerobic treatment of dairy wastewaters: A review. *Process Biochem.* **2005**, *40*, 2583–2595. [CrossRef]
16. Wolfe, S.; Ingold, C.F. Oxidation of organic compounds by zinc permanganate. *J. Am. Chem. Soc.* **1983**, *105*, 7755–7757. [CrossRef]

17. Xu, X.-R.; Li, H.-B.; Wang, W.-H.; Gu, J.-D. Decolorization of dyes and textile wastewater by potassium permanganate. *Chemosphere* **2005**, *59*, 893–898. [CrossRef]
18. Calvosa, L.; Monteverdi, A.; Rindone, B.; Riva, G. Ozone oxidation of compounds resistant to biological degradation. *Water Res.* **1991**, *25*, 985–993. [CrossRef]
19. Stasinakis, A.S. Use of selected advanced oxidation processes (aops) for wastewater treatment—A mini review. *Glob. NEST J.* **2008**, *10*, 376–385.
20. Azbar, N.; Yonar, T.; Kestioglu, K. Comparison of various advanced oxidation processes and chemical treatment methods for COD and color removal from a polyester and acetate fiber dyeing effluent. *Chemosphere* **2004**, *55*, 35–43. [CrossRef]
21. Lamarche, P.; Droste, R.L. Air-stripping mass transfer correlations for volatile organics. *J. Am. Water Works Assoc.* **1989**, *81*, 78–89. [CrossRef]
22. Busca, G.; Berardinelli, S.; Resini, C.; Arrighi, L. Technologies for the removal of phenol from fluid streams: A short review of recent developments. *J. Hazard. Mater.* **2008**, *160*, 265–288. [CrossRef] [PubMed]
23. Andreozzi, R.; Caprio, V.; Insola, A.; Marotta, R. Advanced oxidation processes (AOP) for water purification and recovery. *Catal. Today* **1999**, *53*, 51–59. [CrossRef]
24. Poyatos, J.M.; Muñio, M.M.; Almecija, M.C.; Torres, J.C.; Hontoria, E.; Osorio, F. Advanced oxidation processes for wastewater treatment: State of the art. *Water Air Soil Pollut.* **2010**, *205*, 187–204. [CrossRef]
25. Skoumal, M.; Cabot, P.-L.; Centellas, F.; Arias, C.; Rodríguez, R.M.; Garrido, J.A.; Brillas, E. Mineralization of paracetamol by ozonation catalyzed with Fe^{2+}, Cu^{2+} and UVA light. *Appl. Catal. B Environ.* **2006**, *66*, 228–240. [CrossRef]
26. Rosenfeldt, E.J.; Chen, P.J.; Kullman, S.; Linden, K.G. Destruction of estrogenic activity in water using UV advanced oxidation. *Sci. Total Environ.* **2007**, *377*, 105–113. [CrossRef]
27. Haber, F.; Weiss, J. The catalytic decomposition of hydrogen peroxide by iron salts. *Proc. R. Soc. Lond. Ser. A Math. Phys. Sci.* **1934**, *147*, 332–351.
28. Mahamuni, N.N.; Adewuyi, Y.G. Advanced oxidation processes (aops) involving ultrasound for waste water treatment: A review with emphasis on cost estimation. *Ultrason. Sonochem.* **2010**, *17*, 990–1003. [CrossRef]
29. Herney-Ramirez, J.; Vicente, M.A.; Madeira, L.M. Heterogeneous photo-Fenton oxidation with pillared clay-based catalysts for wastewater treatment: A review. *Appl. Catal. B Environ.* **2010**, *98*, 10–26. [CrossRef]
30. Esteves, B.M.; Rodrigues, C.S.D.; Madeira, L.M. Wastewater treatment by heterogeneous Fenton-like processes in continuous reactors. In *Applications of Advanced Oxidation Processes (AOPs) in Drinking Water Treatment*; Gil, A., Galeano, L.A., Vicente, M.Á., Eds.; Springer International Publishing: Cham, Switzerland, 2019; pp. 211–255.
31. Pawłat, J.; Stryczewska Henryka, D.; Ebihara, K. Sterilization techniques for soil remediation and agriculture based on ozone and AOP. *J. Adv. Oxid. Technol.* **2010**, *13*, 138–145. [CrossRef]
32. Flotron, V.; Delteil, C.; Padellec, Y.; Camel, V. Removal of sorbed polycyclic aromatic hydrocarbons from soil, sludge and sediment samples using the Fenton's reagent process. *Chemosphere* **2005**, *59*, 1427–1437. [CrossRef]
33. Tokumura, M.; Nakajima, R.; Znad, H.T.; Kawase, Y. Chemical absorption process for degradation of voc gas using heterogeneous gas–liquid photocatalytic oxidation: Toluene degradation by photo-Fenton reaction. *Chemosphere* **2008**, *73*, 768–775. [CrossRef]
34. Liu, G.; Ji, J.; Huang, H.; Xie, R.; Feng, Q.; Shu, Y.; Zhan, Y.; Fang, R.; He, M.; Liu, S.; et al. UV/H_2O_2: An efficient aqueous advanced oxidation process for VOCs removal. *Chem. Eng. J.* **2017**, *324*, 44–50. [CrossRef]
35. Domeño, C.; Rodríguez-Lafuente, Á.; Martos, J.; Bilbao, R.; Nerín, C. VOC removal and deodorization of effluent gases from an industrial plant by photo-oxidation, chemical oxidation, and ozonization. *Environ. Sci. Technol.* **2010**, *44*, 2585–2591. [CrossRef]
36. Tokumura, M.; Shibusawa, M.; Kawase, Y. Dynamic simulation of degradation of toluene in waste gas by the photo-Fenton reaction in a bubble column. *Chem. Eng. Sci.* **2013**, *100*, 212–224. [CrossRef]
37. Toor, R.; Mohseni, M. $UV-H_2O_2$ based AOP and its integration with biological activated carbon treatment for dbp reduction in drinking water. *Chemosphere* **2007**, *66*, 2087–2095. [CrossRef] [PubMed]
38. Shannon, M.A.; Bohn, P.W.; Elimelech, M.; Georgiadis, J.G.; Mariñas, B.J.; Mayes, A.M. Science and technology for water purification in the coming decades. *Nature* **2008**, *452*, 301–310. [CrossRef] [PubMed]

39. Comninellis, C.; Kapalka, A.; Malato, S.; Parsons, S.A.; Mantzavinos, I.P.D. Advanced oxidation processes for water treatment: Advances and trends for r&d. *J. Chem. Technol. Biotechnol.* **2008**, *83*, 769–776.
40. Al Momani, F.A. Potential use of solar energy for waste activated sludge treatment. *Int. J. Sustain. Eng.* **2013**, *6*, 82–91. [CrossRef]
41. Krzemieniewski, M.; Dębowski, M.; Janczukowicz, W.; Pesta, J. Effect of sludge conditioning by chemical methods with magnetic field application. *Pol. J. Environ. Stud.* **2003**, *12*, 595–605.
42. Legrini, O.; Oliveros, E.; Braun, A.M. Photochemical processes for water treatment. *Chem. Rev.* **1993**, *93*, 671–698. [CrossRef]
43. Brigda, R.J. Consider Fenton's chemistry for wastewater treatment. *Chem. Eng. Process.* **1995**, *91*, 62–66.
44. Ikehata, K.; Jodeiri Naghashkar, N.; Gamal El-Din, M. Degradation of aqueous pharmaceuticals by ozonation and advanced oxidation processes: A review. *Ozone Sci. Eng.* **2006**, *28*, 353–414. [CrossRef]
45. Rice, R.G.; Netzer, A. *Handbook of Ozone Technology and Applications*; Ann Arbor Science Publishers: Butterworths, UK, 1982; Volume 1.
46. Rodrigues, C.S.D.; Neto, A.R.; Duda, R.M.; de Oliveira, R.A.; Boaventura, R.A.R.; Madeira, L.M. Combination of chemical coagulation, photo-Fenton oxidation and biodegradation for the treatment of vinasse from sugar cane ethanol distillery. *J. Clean. Prod.* **2017**, *142*, 3634–3644. [CrossRef]
47. Inchaurrondo, N.S.; Massa, P.; Fenoglio, R.; Font, J.; Haure, P. Efficient catalytic wet peroxide oxidation of phenol at moderate temperature using a high-load supported copper catalyst. *Chem. Eng. J.* **2012**, *198*, 426–434. [CrossRef]
48. Maciel, R.; Sant'Anna, G.L.; Dezotti, M. Phenol removal from high salinity effluents using Fenton's reagent and photo-Fenton reactions. *Chemosphere* **2004**, *57*, 711–719. [CrossRef] [PubMed]
49. Fenton, H.J.H. Oxidation of tartaric acid in presence of iron. *J. Chem. Soc. Trans.* **1894**, *65*, 899–910. [CrossRef]
50. Walling, C. Fenton's reagent revisited. *Acc. Chem. Res.* **1975**, *8*, 125–131. [CrossRef]
51. Gosu, V.; Dhakar, A.; Sikarwar, P.; Kumar, U.K.A.; Subbaramaiah, V.; Zhang, T.C. Wet peroxidation of resorcinol catalyzed by copper impregnated granular activated carbon. *J. Environ. Manag.* **2018**, *223*, 825–833. [CrossRef]
52. Catrinescu, C.; Teodosiu, C.; Macoveanu, M.; Miehe-Brendlé, J.; Le Dred, R. Catalytic wet peroxide oxidation of phenol over fe-exchanged pillared beidellite. *Water Res.* **2003**, *37*, 1154–1160. [CrossRef]
53. Neyens, E.; Baeyens, J. A review of classic fenton's peroxidation as an advanced oxidation technique. *J. Hazard. Mater.* **2003**, *98*, 33–50. [CrossRef]
54. Ribeiro, R.S.; Silva, A.M.T.; Figueiredo, J.L.; Faria, J.L.; Gomes, H.T. Catalytic wet peroxide oxidation: A route towards the application of hybrid magnetic carbon nanocomposites for the degradation of organic pollutants. A review. *Appl. Catal. B Environ.* **2016**, *187*, 428–460. [CrossRef]
55. Perathoner, S.; Centi, G. Wet hydrogen peroxide catalytic oxidation (WHPCO) of organic waste in agro-food and industrial streams. *Top. Catal.* **2005**, *33*, 207–224. [CrossRef]
56. Melero, J.A.; Martínez, F.; Botas, J.A.; Molina, R.; Pariente, M.I. Heterogeneous catalytic wet peroxide oxidation systems for the treatment of an industrial pharmaceutical wastewater. *Water Res.* **2009**, *43*, 4010–4018. [CrossRef] [PubMed]
57. European Economic Community. *List of Council Directives 76/4647*; European Economic Community: Brussels, Belgium, 1982.
58. Feng, J.; Hu, X.; Yue, P.L. Effect of initial solution ph on the degradation of orange II using clay-based Fe nanocomposites as heterogeneous photo-Fenton catalyst. *Water Res.* **2006**, *40*, 641–646. [CrossRef] [PubMed]
59. Hartmann, M.; Kullmann, S.; Keller, H. Wastewater treatment with heterogeneous Fenton-type catalysts based on porous materials. *J. Mater. Chem.* **2010**, *20*, 9002–9017. [CrossRef]
60. Dantas, T.L.P.; Mendonça, V.P.; José, H.J.; Rodrigues, A.E.; Moreira, R.F.P.M. Treatment of textile wastewater by heterogeneous Fenton process using a new composite Fe_2O_3/carbon. *Chem. Eng. J.* **2006**, *118*, 77–82. [CrossRef]
61. Liou, R.-M.; Chen, S.-H.; Hung, M.-Y.; Hsu, C.-S.; Lai, J.-Y. Fe (III) supported on resin as effective catalyst for the heterogeneous oxidation of phenol in aqueous solution. *Chemosphere* **2005**, *59*, 117–125. [CrossRef]
62. Wang, Y.; Zhao, H.; Zhao, G. Iron-copper bimetallic nanoparticles embedded within ordered mesoporous carbon as effective and stable heterogeneous Fenton catalyst for the degradation of organic contaminants. *Appl. Catal. B Environ.* **2015**, *164*, 396–406. [CrossRef]

63. Subbaramaiah, V.; Srivastava, V.C.; Mall, I.D. Catalytic wet peroxidation of pyridine bearing wastewater by cerium supported SBA-15. *J. Hazard. Mater.* **2013**, *248–249*, 355–363. [CrossRef]
64. Aravindhan, R.; Fathima, N.N.; Rao, J.R.; Nair, B.U. Wet oxidation of acid brown dye by hydrogen peroxide using heterogeneous catalyst Mn-Salen-Y zeolite: A potential catalyst. *J. Hazard. Mater.* **2006**, *138*, 152–159. [CrossRef]
65. Hosseini, S.A.; Davodian, M.; Abbasian, A.R. Remediation of phenol and phenolic derivatives by catalytic wet peroxide oxidation over Co-Ni layered double nano hydroxides. *J. Taiwan Inst. Chem. Eng.* **2017**, *75*, 97–104. [CrossRef]
66. Rodrigues, C.S.D.; Carabineiro, S.A.C.; Maldonado-Hódar, F.J.; Madeira, L.M. Wet peroxide oxidation of dye-containing wastewaters using nanosized Au supported on Al_2O_3. *Catal. Today* **2017**, *280*, 165–175. [CrossRef]
67. Rodrigues, C.S.D.; Carabineiro, S.A.C.; Maldonado-Hódar, F.J.; Madeira, L.M. Orange II degradation by wet peroxide oxidation using Au nanosized catalysts: Effect of the support. *Ind. Eng. Chem. Res.* **2017**, *56*, 1988–1998. [CrossRef]
68. Quintanilla, A.; García-Rodríguez, S.; Domínguez, C.M.; Blasco, S.; Casas, J.A.; Rodriguez, J.J. Supported gold nanoparticle catalysts for wet peroxide oxidation. *Appl. Catal. B Environ.* **2012**, *111–112*, 81–89. [CrossRef]
69. Hassan, H.; Hameed, B.H. Fe–clay as effective heterogeneous Fenton catalyst for the decolorization of reactive blue 4. *Chem. Eng. J.* **2011**, *171*, 912–918. [CrossRef]
70. Han, Y.-F.; Phonthammachai, N.; Ramesh, K.; Zhong, Z.; White, T. Removing organic compounds from aqueous medium via wet peroxidation by gold catalysts. *Environ. Sci. Technol.* **2008**, *42*, 908–912. [CrossRef]
71. Ferentz, M.; Landau, M.V.; Vidruk, R.; Herskowitz, M. Fixed-bed catalytic wet peroxide oxidation of phenol with titania and Au/titania catalysts in dark. *Catal. Today* **2015**, *241*, 63–72. [CrossRef]
72. Domínguez, C.M.; Quintanilla, A.; Casas, J.A.; Rodriguez, J.J. Kinetics of wet peroxide oxidation of phenol with a gold/activated carbon catalyst. *Chem. Eng. J.* **2014**, *253*, 486–492. [CrossRef]
73. Martín, R.; Navalon, S.; Alvaro, M.; Garcia, H. Optimized water treatment by combining catalytic Fenton reaction using diamond supported gold and biological degradation. *Appl. Catal. B Environ.* **2011**, *103*, 246–252. [CrossRef]
74. Navalon, S.; Martín, R.; Alvaro, M.; Garcia, H. Gold on diamond nanoparticles as a highly efficient Fenton catalyst. *Angew. Chem.* **2010**, *122*, 8581–8585. [CrossRef]
75. Sempere, D.; Navalon, S.; Dančíková, M.; Alvaro, M.; Garcia, H. Influence of pretreatments on commercial diamond nanoparticles on the photocatalytic activity of supported gold nanoparticles under natural sunlight irradiation. *Appl. Catal. B Environ.* **2013**, *142–143*, 259–267. [CrossRef]
76. Brust, M.; Walker, M.; Bethell, D.; Schiffrin, D.J.; Whyman, R. Synthesis of thiol-derivatised gold nanoparticles in a two-phase liquid-liquid system. *J. Chem. Soc. Chem. Commun.* **1994**, *7*, 801–802. [CrossRef]
77. Primo, A.; García, H. Chapter 18—Supported gold nanoparticles as heterogeneous catalysts. In *New and Future Developments in Catalysis*; Suib, S.L., Ed.; Elsevier: Amsterdam, The Netherlands, 2013; pp. 425–449.
78. Jiang, G.; Wang, L.; Chen, T.; Yu, H.; Chen, C. Preparation of gold nanoparticles in the presence of poly(benzyl ether) alcohol dendrons. *Mater. Chem. Phys.* **2006**, *98*, 76–82. [CrossRef]
79. Turkevich, J.; Stevenson, P.C.; Hillier, J. A study of the nucleation and growth processes in the synthesis of colloidal gold. *Discuss. Faraday Soc.* **1951**, *11*, 55–75. [CrossRef]
80. Carabineiro, S.A.C.; Thompson, D.T. Catalytic applications for gold nanotechnology. In *Nanocatalysis*; Heiz, U., Landman, U., Eds.; Springer: Berlin/Heidelberg, Germany, 2007; pp. 377–489.
81. Haruta, M. Size- and support-dependency in the catalysis of gold. *Catal. Today* **1997**, *36*, 153–166. [CrossRef]
82. Haruta, M. Gold as a novel catalyst in the 21st century: Preparation, working mechanism and applications. *Gold Bull.* **2004**, *37*, 27–36. [CrossRef]
83. Bond, G.C.; Thompson, D.T. Catalysis by gold. *Catal. Rev.* **1999**, *41*, 319–388. [CrossRef]
84. Hodge, N.A.; Kiely, C.J.; Whyman, R.; Siddiqui, M.R.H.; Hutchings, G.J.; Pankhurst, Q.A.; Wagner, F.E.; Rajaram, R.R.; Golunski, S.E. Microstructural comparison of calcined and uncalcined gold/iron-oxide catalysts for low-temperature CO oxidation. *Catal. Today* **2002**, *72*, 133–144. [CrossRef]
85. Haruta, M.; Yamada, N.; Kobayashi, T.; Iijima, S. Gold catalysts prepared by coprecipitation for low-temperature oxidation of hydrogen and of carbon monoxide. *J. Catal.* **1989**, *115*, 301–309. [CrossRef]
86. Abad, A.; Almela, C.; Corma, A.; García, H. Efficient chemoselective alcohol oxidation using oxygen as oxidant. Superior performance of gold over palladium catalysts. *Tetrahedron* **2006**, *62*, 6666–6672. [CrossRef]

87. Baatz, C.; Decker, N.; Prüße, U. New innovative gold catalysts prepared by an improved incipient wetness method. *J. Catal.* **2008**, *258*, 165–169. [CrossRef]
88. Lorençon, E.; Ferreira, D.C.; Resende, R.R.; Krambrock, K. Amphiphilic gold nanoparticles supported on carbon nanotubes: Catalysts for the oxidation of lipophilic compounds by wet peroxide in biphasic systems. *Appl. Catal. A Gen.* **2015**, *505*, 566–574. [CrossRef]
89. Lin, J.-N.; Wan, B.-Z. Effects of preparation conditions on gold/Y-type zeolite for CO oxidation. *Appl. Catal. B Environ.* **2003**, *41*, 83–95. [CrossRef]
90. Haruta, M.; Tsubota, S.; Kobayashi, T.; Kageyama, H.; Genet, M.J.; Delmon, B. Low-temperature oxidation of CO over gold supported on TiO_2, α-Fe_2O_3, and Co_3O_4. *J. Catal.* **1993**, *144*, 175–192. [CrossRef]
91. Herzing, A.A.; Kiely, C.J.; Carley, A.F.; Landon, P.; Hutchings, G.J. Identification of active gold nanoclusters on iron oxide supports for CO oxidation. *Science* **2008**, *321*, 1331–1335. [CrossRef]
92. Corma, A.; Serna, P. Chemoselective hydrogenation of nitro compounds with supported gold catalysts. *Science* **2006**, *313*, 332–334. [CrossRef] [PubMed]
93. Fu, Q.; Saltsburg, H.; Flytzani-Stephanopoulos, M. Active nonmetallic au and pt species on ceria-based water-gas shift catalysts. *Science* **2003**, *301*, 935–938. [CrossRef] [PubMed]
94. Rodriguez, J.A.; Ma, S.; Liu, P.; Hrbek, J.; Evans, J.; Pérez, M. Activity of CeO_x and TiO_x nanoparticles grown on Au(111) in the water-gas shift reaction. *Science* **2007**, *318*, 1757–1760. [CrossRef]
95. Pérez, P.; Soria, M.A.; Carabineiro, S.A.C.; Maldonado-Hódar, F.J.; Mendes, A.; Madeira, L.M. Application of Au/TiO_2 catalysts in the low-temperature water-gas shift reaction. *Int. J. Hydrog. Energy* **2016**, *41*, 4670–4681. [CrossRef]
96. Scirè, S.; Minicò, S.; Crisafulli, C.; Satriano, C.; Pistone, A. Catalytic combustion of volatile organic compounds on gold/cerium oxide catalysts. *Appl. Catal. B Environ.* **2003**, *40*, 43–49. [CrossRef]
97. Centeno, M.A.; Paulis, M.; Montes, M.; Odriozola, J.A. Catalytic combustion of volatile organic compounds on $Au/CeO_2/Al_2O_3$ and Au/Al_2O_3 catalysts. *Appl. Catal. A Gen.* **2002**, *234*, 65–78. [CrossRef]
98. Scirè, S.; Liotta, L.F. Supported gold catalysts for the total oxidation of volatile organic compounds. *Appl. Catal. B Environ.* **2012**, *125*, 222–246. [CrossRef]
99. Chang, Y.-C.; Chen, D.-H. Catalytic reduction of 4-nitrophenol by magnetically recoverable Au nanocatalyst. *J. Hazard. Mater.* **2009**, *165*, 664–669. [CrossRef]
100. Aprile, C.; Corma, A.; Domine, M.E.; Garcia, H.; Mitchell, C. A cascade aerobic epoxidation of alkenes over Au/CeO_2 and ti-mesoporous material by "in situ" formed peroxides. *J. Catal.* **2009**, *264*, 44–53. [CrossRef]
101. Cojocaru, B.; Neaţu, Ş.; Sacaliuc-Pârvulescu, E.; Lévy, F.; Pârvulescu, V.I.; Garcia, H. Influence of gold particle size on the photocatalytic activity for acetone oxidation of Au/TiO_2 catalysts prepared by dc-magnetron sputtering. *Appl. Catal. B Environ.* **2011**, *107*, 140–149. [CrossRef]
102. Marino, T.; Molinari, R.; García, H. Selectivity of gold nanoparticles on the photocatalytic activity of TiO_2 for the hydroxylation of benzene by water. *Catal. Today* **2013**, *206*, 40–45. [CrossRef]
103. Martínez, F.; Calleja, G.; Melero, J.A.; Molina, R. Heterogeneous photo-Fenton degradation of phenolic aqueous solutions over iron-containing SBA-15 catalyst. *Appl. Catal. B Environ.* **2005**, *60*, 181–190. [CrossRef]
104. Kuznetsova, E.V.; Savinov, E.N.; Vostrikova, L.A.; Parmon, V.N. Heterogeneous catalysis in the Fenton-type system FeZSM-5/H_2O_2. *Appl. Catal. B Environ.* **2004**, *51*, 165–170. [CrossRef]
105. Carabineiro, S.A.C.; Machado, B.F.; Bacsa, R.R.; Serp, P.; Dražić, G.; Faria, J.L.; Figueiredo, J.L. Catalytic performance of Au/ZnO nanocatalysts for CO oxidation. *J. Catal.* **2010**, *273*, 191–198. [CrossRef]
106. Ge, L.; Chen, T.; Liu, Z.; Chen, F. The effect of gold loading on the catalytic oxidation performance of CeO_2/H_2O_2 system. *Catal. Today* **2014**, *224*, 209–215. [CrossRef]
107. Drašinac, N.; Erjavec, B.; Dražić, G.; Pintar, A. Peroxo and gold modified titanium nanotubes for effective removal of methyl orange with CWPO under ambient conditions. *Catal. Today* **2017**, *280*, 155–164. [CrossRef]
108. Yang, X.; Tian, P.-F.; Zhang, C.; Deng, Y.-Q.; Xu, J.; Gong, J.; Han, Y.-F. Au/carbon as Fenton-like catalysts for the oxidative degradation of bisphenol A. *Appl. Catal. B Environ.* **2013**, *134–135*, 145–152. [CrossRef]
109. Alvaro, M.; Cojocaru, B.; Ismail, A.A.; Petrea, N.; Ferrer, B.; Harraz, F.A.; Parvulescu, V.I.; Garcia, H. Visible-light photocatalytic activity of gold nanoparticles supported on template-synthesized mesoporous titania for the decontamination of the chemical warfare agent soman. *Appl. Catal. B Environ.* **2010**, *99*, 191–197. [CrossRef]

110. Rodrigues, C.S.D.; Silva, R.M.; Carabineiro, S.A.C.; Maldonado-Hódar, F.J.; Madeira, L.M. Dye-containing wastewater treatment by photo-assisted wet peroxidation using Au nanosized catalysts. *J. Chem. Technol. Biotechnol.* **2018**, *93*, 3223–3323. [CrossRef]
111. Navalon, S.; Martin, R.; Alvaro, M.; Garcia, H. Sunlight-assisted Fenton reaction catalyzed by goldsupported on diamond nanoparticles as pretreatment forbiological degradation of aqueous phenol solutions. *ChemSusChem* **2011**, *4*, 650–657. [CrossRef] [PubMed]
112. Galindo, C.; Jacques, P.; Kalt, A. Photochemical and photocatalytic degradation of an indigoid dye: A case study of acid blue 74 (AB74). *J. Photochem. Photobiol. A Chem.* **2001**, *141*, 47–56. [CrossRef]
113. Spinks, J.W.T.; Woods, R.J. *An Introduction to Radiation Chemistry*, 3rd ed.; John Wiley & Sons Inc.: New York, NY, USA, 1990.
114. Fida, H.; Zhang, G.; Guo, S.; Naeem, A. Heterogeneous Fenton degradation of organic dyes in batch and fixed bed using la-fe montmorillonite as catalyst. *J. Colloid Interface Sci.* **2017**, *490*, 859–868. [CrossRef]
115. Navalon, S.; de Miguel, M.; Martin, R.; Alvaro, M.; Garcia, H. Enhancement of the catalytic activity of supported gold nanoparticles for the fenton reaction by light. *J. Am. Chem. Soc.* **2011**, *133*, 2218–2226. [CrossRef] [PubMed]
116. Zazo, J.A.; Pliego, G.; Blasco, S.; Casas, J.A.; Rodriguez, J.J. Intensification of the Fenton process by increasing the temperature. *Ind. Eng. Chem. Res.* **2011**, *50*, 866–870. [CrossRef]
117. Huang, C.P.; Dong, C.; Tang, Z. Advanced chemical oxidation: Its present role and potential future in hazardous waste treatment. *Waste Manag.* **1993**, *13*, 361–377. [CrossRef]

© 2019 by the authors. Licensee MDPI, Basel, Switzerland. This article is an open access article distributed under the terms and conditions of the Creative Commons Attribution (CC BY) license (http://creativecommons.org/licenses/by/4.0/).

Article

Condensation By-Products in Wet Peroxide Oxidation: Fouling or Catalytic Promotion? Part I. Evidences of an Autocatalytic Process

Asunción Quintanilla [1], Jose L. Diaz de Tuesta [2,3], Cristina Figueruelo [1], Macarena Munoz [1,*] and Jose A. Casas [1]

1. Chemical Engineering Department, Universidad Autónoma de Madrid, Ctra. Colmenar km 15, 28049 Madrid, Spain; asun.quintanilla@uam.es (A.Q.); cgfigueruelo@gmail.com (C.F.); jose.casas@uam.es (J.A.C.)
2. Centro de Investigação de Montanha (CIMO), Instituto Politécnico de Bragança, 5300-253 Bragança, Portugal; jl.diazdetuesta@ipb.pt
3. Laboratório de Processos de Separação e Reação - Laboratório de Catálise e Materiais (LSRE-LCM), Faculdade de Engenharia, Universidade do Porto, 4200-465 Porto, Portugal
* Correspondence: macarena.munnoz@uam.es; Tel.: 34-91-497-3991; Fax: +34-91497-3516

Received: 22 May 2019; Accepted: 6 June 2019; Published: 11 June 2019

Abstract: The present work is aimed at the understanding of the condensation by-products role in wet peroxide oxidation processes. This study has been carried out in absence of catalyst to isolate the (positive or negative) effect of the condensation by-products on the kinetics of the process, and in presence of oxygen, to enhance the oxidation performance. This process was denoted as oxygen-assisted wet peroxide oxidation (WPO-O_2) and was applied to the treatment of phenol. First, the influence of the reaction operating conditions (i.e., temperature, pH_0, initial phenol concentration, H_2O_2 dose and O_2 pressure) was evaluated. The initial phenol concentration and, overall, the H_2O_2 dose, were identified as the most critical variables for the formation of condensation by-products and thus, for the oxidation performance. Afterwards, a flow reactor packed with inert quartz beads was used to facilitate the deposition of such species and thus, to evaluate their impact on the kinetics of the process. It was found that as the quartz beads were covered by condensation by-products along reaction, the disappearance rates of phenol, total organic carbon (TOC) and H_2O_2 were increased. Consequently, an autocatalytic kinetic model, accounting for the catalytic role of the condensation by-products, provides a well description of wet peroxide oxidation performance.

Keywords: wet peroxide oxidation; wet air oxidation; condensation by-products; fouling; autocatalytic kinetics

1. Introduction

The decline of water quality is a global concern issue as it is crucial to the environment and human health, but also to our social and economic development. Over the last hundred years, the use of water grew by a factor of seven and it is expected to increase by around 55% by 2050, including a 400% rise in industry water demand [1]. This sector is not only one of the major water consumers but also the main group responsible for its pollution by a wide variety of hazardous substances, particularly persistent organic pollutants. These effluents must be treated prior to discharge, or involve recycling in the process. Biological treatment is usually not feasible for industrial wastewater treatment as they commonly contain highly toxic and non-biodegradable substances. Thermal treatments such as incineration can lead to the generation of even more hazardous substances, like dioxins or furans [2]. Adsorption is a non-destructive process and requires dealing with the resulting saturated adsorbent. Compared to these technologies, catalytic wet air oxidation (CWAO) represents an interesting alternative for industrial wastewater treatment. In fact, it is a well-established technology for such goals [3]. A number of commercial CWAO units such as LOPROX, Ciba-Geigy, ORCAN, WPO and ATHOS can be found [3]. This process involves the oxidation of the pollutants using a gaseous source of oxygen, commonly air, in the presence of a catalyst (mainly precious metals) under relatively severe conditions (150–300 °C, 10–200 bar) [3]. The use of H_2O_2 as an oxidant precursor instead of air, in the so-called catalytic wet peroxide oxidation (CWPO), is another well-known oxidation treatment that allows the process to operate under considerably milder operating conditions (25–120 °C, <10 bar), and thus, it represents a more environmentally-friendly alternative, although the high cost of hydrogen peroxide limits its application to highly polluted wastewaters.

The main limitation for a widespread application of CWPO at an industrial scale is related to the relatively fast deactivation of the Fe-bearing solid catalysts usually employed in this process [4–6]. Catalyst deactivation is a complex phenomenon but five main reasons are usually behind it: metal leaching, fouling, poisoning, thermal sintering and mechanical damage. Among them, metal leaching and fouling have been usually identified as the most important catalyst deactivation mechanisms in CWPO [4] while fouling is the major deactivation cause of CWAO catalysts [7]. Metal leaching is an irreversible process associated with the presence of organic acids in the reaction medium, refractory species commonly formed upon the oxidation of organic pollutants, which promote the complex formation and mobilization of iron [5]. Fe supported on Al_2O_3 or robust iron minerals appear as the most resistant catalysts to iron leaching [4,8,9] but, unfortunately, it cannot be completely avoided, especially with highly-polluted wastewaters, where organic acids will be formed in significant amounts. In this context, the use of metal-free catalysts has gained great importance in recent years. In particular, carbon materials like activated carbon, graphite, carbon black, carbon nanotubes and carbon xerogels have been postulated as promising CWAO/CWPO catalysts due to their outstanding donor-acceptor properties [10–13]. In any case, both metal-based and metal-free catalysts can suffer fouling, i.e., blocking of the catalytic active sites by carbonaceous deposits, mainly by condensation/polymerization products formed by oxidative coupling reactions.

The generation of condensation by-products along the oxidation of persistent organic pollutants by both CWPO and CWAO processes has been evidenced in the literature in a number of works [7,14–20]. Poerschmann et al. (2009) [15] deeply characterized for the first time the polymeric species generated along the wet peroxide oxidation of phenol with dissolved iron. They demonstrated the formation of different dimers such as biphenyls and diphenylethers, which ultimately resulted in the formation of a dark brown solid oligomer that remained in suspension in the aquatic environment. When a solid catalyst is used, those condensation by-products are adsorbed onto its surface, this phenomenon is usually described as fouling. Commonly, it is a reversible process and the catalyst can be regenerated by applying more efficient oxidation conditions, i.e., increasing H_2O_2 concentration or operating temperature; or by direct catalyst calcination at temperatures around 350 °C [10], threshold value assigned to the burn-off of these polymeric species, although it can also occur at lower temperatures if supported-metal catalysts are used [21]. The deposition of those condensation by-products has been usually regarded as a catalyst deactivation cause (coking) but recent studies have demonstrated that there are many cases like oxidative dehydrogenation, hydrogenation, isomerization and Fischer-Tropsch reactions where such deposits can even improve the catalytic performance [22]. In particular, it has been proved that coke can promote the activity of the catalyst [23], enhance the selectivity to the desired product [24] or even act as a new reaction site [25,26]. It is clear that a deep understanding of the mechanisms by which carbon deposits cause catalyst deactivation or promote catalytic activity is essential to optimize the efficiency of a given process. Nevertheless, the role of these species on CWPO and CWAO processes remains unclear in the literature. In most studies, the adsorption of such species on the catalyst surface was identified as the main reason to explain the loss of activity of the catalysts [10,21,27–30]. Nevertheless, Delgado et al. (2012) [20] found that the covering of the catalyst by carbonaceous deposits was actually the preliminary step for the CWAO of phenol. On the other hand, little is known about the nature of the carbonaceous deposits although it is widely accepted that they are formed through condensation reactions of aromatic radical species [20,30]. Only in a few works, catalyst regeneration by thermal oxidation treatments within the range 200–350 °C in air atmosphere was investigated [10,20,27,29].

Inspired by the recent review of Collet et al. (2016) [22], where the beneficial role of carbonaceous deposits in a number of catalytic processes has been summarized, this work aims to clarify the role of those condensation by-products in wet oxidation. As industrial wastewater treatment is the target of this study and given the wide commercial application of CWAO for such goal, the oxygen-assisted wet peroxide oxidation process (WPO-O_2) has been investigated in order to maximize the efficiency of the system while operating under mild conditions. Phenol has been chosen as the target pollutant given its toxicity and persistency as well as its regular use in research, which facilitates comparison purposes. A complete operating condition study was initially performed in the absence of catalyst to clarify the possible synergic effect between H_2O_2 and O_2 as well as to learn on the formation of condensation by-products in the homogeneous phase. Once selected the optimum operating conditions, a flow reactor packed with inert quartz beads was used in order to facilitate the deposition of those condensation by-products and thus, to evaluate their impact on the oxidation performance. On the basis of the obtained results, the reaction pathway for phenol oxidation was proposed and a complete kinetic model accounting for the activity promoted by the condensation by-products was developed to describe the experimental data (temporal concentration of phenol, total organic carbon (TOC) and H_2O_2).

2. Results and Discussion

2.1. Preliminary Study

The first set of experiments was focused on the comparison between wet air oxidation (WAO), wet peroxide oxidation (WPO) and oxygen-assisted wet peroxide oxidation (WPO-O_2) in the degradation of phenol ([Phenol]$_0$ = 1000 mg L^{-1}) under moderate operating conditions ([H_2O_2]$_0$ = 5000 mg L^{-1}, T = 127 °C, P_{O_2} = 8 bar, Q_{O_2} = 92 NmL min^{-1}, natural pH$_0$). Figure 1 shows the evolution of the target pollutant, TOC and H_2O_2 reactions, with the three treatment systems. It is clear that the presence of H_2O_2 played a key role in the oxidation process. In this sense, both WPO-O_2 and WPO systems allowed to reach the complete conversion of phenol in 10 min reaction time, whereas only 20% phenol removal was achieved upon WAO at the same reaction time. Accordingly, the initial phenol oxidation rates with the former processes were around 100 mg$_{phenol}$ L^{-1} min^{-1} and that of the latter was 60 mg$_{phenol}$ L^{-1} min^{-1}. The synergic effect between oxygen and hydrogen peroxide was more clearly evidenced in the mineralization of TOC. The mineralization yield, i.e., the percentage of TOC completely oxidized to CO_2, was calculated by the difference between the TOC measured in the final reaction samples and the initially present in the starting solution (1000 mg L^{-1} phenol). Up to 80% mineralization yield was reached with the WPO-O_2 process, whereas only 30% and 15% were achieved with WPO and WAO, respectively. The initial TOC abatement rates were 44, 15 and 4 mg$_{TOC}$ L^{-1} min^{-1} for WPO-O_2, WPO and WAO, respectively. On the other hand, the consumption of H_2O_2 along reaction was slightly slower in the presence of oxygen (WPO-O_2), giving rise to a higher efficiency on the use of this reagent (η) –calculated as the ratio between the conversion of TOC and that of H_2O_2 (82% and 32% for WPO and WPO-O_2, respectively).

The kind of oxidation treatment system used also showed a significant impact on the distribution of intermediates generated along the process. As can be seen in Figure 2, the WPO-O_2 process warranted the complete removal of the cyclic compounds, being the final oxidation products short-chain organic acids, which represented 18% of the initial TOC. The remaining 82% of the initial TOC was mineralized. The amount of CO_2 was calculated as the difference between the TOC of the target solution (1000 mg L^{-1} phenol) and the TOC of the final reaction effluent. Therefore, in this case, the carbon balance was closed at almost 100% after 3 h reaction time. On the contrary, WPO and WAO processes, which showed significantly lower mineralization yields (30% and 15%, respectively), led to a significant amount of aromatic intermediates (9% and 27%, respectively) and non-identified products (47% and 53%, respectively), calculated by the difference between the TOC measured and the calculated from the identified compounds, usually associated with condensation/polymerization compounds [14,15]. Consequently, the presence of short-chain organic acids was of very low relevance (9% and 7%, respectively). The evolution of the cyclic compounds (hydroquinone, resorcinol, catechol and benzoquinone) along reaction as well as the amount of short-chain organic acids (maleic, malonic, oxalic, acetic and formic) present in the final effluents are provided in Figures S1 and S2 of the Supplementary Material, respectively.

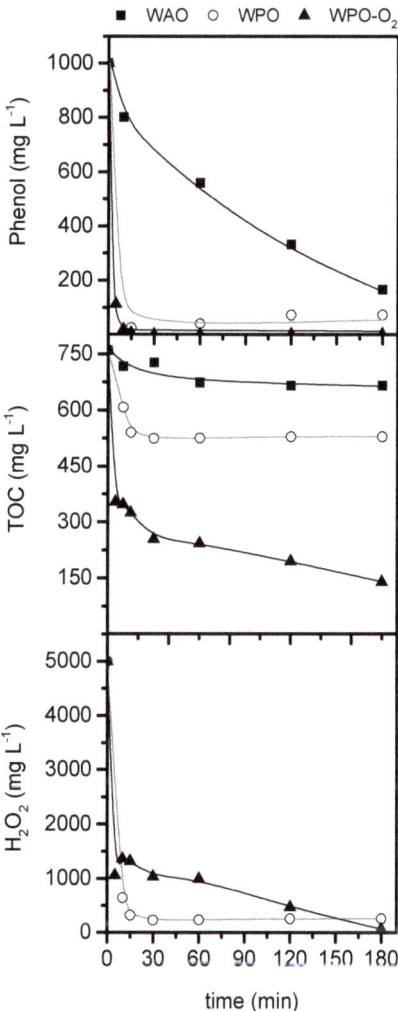

Figure 1. Temporal concentration profiles of H_2O_2, phenol and TOC upon the treatment of phenol by WAO, WPO and WPO-O_2 processes. Operating conditions: [Phenol]$_0$ = 1000 mg L^{-1}, [H_2O_2]$_0$ = 5000 mg L^{-1}, P$_{O2}$ = 8 bar (92 N mL$_{O2}$ min^{-1}), T = 127 °C and natural pH$_0$.

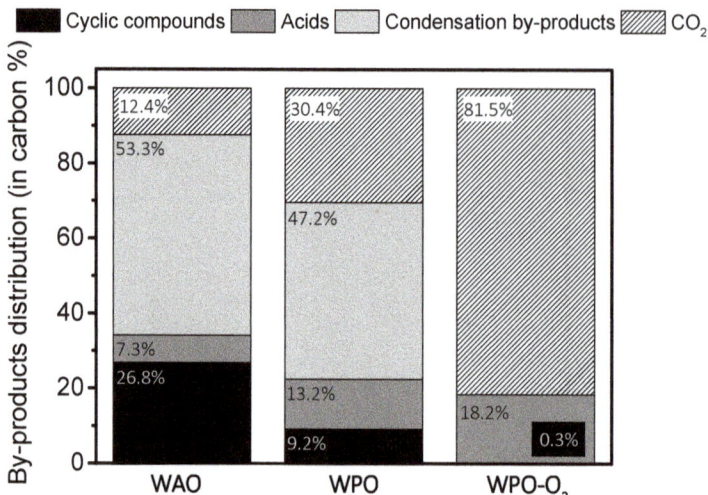

Figure 2. By-product distribution upon the WAO, WPO and WPO-O_2 of phenol after 3 h of reaction. Operating conditions: [Phenol]$_0$ = 1000 mg L^{-1}, [H$_2$O$_2$]$_0$ = 5000 mg L^{-1}, P$_{O2}$ = 8 bar (92 N mL$_{O2}$ min^{-1}), T = 127 °C and natural pH$_0$.

According to the obtained results, it is clear that the WPO-O_2 process operated under moderate conditions represents an interesting non-catalytic oxidation process in terms of degradation rate, mineralization yields, H_2O_2 consumption efficiency and products distribution. Therefore, it represents a suitable starting point to evaluate the role of condensation by-products on the oxidation performance.

2.2. Operating conditions study: Formation of condensation by-products

A complete operating condition study was performed, analyzing the effect of pH$_0$ (3–9), initial phenol concentration (1000–5000 mg L^{-1}), H_2O_2 dose (20–100% of the stoichiometric amount), O_2 pressure (5–10 bar) and operating temperature (100–140 °C). A summary of the obtained results in terms of TOC mineralization can be seen in Figure 3 (see Figures S3–S7 of the Supplementary Material for all experimental data). As observed, the major impact on TOC mineralization was related to the H_2O_2 dose, an essential variable to be optimized in the WPO-O_2 process given its significant weight on the economy of the process. On the other hand, the pressure of O_2 did not show a relevant influence on the oxidation performance, demonstrating the prominent role of hydroxyl radicals from H_2O_2 as oxidant. Nevertheless, it should be noted that mineralization yields higher than the expected by only H_2O_2-promoted oxidation were obtained with substoichiometric H_2O_2 doses, which confirms that oxygen acted as supplemental oxidant. A neutral pH$_0$ value led to a higher extension of the reaction. At higher pH$_0$, the decomposition of H_2O_2 follows another pathway and only a small fraction of H_2O_2 leads to oxidants able to degrade the pollutants [31,32]. In the same line, at acidic conditions, hydroxyl radicals interact with protons, leading to lower reactive species [33]. With regard to the temperature, it presented a significant effect on the oxidation rate (initial phenol abatement rates of 100, 150 and 180 mg$_{Phenol}$ L^{-1} min^{-1} were obtained at 100, 127 and 140 °C, respectively) but it did not increase remarkably the extension of the reaction, which depends on the amount of oxidant (see Figure S8 of the Supplementary Material for experimental data). Finally, the initial concentration of phenol also showed an important influence on the mineralization yield, which can be explained by the different efficiency on the consumption of H_2O_2 achieved with increasing phenol concentration. The presence of higher H_2O_2 amounts from the beginning of the reaction unavoidably led to the appearance of parasitic reactions, being especially important with an initial phenol concentration of 5000 mg L^{-1}. Furthermore, the presence of high amounts of phenol also led to the formation of high quantities of organic radicals, leading to the formation of higher concentrations of condensation products. In this case, the distribution of the H_2O_2 dose along reaction would be required to achieve higher mineralization yields. On the basis of these results, a neutral pH_0 value, an initial phenol concentration of 1000 mg L^{-1}, an O_2 pressure of 8 bar and 127 °C were the operating conditions selected for further experiments while the effect of H_2O_2 was evaluated in more detail.

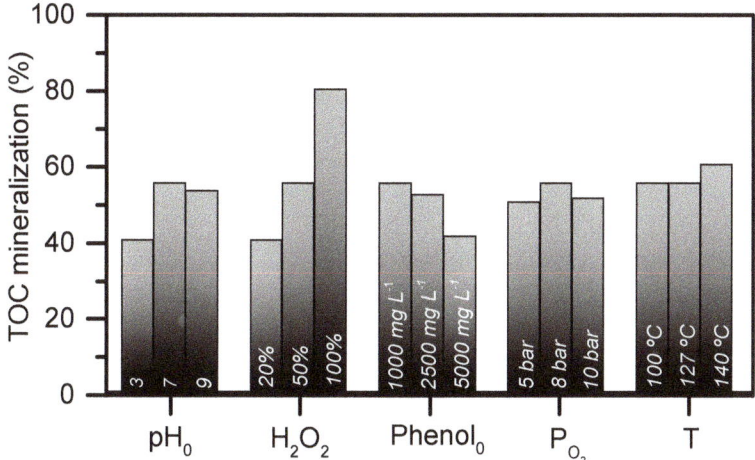

Figure 3. Effect of the operating conditions on the phenol mineralization after 3 h of reaction upon WPO-O_2 process. Standard conditions: $[Phenol]_0$ = 1000 mg L^{-1}, $[H_2O_2]_0$ = 2500 mg L^{-1}, P_{O2} = 8 bar, T = 127 °C and natural pH_0.

The operating condition study allowed to confirm the critical role of H_2O_2 on the performance of the WPO-O_2 process. On the other hand, taking into account previous works [14,15,34], it is expected that this variable will be intimately related to the formation of condensation by-products. To better understand the effect of H_2O_2 on the reaction, Figure 4 shows the evolution of this reagent as well as the resulting mineralization yield along the treatment using different H_2O_2 doses. The evolution of color, was also evaluated. As observed, after 3 h of reaction, the effluent color varied from dark-brown to colorless depending on the H_2O_2 dose and therefore on the TOC conversion achieved (from 40 to 80%, respectively). The change in color upon phenol oxidation have been previously described in the literature [12,14,15,20,34,35], not being attributed to low significance trace components but to the main reaction intermediates, mostly condensation products. These species are formed by oxidative coupling reactions of the highly reactive phenolic radical derivatives [36].

The results allowed the confirmation that the formation of condensation by-products was actually a primary step on the oxidation reaction and that they could be completely eliminated along the process when the stoichiometric amount of H_2O_2 was used. Accordingly, the color of the reaction mixture can be seen as a direct indication of the oxidation level achieved. In fact, almost 40% TOC mineralization was achieved in only a three minute reaction time (brown solution color), when the stoichiometric amount of H_2O_2 was used. That mineralization yield was similar to the achieved when 20% of the stoichiometric amount of H_2O_2 was employed. In the same line, the light brown color was associated with a mineralization yield around 55% while the colorless effluent showed a TOC reduction above 80%.

In addition to the condensation by-products, aromatic species such as hydroquinone, catechol and in lower amounts p-benzoquinone and resorcinol were detected upon the WPO-O_2 at the operating conditions of Figure 4b. Acids such as maleic, fumaric, oxalic, malonic, acetic and formic were also identified. The aromatic species were completed oxidized after 3 h of reaction at the 100% H_2O_2

stoichiometric dose. Only, oxalic, acetic and formic acids were detected in the reactor effluent at those operating conditions. Based on these findings, the reaction pathway of Figure 5 is proposed for the WPO-O_2 of phenol.

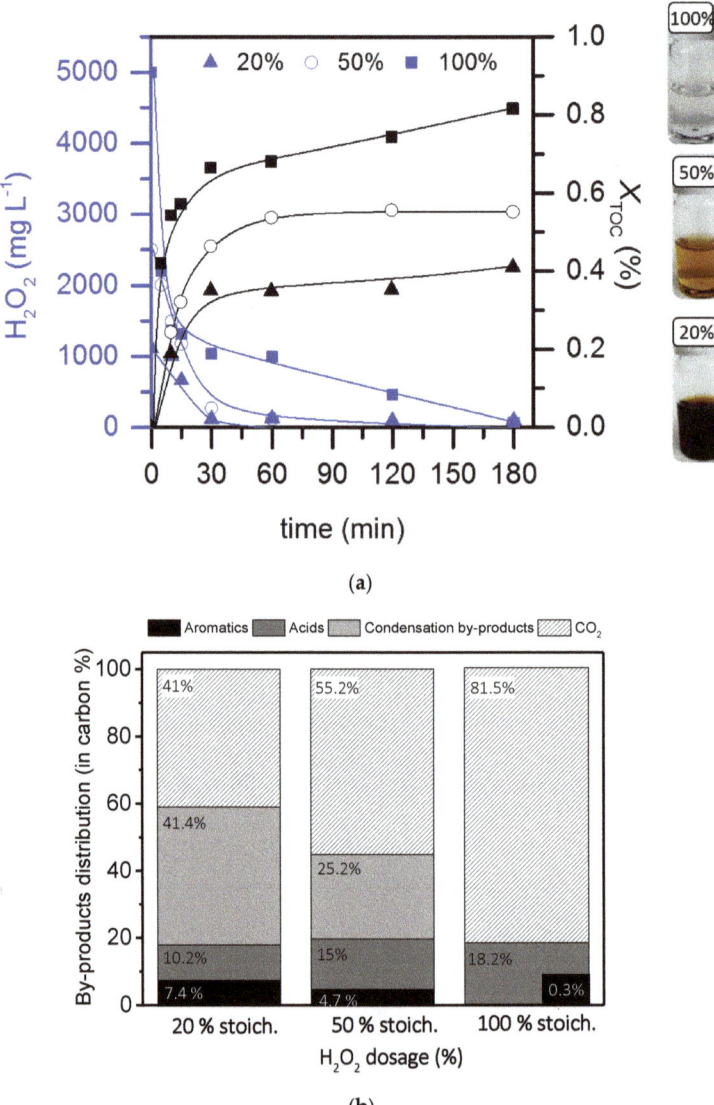

Figure 4. Effect of the H_2O_2 dose on the H_2O_2 consumption and mineralization (**a**) and by-product distribution at 3 h of reaction (**b**) upon phenol WPO-O_2. Operating conditions: [Phenol]$_0$ = 1000 mg L^{-1}, P$_{O2}$ = 8 bar (92 N mL$_{O2}$ min^{-1}), T = 127 °C and natural pH$_0$. The inset images in (**a**) show the aspect of the liquid effluent after 3 h of reaction.

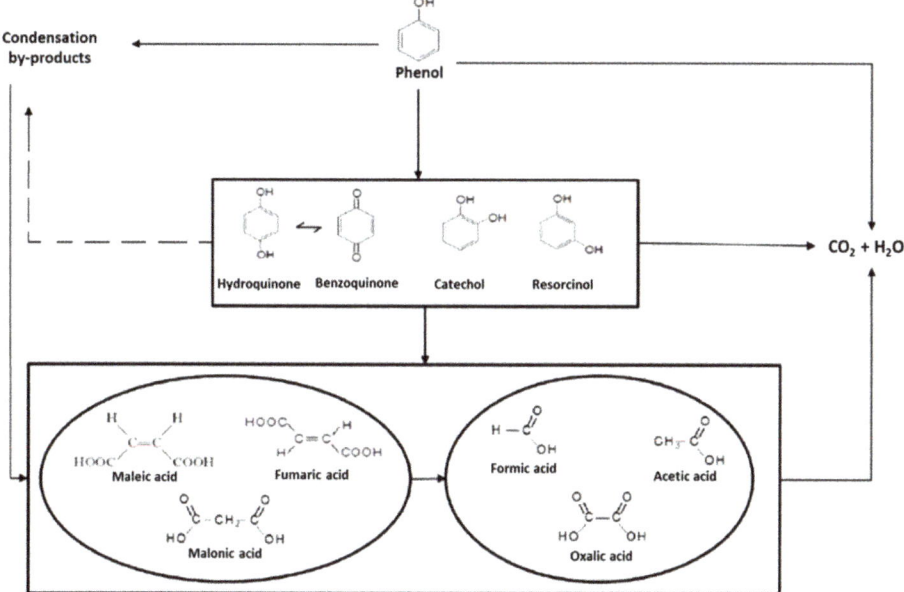

Figure 5. Proposed reaction pathway for the degradation of phenol by WPO-O_2.

2.3. Role of condensation by-products on the kinetics of the process

In order to elucidate the role of the condensation by-products upon the WPO-O_2 process, the kinetics were investigated in a flow reactor. The reactor was packed with inert quartz beads (mass: W = 29 g, particle diameter: ϕ_{beads} = 2 mm, and bed porosity: ε_L = 0.32) to provide an available surface for the deposition of condensation by-products. The oxidation runs were performed in concurrent up-flow under the operating conditions previously selected: [Phenol]$_0$ = 1000 mg L^{-1}, P_{O2} = 8 bar and neutral pH$_0$. The dose of H_2O_2 was established at 50% of the stoichiometric amount (2500 mg L^{-1}) to warrant the presence of condensation by-products. The temperature was varied within the range of 100–140 °C.

The concentration profiles of phenol, TOC and H_2O_2 vs. residence time (t_r) at the different operating conditions are shown in Figure 6 (in symbols).

Assuming isothermal plug-flow reactor, the mass balance of a given compound in liquid-phase (i) can be expressed as:

$$-Q_L \cdot dC_i = (-r_i) \cdot dV_L \tag{1}$$

where C_i is the molar concentration of the i species in mol L^{-1}, $(-r_i)$ the reaction rate of i reactant in mol L^{-1} s^{-1} and V_L the liquid volume in the reactor, in L, calculated as $\varepsilon_L \cdot V_R$.

Considering that the flow rate (Q_L) remains constant, the residence time (t_r) is defined as V_L/Q_L, then Equation (2) can be obtained from Equation (1):

$$-\frac{dC_i}{dt_r} = (-r_i) \tag{2}$$

Accordingly, the corresponding mass balances for phenol disappearance, H_2O_2 consumption and TOC abatement can be expressed as:

$$-\frac{dC_{H_2O_2}}{dt_r} = (-r_{H_2O_2}) \tag{3}$$

$$-\frac{dC_{Phenol}}{dt_r} = (-r_{phenol}) \tag{4}$$

$$-\frac{dC_{TOC}}{dt_r} = (-r_{TOC}) \tag{5}$$

Different reaction rate equations have been proposed to fit the experimental data (Figure 6), viz. pseudo-first order for each species, Langmuir-Hinshelwood-Hougen-Watson (LHHW) and autocatalytic rate equations. The numerical integration of the rate equations in the plug-flow reactor with the initial conditions $C_{Phenol} = C_{Phenol,0}$, $C_{H2O2} = C_{H2O2,0}$ at t = 0 was solved by using the Microsoft Excel Solver (Microsoft Office, 2010, MicrosoftCorp., Redmond, WA, USA) based on the Generalized Reduced Gradient (GRG) algorithm for least squares minimization. The equations were fitted at all temperatures simultaneously considering the Arrhenius equation to determine activation energy (Ea), the pre-exponential factor (k_0) and hence the kinetic constants (k).

The pseudo-first order and LHHW models failed to describe the experimental data (see Figures S9 and S10, respectively, of Supplementary Material to compare the experimental and predicted values). None of them consider the presence of an induction period in the concentration profiles of the three species (phenol, H_2O_2 and TOC), which was particularly significant at 100 °C. On the contrary, the autocatalytic model takes into account this phenomenon. This model considers that the catalyst is formed during the reaction, as in the case of the condensation by-products. As it has been above demonstrated, they are formed in the first stages of the reaction and afterwards, they are progressively degraded. Therefore, it is proposed that the increase in the concentration of condensation by-products in the beginning of the reaction enhances the decomposition of H_2O_2, and thus, the oxidation of phenol and TOC. The scheme proposed to describe the autocatalysis of H_2O_2 decomposition into radical species and the reaction of those radical species with phenol and TOC is the following:

$$H_2O_2 \xrightarrow{O_2/T} HO + OH^- \tag{6}$$

$$Phenol + HO \rightarrow Phenol + H_2O \tag{7}$$

$$Phenol + Phenol \rightarrow CP \tag{8}$$

$$H_2O_2 \xrightarrow{CP} HO + OH^- \tag{9}$$

where CP represents the condensation by-products.

The corresponding rate equations are:

$$(-r_{H_2O_2}) = k_{H_2O_2} \cdot C_{H_2O_2} \cdot C_{CP} = k\prime_{H_2O_2} \cdot C_{H_2O_2} \cdot (C_{H_2O_2,0} - C_{H_2O_2}) \tag{10}$$

$$\begin{aligned}(-r_{Phenol}) &= k_{Phenol} \cdot C_{Phenol} \cdot C_{CP} \\ &= k\prime_{Phenol} \cdot C_{Phenol} \cdot (C_{H_2O_2,0} - C_{H_2O_2})\end{aligned} \tag{11}$$

$$(-r_{TOC}) = k_{TOC} \cdot C_{TOC} \cdot C_{CP} = k\prime_{TOC} \cdot C_{TOC} \cdot (C_{H_2O_2,0} - C_{H_2O_2}) \tag{12}$$

where the CP concentration is proportional to the decomposed H_2O_2 amount.

The resulting differential equations of this autocatalytic model are:

$$(-r_{H_2O_2}) = 925 \cdot \exp\left(\frac{-4100}{T}\right) \cdot C_{H_2O_2} \cdot (0.074 - C_{H_2O_2}) \tag{13}$$

$$(-r_{Phenol}) = 8.5 \cdot 10^4 \cdot \exp\left(\frac{-9500}{T}\right) \cdot C_{Phenol} \cdot (0.074 - C_{H_2O_2}) \tag{14}$$

$$(-r_{TOC}) = 1.5 \cdot 10^4 \cdot \exp\left(\frac{-5700}{T}\right) \cdot C_{H_2O_2} \cdot (0.074 - C_{H_2O_2}) \tag{15}$$

The good agreement found between the experimental and calculated values (in Figure 6) allows to validate the autocatalytic model proposed and to demonstrate the catalytic role of condensation by-products in the WPO process. The parity plot in Figure 7 finally compare the calculated and measured concentrations of phenol, H_2O_2 and TOC evolution upon WPO-O_2 of phenol over quartz beads within the temperature range 100–140 °C using the autocatalytic kinetic model. The initial low-activity period can be then explained by the formation of such species in the reaction medium with the consequent coverage of the quartz beads, which is considerably slower at low temperatures. For instance, at 10 min of residence time and 100 °C, no significant phenol conversion was observed whereas up to 90% phenol removal was achieved at 140 °C. It is important to mention that only 50% of the stoichiometric amount of H_2O_2 was used and thus, complete removal of the condensation by-products was not completely achieved, as evidenced by the brown color of the used quartz beads (Figure 8). Nevertheless, a higher degradation of these species at higher reaction temperatures is expected.

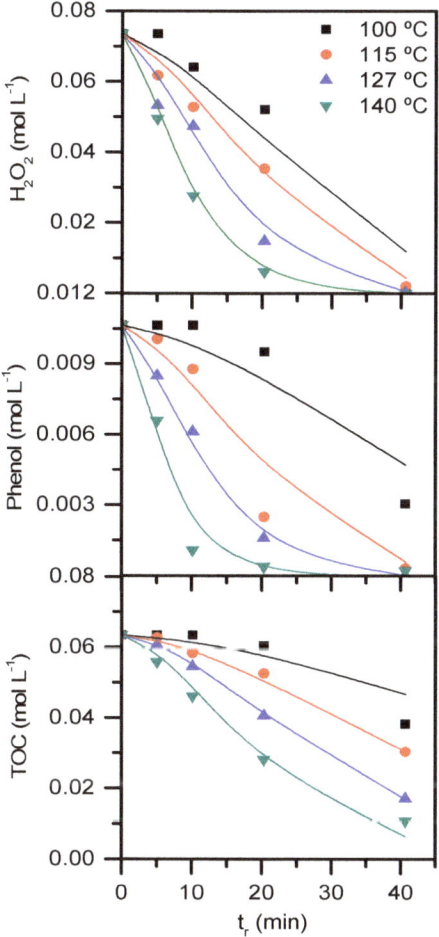

Figure 6. Experimental (symbols) and predicted (lines) concentrations of H_2O_2, phenol and TOC by the autocatalytic kinetic model of the phenol WPO-O_2 over quartz beads at different temperatures. Operating conditions: [Phenol]$_0$ = 1000 mg L^{-1}, [H_2O_2]$_0$ = 2500 mg L^{-1}, P_{O2} = 8 bar, natural pH$_0$ and W$_{quartz}$ = 29 g.

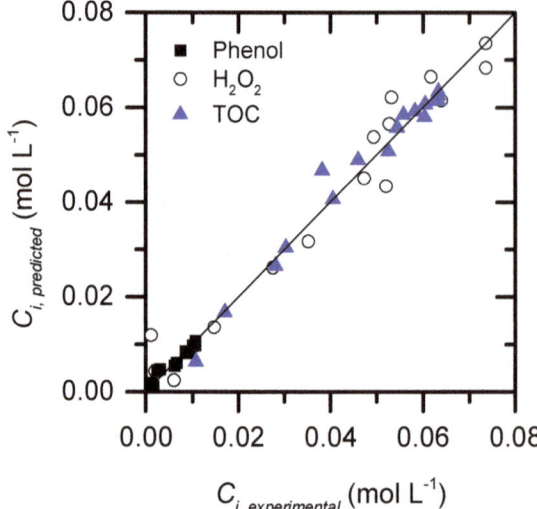

Figure 7. Parity plot for phenol, H_2O_2 and TOC evolution upon WPO-O_2 of phenol over quartz beads within the temperature range 100 – 140 °C using the autocatalytic kinetic model. Operating conditions: [Phenol]$_0$ = 1000 mg L^{-1}, [H_2O_2]$_0$ = 2500 mg L^{-1}, P_{O2} = 8 bar, natural pH$_0$ and W_{quartz} = 29 g.

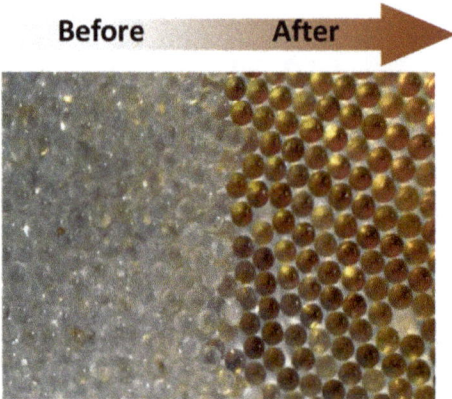

Figure 8. Image of the quartz beads before and after the WPO-O_2 process under the operating conditions of Figure 6.

The results obtained in this work allow us to demonstrate that the condensation by-products formed with wet oxidation can catalyze the process. These results are in line of other studies where the beneficial role of carbon deposits has been also proved in different reactions [22]. For instance, Teschner et al. (2006) [24] found that the palladium-carbon surface formed in the early stages of alkyne hydrogenation was actually the active phase responsible for the highly selective hydrogenation of 1-pentyne to 1-pentene. In the same line, Gornay et al. (2010) [23] demonstrated that the coke formed on the walls of the reactor had an "accelerating" effect on the octanoic acid pyrolisis. All in all, the catalytic effect promoted by the condensation by-products in the wet oxidation of phenol was not previously demonstrated in the literature to the best of our knowledge. In fact, the coverage of the catalyst by those species was usually associated to its loss of activity. For that reason, this study was performed in the absence of catalyst. In fact, the obtained results are of the same order of magnitude as those achieved with metal-free catalysts of relatively low-activity, such as carbon black [12]. An initial TOC abatement rate of 8 mg$_{TOC}$ L^{-1} min^{-1} was obtained in the CWPO of phenol under similar operating conditions ([Phenol]$_0$ = 1000 mg L^{-1}; [H$_2$O$_2$]$_0$ = 5000 mg L^{-1}; 130 °C), using a catalyst concentration of 5 g L^{-1}, while 12 mg$_{TOC}$ L^{-1} min^{-1} was obtained in the current work. Obviously, the activity of metal-free and metal-based catalysts cannot lead to comparable yields. In fact, when a metal-based catalyst such as Fe/AC or Fe/Al$_2$O$_3$ is used, the activity is much higher (up to one order of magnitude). In this sense, initial TOC abatement rates of 72 mg$_{TOC}$ L^{-1} min^{-1} and 32 mg$_{TOC}$ L^{-1} min^{-1}, were obtained with those catalysts, respectively, in the CWPO of phenol under similar operating conditions [37]. It is clear that in these cases, the coverage of the active sites even with catalytically active condensation by-products would unavoidably lead to lower catalytic activities.

3. Experimental

3.1. Oxidation experiments

Oxidation experiments were carried out in a 75 mL autoclave reactor (Berghof). The reactor consisted of a stainless-steel pressure vessel (PTFE) placed on a magnetic stirrer and surrounded by an electric resistance heating block run by the corresponding control system. Pressure was measured by a transducer. Six ports Valco valve VICI with two positions allowing the gas flow to pass through the reactor or bypass it. The inlet gas flow-rate (92 NmL min^{-1} of pure N$_2$ or O$_2$) bubbling into the liquid was adjusted by mass flow controllers (Hi-Tec Bronkhorst, Ruurlo, the Netherlands). In a typical oxygen-assisted wet peroxide oxidation (WPO-O$_2$), 70 mL of phenol solution at natural pH was charged to the vessel. Then, the reactor was stoppered, heated and pressurized under nitrogen atmosphere and a stirring rate of 750 rpm. After stabilization, the N$_2$ flow was switched into O$_2$. The O$_2$ stream flushed 5 mL of H$_2$O$_2$ aqueous solution of the appropriate concentration into the reactor. This was considered the starting reaction time. In the wet peroxide oxidation experiments (WPO), H$_2$O$_2$ was flushed with the N$_2$ stream, the only gas employed in the experiment. In the wet air oxidation (WAO), the experimental procedure was similar to the first described but H$_2$O$_2$ was not flushed to the reactor and 75 mL of phenol solution was charged from the beginning.

The standard selected conditions for WPO-O$_2$, WPO and WAO were natural pH$_0$, 1000 mg L^{-1} of phenol, 127 °C and 0.8 bar. If necessary, 5000 mg L^{-1} H$_2$O$_2$ was added (100% of the stoichiometric amount relative to the initial phenol concentration). Also, WPO-O$_2$ was performed in a wide range of operating conditions: pH$_0$ = 3–9, 1000–5000 mg L^{-1} of initial phenol, 20–100% stoichiometric dose of H$_2$O$_2$, 100–140 °C and 0.5–1 bar to study the formation of condensation by-products.

WPO-O$_2$ experiments were also conducted in a fixed-bed reactor consisting of a titanium tube with a 0.91 cm internal diameter and 30 cm long (reactor volume, V$_R$ = 4 cm^3), loaded with 29 g of quartz beads (ϕ_{beads} = 2mm and ε_L = 0.32). The temperature was measured by a thermocouple located in the bed. The liquid and gas phases were passed through the bed in concurrent up flow. Detailed information about the components and operation procedure of this setup has been reported elsewhere [38]. A 92 NmL min^{-1} pure oxygen was continuously passed in all the experiments. The oxidation runs were performed at different temperatures (100–140 °C) and 0.8 bar of total pressure. An aqueous solution of phenol at natural pH$_0$ was continuously fed to the reactor at 1000 mg L^{-1} and 50% stoichiometric dose of H$_2$O$_2$ at different flow rates (Q$_L$) = 1–0.125 mL min^{-1} to cover the experimental range of residence time (t$_r$) = 5–41 min, calculated as the ratio of the liquid volume (V$_L$) and flow rate (Q$_L$). V$_L$ was equal to $\varepsilon_L \bullet V_R$.

Liquid samples were periodically withdrawn from the reactors and immediately injected in a vial (submerged in crushed ice) containing a known volume of cold distilled water. The diluted samples were filtered (0.45 µm Nylon filter) and subsequently analyzed by different techniques.

3.2. Analytical methods

Phenol and aromatic intermediates evolution along oxidation reactions was followed by high performance liquid chromatography (Ultimate 3000, Thermo Scientific, Waltham, MA, USA) using a C18 5 μm column (Kinetex from Phenomenex, 4.6 mm diameter, 15 cm long) and a 4 mM H_2SO_4 aqueous solution as stationary and mobile phases, respectively. The quantification was performed at wavelengths of 210 and 246 nm. The measurement of short-chain organic acids was carried out by means of chromatography (Personal IC mod. 790, Metrohm). A mixture of Na_2CO_3 (3.2 mM) and $NaHCO_3$ (1 mM) aqueous solutions and a Metrosep A sup 5–250 column (4 mm diameter, 25 cm long) were used. Total organic carbon (TOC) was quantified with a TOC analyzer (Shimadzu TOC VSCH, Kyoto, Japan). H_2O_2 concentration was obtained by colorimetric titration $TiOSO_4$ method [39] using a UV2100 Shimadzu UV–vis spectrophotometer.

4. Conclusions

The results obtained in this work allowed us to demonstrate that the condensation by-products formed along wet peroxide oxidation can act as catalytic promoters. It was found that as reactions proceeded, the inert quartz beads were covered with condensation by-products and the disappearance rates of phenol, TOC and H_2O_2 were significantly increased. An initial low-period activity was clearly observed at low operating temperatures, especially 100 °C. The experimental data, obtained in the range of 100–140 °C, could not be described by conventional kinetic models like first order or Langmuir-Hinshelwood-Hougen-Watson. On the opposite, a good fit was reached by an autocatalytic kinetic model, which accounted for the activity promoted by the condensation by-products formed during the reaction. Interestingly, the activity of these species was found to be of the same order of magnitude as the reported with metal-free catalysts such as carbon black. Nevertheless, it was clearly lower than the achieved with metal-based catalysts, which explains that condensation by-products are usually seen as catalyst deactivators in the literature.

Supplementary Materials: The following are available online at http://www.mdpi.com/2073-4344/9/6/516/s1, Figure S1. Temporal concentration profiles of hydroquinone, resorcinol, catechol and benzoquinone upon the treatment of phenol by WPO-O_2. Operating conditions: [Phenol]$_0$ = 1000 mg L^{-1}, [H_2O_2]$_0$ = 5000 mg L^{-1}, P_{O2} = 8 bar (92 N mL$_{O2}$ min^{-1}), T = 127 °C and natural pH$_0$.; Figure S2. Short-chain organic acid distribution upon the WAO, WPO and WPO-O2 of phenol after 3 h of reaction. Operating conditions: [Phenol]$_0$ = 1000 mg L^{-1}, [H_2O_2]$_0$ = 5000 mg L^{-1}, P_{O2} = 8 bar (92 N mL$_{O2}$ min^{-1}), T = 127 °C and natural pH$_0$; Figure S3. Effect of the pH$_0$ on the phenol WPO-O_2. Operating conditions: [Phenol]$_0$ = 1000 mg L^{-1}, [H_2O_2]$_0$ = 2500 mg L^{-1}, T = 127 °C and P_{O2} = 8 bar; Figure S4. Effect of the H_2O_2 dose on the phenol WPO-O_2. Operating conditions: [Phenol]$_0$ = 1000 mg L^{-1}, T = 127 °C, P_{O2} = 8 bar and natural pH$_0$; Figure S5. Effect of the initial phenol concentration on the phenol WPO-O_2. Operating conditions: T = 127 °C, P_{O2} = 8 bar and natural pH$_0$; Figure S6. Effect of the oxygen pressure on the phenol WPO-O_2. Operating conditions: [Phenol]$_0$ = 1000 mg L^{-1}, [H_2O_2]$_0$ = 2500 mg L^{-1}, T = 127 °C, P_{O2} = 8 bar and natural pH$_0$; Figure S7. Effect of the temperature on the phenol WPO-O_2. Operating conditions: [Phenol]$_0$ = 1000 mg L^{-1}, [H_2O_2]$_0$ = 2500 mg L^{-1}, P_{O2} = 8 bar and natural pH$_0$; Figure S8. Effect of the operating conditions on the initial TOC abatement rate upon WPO-O_2 process. Standard conditions: [Phenol]$_0$ = 1000 mg L^{-1}, [H_2O_2]$_0$ = 2500 mg L^{-1}, P_{O2} = 8 bar, T = 127 °C and natural pH$_0$; Figure S9. Experimental (symbols) and predicted (lines) concentrations of H_2O_2, phenol and TOC by the pseudo-first order kinetic model of the phenol WPO-O_2 over quartz beads at different temperatures. Operating conditions: [Phenol]$_0$ = 1000 mg L^{-1}, [H_2O_2]$_0$ = 2500 mg L^{-1}, P_{O2} = 8 bar, natural pH$_0$ and W$_{quartz\ beads}$ = 29 g; Figure S10. Experimental (symbols) and predicted (lines) concentrations of H_2O_2, phenol and TOC by the LHHW kinetic model of the phenol WPO-O_2 over quartz beads at different temperatures. Operating conditions: [Phenol]$_0$ = 1000 mg L^{-1}, [H_2O_2]$_0$ = 2500 mg L^{-1}, P_{O2} = 8 bar, natural pH$_0$ and W$_{quartz\ beads}$ = 29 g.

Author Contributions: Conceptualization, J.A.C. and A.Q.; Data curation, J.L.D.d.T. and C.F.; Formal analysis, A.Q., J.L.D.d.T., C.F. and M.M.; Funding acquisition, J.A.C.; Investigation, A.Q., J.L.D.d.T., C.F. and M.M.; Methodology, A.Q., J.L.D.d.T. and C.F.; Project administration, J.A.C.; Supervision, J.A.C. and A.Q.; Validation, J.L.D.d.T. and C.F.; Writing – original draft, A.Q. and M.M.; Writing – review & editing, A.Q., M.M. and J.A.C.

Funding: This research was funded by the Spanish MINECO, through the project CTM-2016-76454-R and by the CM, through the project P2018/EMT-4341. M. Munoz thanks the Spanish MINECO for the Ramón y Cajal postdoctoral contract (RYC-2016-20648).

Conflicts of Interest: The authors declare no conflict of interest.

References

1. UN-Water 2015, Compendium of Water Quality Regulatory Frameworks: Which Water for Which Use? Available online: http://www.unwater.org/app/uploads/2017/05/Compendium-of-Water-Quality-Main-Report_4.pdf (accessed on 16 February 2019).
2. Ryu, J.Y. Formation of chlorinated phenols, dibenzo-p-dioxins, dienzofurans, benzenes, benzoquinones and perchloroethylenes from phenols in oxidative and copper(II) chloride-catalyzed thermal process. *Chemosphere* **2008**, *71*, 1100–1109. [CrossRef] [PubMed]
3. Levec, J.; Pintar, A. Catalytic wet-air oxidation processes: A review. *Catal. Today* **2007**, *124*, 172–184. [CrossRef]
4. Munoz, M.; de Pedro, Z.M.; Casas, J.A.; Rodriguez, J.J. Preparation of magnetite-based catalysts and their application in heterogeneous Fenton oxidation – A review. *Appl. Catal. B* **2015**, *176–177*, 249–265. [CrossRef]
5. Zazo, J.A.; Casas, J.A.; Mohedano, A.F.; Rodríguez, J.J. Catalytic wet peroxide oxidation of phenol with a Fe/active carbon catalyst. *Appl. Catal. B* **2006**, *65*, 261–268. [CrossRef]
6. Rey, A.; Faraldos, M.; Casas, J.A.; Zazo, J.A.; Bahamonde, A.; Rodríguez, J.J. Catalytic wet peroxide oxidation of phenol over Fe/AC catalysts: Influence of iron precursor and activated carbon surface. *Appl. Catal. B* **2009**, *86*, 69–77. [CrossRef]
7. Keav, S.; de los Monteros, A.E.; Barbier, J.; Duprez, D. Wet Air Oxidation of phenol over Pt and Ru catalysts supported on cerium-based oxides: Resistance to fouling and kinetic modelling. *Appl. Catal. B* **2014**, *150–151*, 402–410. [CrossRef]
8. Bautista, P.; Mohedano, A.F.; Menéndez, N.; Casas, J.A.; Zazo, J.A.; Rodriguez, J.J. Highly stable Fe/γAl$_2$O$_3$ catalyst for catalytic wet peroxide oxidation. *J. Chem. Technol. Biotechnol.* **2011**, *86*, 497–504. [CrossRef]
9. Munoz, M.; Domínguez, P.; de Pedro, Z.M.; Casas, J.A.; Rodriguez, J.J. Naturally-occurring iron minerals as inexpensive catalysts for CWPO. *Appl. Catal. B* **2017**, *203*, 166–173. [CrossRef]
10. Domínguez, C.M.; Ocón, P.; Quintanilla, A.; Casas, J.A.; Rodriguez, J.J. Highly efficient application of activated carbon as catalyst for wet peroxide oxidation. *Appl. Catal. B* **2013**, *140–141*, 663–670. [CrossRef]
11. Domínguez, C.M.; Ocón, P.; Quintanilla, A.; Casas, J.A.; Rodriguez, J.J. Graphite and carbon black materials as catalysts for wet peroxide oxidation. *Appl. Catal. B* **2014**, *144*, 599–606. [CrossRef]
12. Diaz de Tuesta, J.L.; Quintanilla, A.; Casas, J.A.; Rodriguez, J.J. Kinetic modeling of wet peroxide oxidation with a carbon black catalyst. *Appl. Catal. B* **2017**, *209*, 701–710. [CrossRef]
13. Ribeiro, R.S.; Fathy, N.A.; Attia, A.A.; Silva, A.M.T.; Faria, J.L.; Gomes, H.T. Activated carbon xerogels for the removal of the anionic azo dyes Orange II and Chromotrope 2R by adsorption and catalytic wet peroxide oxidation. *Chem. Eng. J.* **2012**, *195–196*, 112–121. [CrossRef]
14. Munoz, M.; de Pedro, Z.M.; Casas, J.A.; Rodriguez, J.J. Assessment of the generation of chlorinated byproducts upon Fenton-like oxidation of chlorophenols at different conditions. *J. Hazard. Mater.* **2011**, *190*, 993–1000. [CrossRef] [PubMed]
15. Poerschmann, J.; Trommler, U.; Górecki, T.; Kopinke, F.D. Formation of chlorinated biphenyls, diphenyl ethers and benzofurans as a result of Fenton-driven oxidation of 2-chlorophenol. *Chemosphere* **2009**, *75*, 772–780. [CrossRef] [PubMed]
16. Poerschmann, J.; Trommler, U.; Górecki, T. Aromatic intermediate formation during oxidative degradation of Bisphenol A by homogeneous sub-stoichiometric Fenton reaction. *Chemosphere* **2010**, *79*, 975–986. [CrossRef] [PubMed]
17. Vallejo, M.; Fernández-Castro, P.; San Román, M.F.; Ortiz, I. Assessment of PCDD/Fs formation in the Fenton oxidation of 2-chlorophenol: Influence of the iron dose applied. *Chemosphere* **2015**, *137*, 135–141. [CrossRef]
18. Vallejo, M.; Fresnedo San Román, M.; Ortiz, I.; Irabien, A. Overview of the PCDD/Fs degradation potential and formation risk in the application of advanced oxidation processes (AOPs) to wastewater treatment. *Chemosphere* **2015**, *118*, 44–56. [CrossRef]
19. Delgado, J.J.; Pérez-Omil, J.A.; Rodríguez-Izquierdo, J.M.; Cauqui, M.A. The role of the carbonaceous deposits in the Catalytic Wet Oxidation (CWO) of phenol. *Catal. Comm.* **2006**, *7*, 639–643. [CrossRef]

20. Delgado, J.J.; Chen, X.; Pérez-Omil, J.A.; Rodríguez-Izquierdo, J.M.; Cauqui, M.A. The effect of reaction conditions on the apparent deactivation of Ce–Zr mixed oxides for the catalytic wet oxidation of phenol. *Catal. Today* **2012**, *180*, 25–33. [CrossRef]
21. Lee, D.K.; Kim, D.S.; Kim, T.H.; Lee, Y.K.; Jeong, S.E.; Le, N.T.; Cho, M.J.; Henam, S.D. Deactivation of Pt catalysts during wet oxidation of phenol. *Catal. Today* **2010**, *154*, 244–249. [CrossRef]
22. Collett, C.H.; McGregor, J. Things go better with coke: the beneficial role of carbonaceous deposits in heterogeneous catalysis. *Catal. Sci. Technol.* **2016**, *6*, 363–378. [CrossRef]
23. Gornay, J.; Coniglio, L.; Billaud, F.; Wild, G. Octanoic acid pyrolysis in a stainless-steel tube: What is the role of the coke formed on the wall? *J. Anal. Appl. Pyrol.* **2010**, *87*, 78–84. [CrossRef]
24. Teschner, D.; Vass, E.; Hävecker, M.; Zafeiratos, S.; Schnörch, P.; Sauer, H.; Knop-Gericke, A.; Schlögl, R.; Chamam, M.; Wootsch, A.; et al. Alkyne hydrogenation over Pd catalysts: A new paradigm. *J. Catal.* **2006**, *242*, 26–37. [CrossRef]
25. Vrieland, G.E.; Menon, P.G. Nature of the catalytically active carbonaceous sites for the oxydehydrogenation of ethylbenzene to styrene: A brief review. *Appl. Catal.* **1991**, *77*, 1–8. [CrossRef]
26. Fiedorow, R.; Przystajko, W.; Sopa, M.; Dalla Lana, I.G. The nature and catalytic influence of coke formed on alumina: Oxidative dehydrogenation of ethylbenzene. *J. Catal.* **1981**, *68*, 33–41. [CrossRef]
27. Hamoudi, S.; Larachi, F.; Adnot, A.; Sayari, A. Characterization of Spent MnO2/CeO2 Wet Oxidation Catalyst by TPO–MS, XPS, and S-SIMS. *J. Catal.* **1999**, *185*, 333–344. [CrossRef]
28. Kim, S.; Ihm, S. Nature of carbonaceous deposits on the alumina supported transition metal oxide catalysts in the wet air oxidation of phenol. *Top. Catal.* **2005**, *33*, 171–179. [CrossRef]
29. Catrinescu, C.; Teodosiu, C.; Macoveanu, M.; Miehe-Brendlé, J.; Le Dred, R. Catalytic wet peroxide oxidation of phenol over Fe-exchanged pillared beidellite. *Water Res.* **2003**, *37*, 1154–1160. [CrossRef]
30. Yadav, A.; Verma, N. Carbon bead-supported copper-dispersed carbon nanofibers: An efficient catalyst for wet air oxidation of industrial wastewater in a recycle flow reactor. *J. Ind. Eng. Chem.* **2018**, *67*, 448–460. [CrossRef]
31. Huang, H.; Lu, M.; Chen, J. Catalytic Decomposition of hydrogen peroxide and 2-chlorophenol with iron oxides. *Water Res.* **2001**, *35*, 2291–2299. [CrossRef]
32. Pham, A.L.; Lee, C.; Doyle, F.M.; Sedlak, D.L. A silica-supported iron oxide catalyst capable of activating hydrogen peroxide at neutral pH values. *Environ. Sci. Technol.* **2009**, *43*, 8930–8935. [CrossRef]
33. Kwon, B.G.; Lee, D.S.; Kang, N.; Yoon, J. Characteristics of p-chlorophenol oxidation by Fenton's reagent. *Water Res.* **1999**, *33*, 2110–2118. [CrossRef]
34. Munoz, M.; de Pedro, Z.M.; Casas, J.A.; Rodriguez, J.J. Combining efficiently catalytic hydrodechlorination and wet peroxide oxidation (HDC–CWPO) for the abatement of organochlorinated water pollutants. *Appl. Catal. B* **2014**, *150–151*, 197–203. [CrossRef]
35. Mijangos, F.; Varona, F.; Villota, N. Changes in solution color during phenol oxidation by Fenton reagent. *Environ. Sci. Technol.* **2006**, *40*, 5538–5543. [CrossRef] [PubMed]
36. McDonald, P.D.; Hamilton, G.A. CHAPTER II - Mechanisms of phenolic oxidative coupling reactions. *Org. Chem.* **1973**, *5*, 97–134.
37. Garcia-Costa, A.L.; Zazo, J.A.; Rodriguez, J.J.; Casas, J.A. Microwave-assisted catalytic wet peroxide oxidation. Comparison of Fe catalysts supported on activated carbon and -alumina. *Appl. Catal. B* **2017**, *218*, 637–642. [CrossRef]
38. Quintanilla, A.; Casas, J.A.; Zazo, J.A.; Mohedano, A.F.; Rodríguez, J.J. Wet air oxidation of phenol at mild conditions with a Fe/activated carbon catalyst. *Appl. Catal. B* **2006**, *62*, 115–120. [CrossRef]
39. Eisenberg, G.M. Colorimetric determination of hydrogen peroxide. *Ind. Eng. Chem. Res.* **1943**, *15*, 327–328. [CrossRef]

© 2019 by the authors. Licensee MDPI, Basel, Switzerland. This article is an open access article distributed under the terms and conditions of the Creative Commons Attribution (CC BY) license (http://creativecommons.org/licenses/by/4.0/).

Article

Condensation By-Products in Wet Peroxide Oxidation: Fouling or Catalytic Promotion? Part II: Activity, Nature and Stability

Asunción Quintanilla [1], Jose L. Diaz de Tuesta [2,3], Cristina Figueruelo [1], Macarena Munoz [1,*] and Jose A. Casas [1]

[1] Chemical Engineering Department, Universidad Autónoma de Madrid, Ctra. Colmenar km 15, 28049 Madrid, Spain; asun.quintanilla@uam.es (A.Q.); cgfigueruelo@gmail.com (C.F.); jose.casas@uam.es (J.A.C.)
[2] Centro de Investigação de Montanha (CIMO), Instituto Politécnico de Bragança, 5300-253 Bragança, Portugal; jl.diazdetuesta@ipb.pt
[3] Laboratório de Processos de Separação e Reação-Laboratório de Catálise e Materiais (LSRE-LCM), Faculdade de Engenharia, Universidade do Porto, 4200-465 Porto, Portugal
* Correspondence: macarena.munnoz@uam.es; Tel.: 34-91-497-3991; Fax: +34-91497-3516

Received: 29 April 2019; Accepted: 6 June 2019; Published: 11 June 2019

Abstract: The deposition of condensation by-products onto the catalyst surface upon wet peroxide and wet air oxidation processes has usually been associated with catalyst deactivation. However, in Part I of this paper, it was demonstrated that these carbonaceous deposits actually act as catalytic promoters in the oxygen-assisted wet peroxide oxidation (WPO-O_2) of phenol. Herein, the intrinsic activity, nature and stability of these species have been investigated. To achieve this goal, an up-flow fixed bed reactor packed with porous Al_2O_3 spheres was used to facilitate the deposition of the condensation by-products formed in the liquid phase. It was demonstrated that the condensation by-products catalyzed the decomposition of H_2O_2 and a higher amount of these species leads to a higher degree of oxidation degree The reaction rates, conversion values and intermediates' distribution were analyzed. The characterization of the carbonaceous deposits on the Al_2O_3 spheres showed a significant amount of condensation by-products (~6 wt.%) after 650 h of time on stream. They are of aromatic nature and present oxygen functional groups consisting of quinones, phenols, aldehydes, carboxylics and ketones. The initial phenol concentration and H_2O_2 dose were found to be crucial variables for the generation and consumption of such species, respectively.

Keywords: wet peroxide oxidation; wet air oxidation; condensation by-products nature; fouling; autocatalytic kinetics

1. Introduction

The treatment of wastewater containing aromatic compounds by oxidizing radicals, like those generated in catalytic wet peroxide oxidation (CWPO) and catalytic wet air oxidation (CWAO) processes, unavoidably involve oxidative coupling or polymerization reactions [1–4]. In these reactions, the aromatic molecules polymerize leading to a complex mixture of intermediates usually known as condensation by-products, which remain adsorbed onto the porous catalyst surface. By using the appropriate operating conditions, such as H_2O_2 dose [5] or temperature [1,6], and also by the integration of different treatments [7], these species can be further oxidized to CO_2.

The presence of such carbonaceous deposits or "coke" on the catalyst has been recognized as a major cause of its deactivation as these deposits can cover the catalytic active sites and thus, block the redox cycle [1,3,8–13]. However, recent studies have shown that the presence of those species enhances the catalytic performance in a number of processes such as oxidative dehydrogenation, hydrogenation, isomerization and Fischer-Tropsch [14]. In fact, the poly-aromatic structure of condensation by-products could also have a positive effect in CWPO and CWAO processes since it could facilitate the electron transfer to the oxidants initiating the radical reaction mechanism, as if it was a carbocatalyst surface. However, to the best of our knowledge, this theory has not been addressed in the literature so far.

Table 1 presents a brief summary of the studies which have considered the formation of condensation by-products during aromatic-polluted wastewater treatment using both CWAO and CWPO processes. As can be seen, in all studies the deposition of carbonaceous deposits was identified as the main reason for the deactivation of the catalysts. However, it is not possible to generalize the results since this mechanism depends on the nature of the catalyst and the operating conditions, in particular temperature, initial pollutant concentration and oxidant dose. On the other hand, it must be noted that the deactivation was significantly more pronounced when metal-based catalysts were used [1,11,12], whereas carbocatalysts were much more resistant [3]. In the same line, the regeneration of solids by calcination treatments allowed the complete recovery of the initial activity of the carbocatalysts [3]. Although the regeneration of some metal-based catalysts by calcination has also been demonstrated [1,15], it is not so straightforward in this case because the metal nanoparticles would also be oxidized, which could be counterproductive for the process, for instance, if the popular magnetic catalysts are used. On the other hand, as shown in Table 1, the nature of condensation by-products deposited onto the catalyst surface upon CWAO and CWPO reactions has not been deeply investigated and many questions remain.

In our previous contribution, Part I of this study [16], it was demonstrated that the carbonaceous materials deposited on the surface of inert quartz beads during the WPO-O_2 of phenol were able to catalyze the oxidation process. That is, the disappearance rates of phenol, total organic carbon (TOC) and H_2O_2 were progressively increased during the reaction as the quartz beads were covered by the condensation by-products, which was successfully described by an autocatalytic kinetic model. Here, the activity, nature and stability of those species have been more deeply investigated. To achieve this goal, inert Al_2O_3 spheres were used instead of quartz beads in the fixed bed reactor as they provide a significantly higher surface area for the deposition of the condensation by-products. The effect of these deposits on the kinetics of the process were investigated in the decomposition of H_2O_2 and also in the WPO-O_2 of phenol. The amount of condensation by-products deposited onto the Al_2O_3 spheres were measured and their nature was fully characterized by several techniques. Finally, the stability of these species was evaluated under different process operating conditions.

Table 1. A brief summary of CWAO and CWPO studies in which the formation of condensation by-products upon wastewater treatment was considered.

Process	Catalyst	Target Pollutant	Operating Conditions	Condensation Products	Reference
CWAO	MnO_2/CeO_2	Phenol	$T = 80–130\ °C$ $P_{O_2} = 5\ bar$ $C_{cat} = 1–5\ g\ L^{-1}$ $C_{cont} = 1–10\ g\ L^{-1}$	*Formation*: A carbonaceous overlayer was formed upon CWAO. The amount of surface carbon increased with reaction time, whereas its composition was reaction time and temperature dependent. *Characterization*: TPO-MS, XPS and S-SIMS. Low polycondensation (4 condensed aromatic rings) deposits were formed. The presence of aliphatic species as well as oxygen-containing compounds from alcohol/ether origin was also evidenced. *Regeneration*: oxidation in air (200–300 °C).	[8]
CWAO	Mn/Al_2O_3 Fe/Al_2O_3 Co/Al_2O_3 Ni/Al_2O_3 Cu/Al_2O_3	Phenol	$T = 150\ °C$ $P_{air} = 20\ bar$ $C_{cat} = 3\ g\ L^{-1}$ $C_{cont} = 1\ g\ L^{-1}$	*Formation*: Used catalysts showed a darker color due to the formation of carbonaceous deposits. Mn/Al_2O_3 showed the highest amount of deposits. *Characterization*: Specific surface area, NMR and FTIR. The carbonaceous deposits showed a microporous structure, increasing the surface area of the catalyst. They were mostly of aromatic nature, containing also oxygen-bearing groups such as carboxylic acids and alcohols. *Regeneration*: N/A	[9]
CWAO	Pt/Al_2O_3 Pt/CeO_2	Phenol	$T = 150\ °C$ $P_{O_2} = 14\ bar$ $C_{cat} = 1–3\ g\ L^{-1}$ $C_{cont} = 0.5–10\ g\ L^{-1}$	*Formation*: Along reaction, carbonaceous deposits were formed on the catalyst surface, leading to its deactivation, especially with Pt/Al_2O_3. *Characterization*: TEM, EDX, XPS and TPO. The carbonaceous deposits were predominantly formed on the surface of Pt particles. The deposits were burned at 250–700 °C in Pt/Al_2O_3 and 250–450 °C in Pt/CeO_2. *Regeneration*: N/A	[10]
CWAO	Ce-Zr mixed oxides	Phenol	$T = 120–160\ °C$ $P_{O_2} = 5–20\ bar$ $C_{cat} = 1.9–5.8\ g\ L^{-1}$ $C_{cont} = 0.25–3.8\ g\ L^{-1}$	*Formation*: Phenolic polymers are formed and further degraded along reaction depending on the operating conditions. At 120 °C, their formation is favored, leading to a deactivation of the catalyst. At 160 °C, the deposits are completely oxidized to CO_2. *Characterization*: Quantification by carbon-mass balances, TPO and specific surface area. The carbon deposits amounted from 2.9% to 30% wt. They were burn-off at mild temperatures (<350 °C). The specific surface area was slightly decreased due to carbon deposits coverage (1%–13%). *Regeneration*: Calcination at 350 °C.	[1]
CWAO	Ru/Ce, ZrCe or ZrCePr Pt/Ce, ZrCe or ZrCePr	Phenol	$T = 160\ °C$ $P_{O_2} = 20\ bar$ $C_{cat} = 4\ g\ L^{-1}$ $C_{cont} = 2.1\ g\ L^{-1}$	*Formation*: Heavy polymer species are adsorbed on the catalytic surface along reaction, leading to the blockage of the redox cycle by limiting the reoxidation of the catalytic sites. The increase in the number of catalytic oxidation sites increased the accumulation of adsorbed species, while high metal dispersion allowed their degradation. *Characterization*: N/A *Regeneration*: N/A	[11]

Table 1. Cont.

Process	Catalyst	Target Pollutant	Operating Conditions	Condensation Products	Reference
CWAO	Cu-PhB-CNF-K	Phenol Industrial wastewater	T = 180–230 °C P_{O_2} = 12–16 bar C_{cat} = 0.25–1 g L^{-1} C_{cont} (phenol) = 1–1.5 g L^{-1} COD = 111–120 go_2 L^{-1}	*Formation*: Carbonaceous deposits formation due to copolymerization of phenol, leading to coverage of catalytic active sites and the deactivation of the catalyst. *Characterization*: N/A *Regeneration*: N/A	[12]
CWPO	Dissolved iron (Fe^{3+})	4-chlorophenol	T = 50 °C pH_0 = 3 C_{cat} = 1–20 mg L^{-1} $C_{H_2O_2}$ = 20–100% stoich. dose C_{cont} = 0.1–2 g L^{-1}	*Formation*: At short reaction times and/or using substoichiometric doses of H_2O_2, the development of a brownish color in the reaction medium and the existence of a dark solid residue after evaporation were observed. Final colorless effluents were obtained when the stoichiometric dose of H_2O_2 was used. *Characterization*: GC-MS of homogeneous condensation by-products and elemental analysis of the solid. Different dimers such as chlorinated diphenyl ethers, PCBs and even dioxins were identified in the liquid phase. The solid mainly contained carbon, oxygen and chlorine, including also iron (46.4% C, 2.6% H, 12.6% Cl, 3.6% Fe and 34.8% O). *Regeneration*: N/A	[5]
CWPO	Fe/pillared beidellite	Phenol	T = 50 °C pH_0 = 3.5 C_{cat} =0.5 g L^{-1} $C_{H_2O_2}$ = 1.25 g L^{-1} (stoich. dose) C_{cont} = 0.25 g L^{-1}	*Formation*: Organic species were deposited in the used catalyst, leading to a loss of activity upon successive runs. It recovered its initial activity after a calcination step. *Characterization*: N/A *Regeneration*: Calcination.	[13]
CWPO	Fe_2O_3/γ-Al_2O_3	2-chlorophenol 2,4-dichlorophenol 2,4,6-trichlorophenol	T = 50 °C pH_0 = 3 C_{cat} = 2 g L^{-1} $C_{H_2O_2}$ = 20–100% stoich. dose C_{cont} = 0.1–2 g L^{-1}	*Formation*: Condensation by-products were formed in high amounts at initial reaction times and/or substoichiometric doses of H_2O_2. A dramatic color change of the reaction medium was observed. The used catalysts showed a brown color when doses below 75% of H_2O_2 were used. *Characterization*: Specific surface area and elemental analysis. The former decreased due to the presence of carbonaceous deposits at H_2O_2 doses below 75% of the stoichiometric amount (up to 33% area reduction with a 20% H_2O_2 stoichiometric dose). The used catalyst with 20% H_2O_2 stoichiometric dose showed 7.7% C, 0.99% H and 2.74% Cl. *Regeneration*: N/A	[7]
CWPO	Activated carbon	Phenol	T = 80 °C pH_0 = 3.5 C_{cat} = 2.5 g L^{-1} $C_{H_2O_2}$ = 100% stoichiometric dose C_{cont} = 5 g L^{-1}	*Formation*: Condensation by-products were adsorbed onto the carbon surface, causing a progressive deactivation upon successive uses. *Characterization*: TGA and cyclic voltammetry. The burn-off of condensation products occurs at 350 °C. An important decrease of the electrochemical properties of the catalyst due to the presence of carbon deposits was observed. *Regeneration*: calcination (350 °C, 24 h).	[3]

N/A, not analyzed.

2. Results and Discussion

2.1. Catalytic Activity Promoted by Condensation By-Products Deposits

In a first approach to directly assess the activity promoted by the carbonaceous by-products deposited on the surface of inert solids in a fixed bed reactor during the WPO-O_2 of phenol, their ability to decompose H_2O_2 was investigated. The decomposition of H_2O_2 over fresh Al_2O_3 spheres and those used in the WPO-O_2 of phenol during 145 h on stream were assessed. Al_2O_3 spheres were used instead of quartz beads in this work because they are also catalytically inert particles for the decomposition of H_2O_2 but provide higher available specific surface area (S_{BET} = 275 m^2 g^{-1}) for the deposition of condensation by-products formed in the liquid phase. The concentration of H_2O_2 tested was established at the same amount used in the WPO-O_2 of phenol (2500 mg L^{-1}) [16]. The rest of the operating conditions were fixed at P_{O2} = 8 bar (100 N mL$_{O2}$ min^{-1}), T = 150 °C and neutral pH$_0$.

The obtained results can be seen in Figure 1. It should be noted that the used Al_2O_3 spheres showed a dark brown color according to the presence of carbonaceous deposits on their surface [16]. Although H_2O_2 was decomposed in both cases, significant differences were found in the oxidation rate. The resulting pseudo-first order rate constant values for H_2O_2 decomposition were 0.181 and 0.345 L g$_{cat}^{-1}$ h^{-1} with fresh and used Al_2O_3 spheres, respectively. The C-coated Al_2O_3 solid (Al_2O_3 covered by condensation by-products) decomposed H_2O_2 about two times faster than the fresh one, confirming the important role of condensation by-products on the kinetics of the process. These results are consistent with those of Part I of this study [16], the condensation by-products are catalytic species able to promote the redox reactions to decompose H_2O_2 into hydroxyl and hydroperoxyl radicals. Therefore, the formation of condensation by-products is the preliminary step in the phenol wet peroxide oxidation. To the best of our knowledge, the activity promoted by the typical carbonaceous deposits formed upon the CWPO, and also CWAO of organic pollutants has not been previously reported in the literature. On the contrary, the catalyst coverage by those species has commonly been identified as an important reason for the loss of its activity [9–13]. Clearly, when highly active metal-based catalysts are covered by condensation by-products, the activity of the solid is unavoidably decreased as the activity promoted by those species cannot compete with that of metal particles. Accordingly, to prove the intrinsic activity of the carbonaceous deposits, the process must be carried out in the absence of catalyst with an inert solid, as demonstrated in this work.

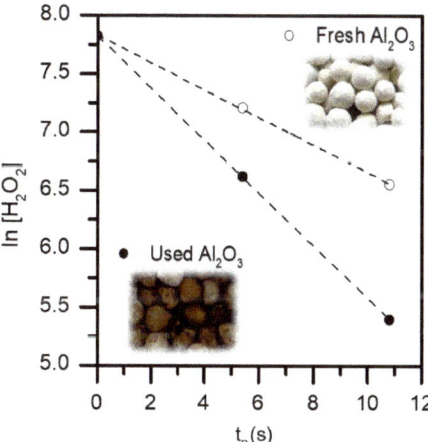

Figure 1. H_2O_2 decomposition over fresh and 145 h-used Al_2O_3 upon the phenol WPO-O_2. Operating conditions: [H_2O_2]$_0$ = 2500 mg L^{-1}, P_{O2} = 8 bar (92 N mL$_{O2}$ min^{-1}), T = 150 °C, natural pH$_0$ and W_{Al2O3} = 2 g. The inset images show the Al_2O_3 spheres before and after being used in the WPO-O_2 of phenol.

To get a better insight into the catalytic performance of condensation by-products, the WPO-O_2 performance was studied over a Al_2O_3 fixed bed, and the results were compared to those obtained over quartz beads, as reported in Part I of this work [16]. It was expected that the amount of the condensation by-products deposited would be different in both beds due to the highest specific surface area of the porous Al_2O_3. The results in terms of phenol, TOC and H_2O_2 conversion at the following operating conditions: [phenol]$_0$ = 1000 mg L^{-1}, [H_2O_2]$_0$ = 2500 mg L^{-1}, T = 100–140 °C, P_{O2} = 8 bar, natural pH$_0$ and t_R = 7 min, are summarized in Figure 2. The resulting initial oxidation rate values (g_i g_{solid}^{-1} h^{-1}) are also included. In all cases, the conversion values were significantly higher in the presence of Al_2O_3 spheres. Accordingly, the reaction proceeded much faster with Al_2O_3, obtaining rate values around one order of magnitude higher. Also, higher extension of the reaction was reached, as indicated in the intermediate distribution obtained with both systems at given operating conditions (Figure 3). In this sense, the mineralization yield (or TOC conversion) was around 72% with the Al_2O_3 fixed bed while only 5% was obtained with the quartz one. In the same line, dissolved aromatic compounds (mainly p-benzoquinone, catechol and hydroquinone) were almost completely eliminated in the presence of Al_2O_3 while they were the main reaction products (representing 84% of the initial TOC) with quartz beads. Taking into account that both quartz and Al_2O_3 are inert materials, the only difference among them can be attributed to the available surface area provided for the deposition of condensation by-products, which is remarkably higher in the case of Al_2O_3 area (S_{BET} = 275 m^2 g^{-1}). Accordingly, the catalytic activity promoted by condensation by-products is significantly stronger with this solid.

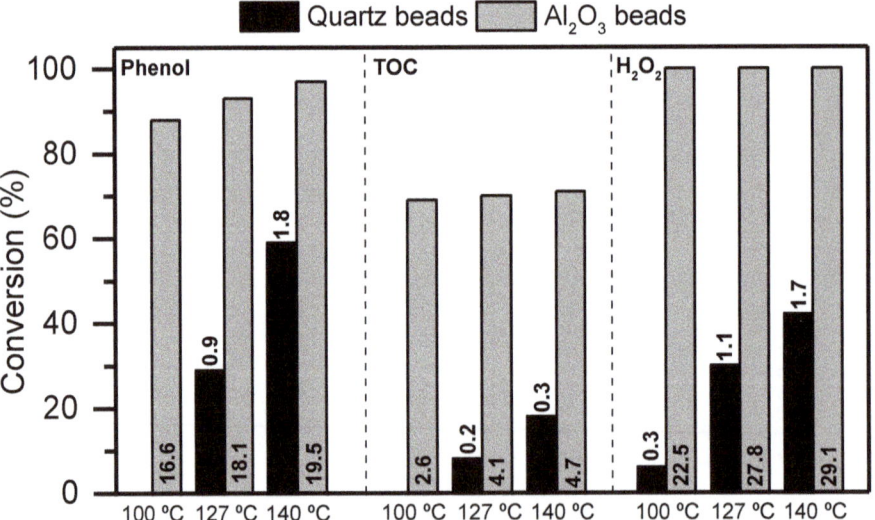

Figure 2. Phenol disappearance, TOC removal and H_2O_2 consumption upon the phenol WPO-O_2 over quartz and Al_2O_3 beads (W_{quarzt} = 29 g and W_{Al2O3} = 2 g, respectively) at different temperatures. Operating conditions: [phenol]$_0$ = 1000 mg L^{-1}, [H_2O_2]$_0$ = 2500 mg L^{-1}, P_{O2} = 8 bar (92 N mL$_{O2}$ min^{-1}), natural pH$_0$ and t_R = 7 min. The inset numbers correspond to the initial rate values calculated in g_i g_{solid}^{-1} h^{-1}.

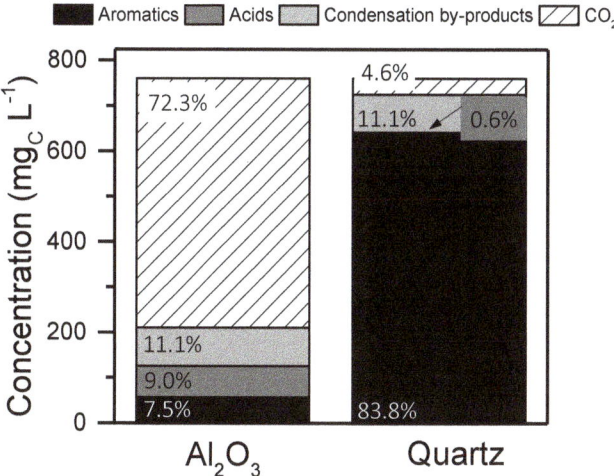

Figure 3. Comparison of the reaction products distribution using quartz and Al_2O_3 beads at 7 min of reaction. Operating conditions are the same as Figure 2.

2.2. Nature of the Condensation By-Product Deposits

Once it was demonstrated that the condensation by-products deposited on the surface of an inert solid (Al_2O_3 or quartz beads) were able to decompose H_2O_2 and thus, to autocatalyze the WPO-O_2 of phenol, an in-depth characterization of the solids before and after being used in the reaction was carried out. Given the significantly higher specific surface area of Al_2O_3 for the deposition of the by-product species, this solid was selected for the characterization study. The main goals were to learn about the amount, composition and nature of the deposited carbonaceous by-products.

The textural properties of both fresh and used Al_2O_3 spheres (after 650 h in stream) were first determined. Up to 70% fall on the S_{BET} was observed, from 275 to 84 m^2 g^{-1}, which seems to be due to the presence of carbonaceous deposits, although the possible hydration of Al_2O_3 under the WPO-O_2 operating conditions should also be considered, as was further confirmed by thermogravimetric analysis (TGA) and Fourier transform infrared spectroscopy (FTIR). Also, the mesopore volume was reduced from 0.32 to 0.15 cm^3 g^{-1}. Scanning Electron Microscope (SEM) images of the Al_2O_3 spheres before and after being used in the WPO-O_2 of phenol can be seen in Figure 4. As observed, the fresh sphere presents a clearly rough surface, consistent with the porosity of the solid, while the used one shows a smooth aspect, which could be indicative of the coverage of the material by carbonaceous deposits. This effect can be better appreciated in the images taken at high resolution (x1K, x10K). These results are consistent with the decrease of S_{BET} and porosity previously described. The bright areas that are observed in some images, i.e., Figure 4e, correspond to the presence of metals, mainly iron (Fe) and titanium (Ti), as was confirmed by Energy Dispersive X-ray spectroscopy (EDX) (Figure S1 in the Supplementary Material shows the EDX spectra). These metals come from the leaching of the tubes and reactor, respectively. Their total content is less than 0.2 wt.% (measured by inductively coupled plasma mass spectrometry, ICP-MS), therefore their effect on the reaction can be considered negligible.

The presence of carbonaceous deposits on the Al_2O_3 spheres was further confirmed by thermogravimetric analysis (TGA) of the solids in air atmosphere. The obtained results are depicted in Figure 5. Al_2O_3 suffered dehydration and a consequent phase transformation at temperatures around 450–500 °C [17–19], which can be clearly observed in both fresh and used samples. Nevertheless, in the latter, there was a shoulder at 350 °C and the weight loss from 300 to 600 °C was significantly higher than that observed in the fresh material. This peak at 350 °C was ascribed to the carbonaceous deposits, which were removed during the thermal treatment as indicated by the white color of the resulting Al$_2$O$_3$ spheres. By integrating the curves corresponding to the fresh and used samples, the weight loss due to the condensation by-products burned-off was calculated to be around 6% within the temperature range of 30–900 °C, which is consistent with the results of the ICP (provided above).

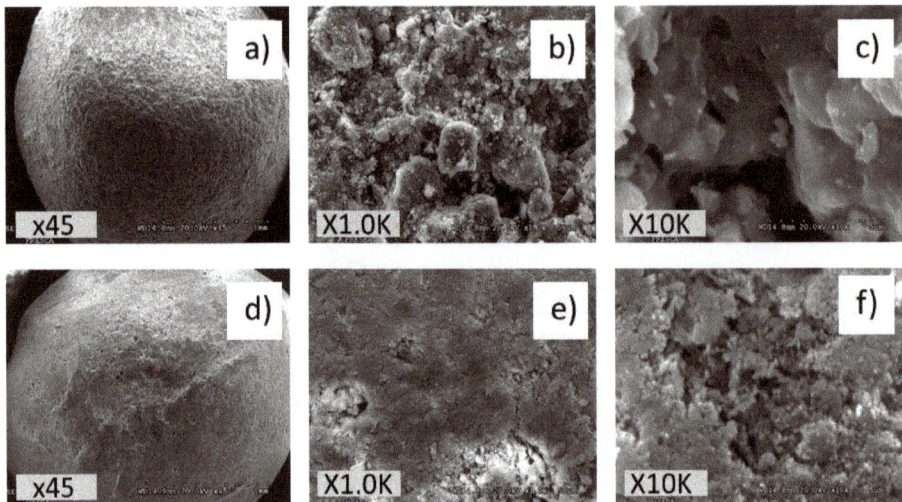

Figure 4. SEM micrographs of the Al$_2$O$_3$ spheres before (**a**,**b**,**c**) and after (**d**,**e**,**f**) being used in the WPO-O$_2$ of phenol.

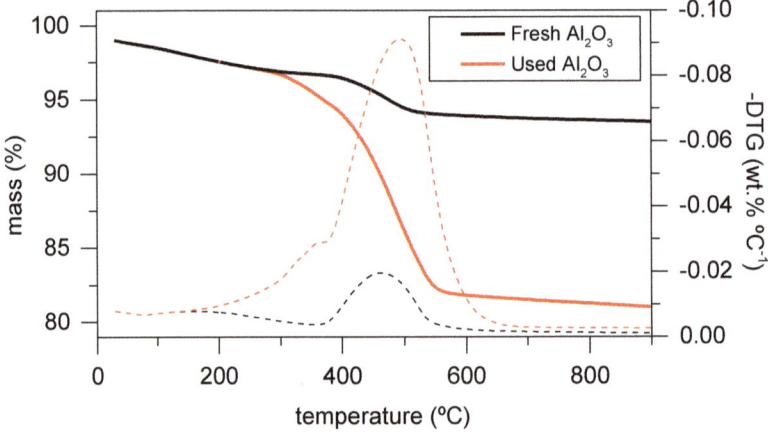

Figure 5. Differential thermogravimetric profiles (wt.%·°C^{-1}) of the fresh and used Al$_2$O$_3$ in helium atmosphere at 10 °C min^{-1}.

The characterization of the fresh and used Al$_2$O$_3$ spheres by different techniques such as elemental analysis (EA) and ICP-MS allowed us to get an insight on the composition of the condensation. The content of carbon and hydrogen increased from 0.11 to 2.90 wt.% and 1.61 to 2.07 wt.%, for the fresh and spent Al$_2$O$_3$, respectively. The high content of hydrogen in the fresh Al$_2$O$_3$ spheres (1.61 wt.%) could indicate the presence of bohemite phase AlO(OH). In fact, X-Ray Diffraction (XRD) confirmed the coexistence of both phases, bohemite and γ-Al$_2$O$_3$, the spectrum is shown in Figure S2 in the Supplementary Material. Taking into account that the sum of carbon and hydrogen content in the used Al$_2$O$_3$ spheres is lower than the 6 wt.% of reaction products, identified as the condensation by TGA, it is expected that the carbonaceous by-products contain oxygen in their structure. The amount of Al (measured by IPC-MS) decreased from 46.1 to 43.1 wt.%, which corresponds to the 6 wt.% of the condensation deposits.

To learn more about the functionalization of the condensation by-products deposited on the Al_2O_3 spheres, FTIR and X-Ray photoelectron spectroscopy (XPS) analyses were performed on the fresh and used Al_2O_3. FTIR spectra are given in Figure 6. The characteristic bands of fresh Al_2O_3 appear in the range of 1070–1150 cm^{-1}, which are typical of bohemite (AlO(OH)) [20,21], and also in the range of 3093–3295 cm^{-1}, which correspond to the vibration of OH bonds. These bands were of greater intensity in the used Al_2O_3 sample, in line with the hydration of the alumina upon the oxidation process, as observed by TGA. Nevertheless, the band at around 3100 cm^{-1} can also be assigned to the = C–H stretch in the aromatics [9], which could indicate that carbonaceous deposits show an aromatic character. On the other hand, by comparing both spectra, the appearance of some new bands in the used solid can be clearly seen. In fact, the new band in the range between 1025 and 1250 cm^{-1} can also be assigned to the C–H in-plane banding of the phenyl group [9]. The aromatic nature of the carbonaceous deposits is consistent with previous works where the precipitation of a solid polymer following the CWPO of phenol was observed [4,5]. Poerschmann et al. (2009) [4] postulated that carbonaceous solids showed features of humic acids, as they were formed by the oxidative coupling reactions of phenolic rings. The band at 1427 cm^{-1} revealed the presence of C–O and OH groups in carboxylic acids, while those at 1698 and 1718 cm^{-1} indicated the occurrence of C=O groups and C=O from quinones, aldehydes and ketones, respectively. The bands in the range of 2100–2300 cm^{-1} could be due to C≡C stretch [22]. Therefore, the carbonaceous deposits exhibit a complex composition of phenols, carboxylic acids, aldehydes, ketones, aromatic carbon rings and aliphatic carbons.

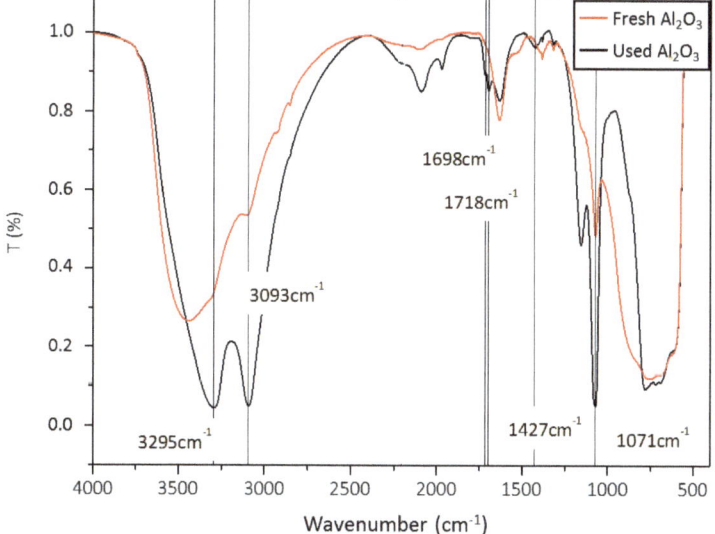

Figure 6. FTIR spectra of the Al_2O_3 spheres before and after being used in the phenol WPO-O_2.

In line with the previous characterization analyses, XPS spectrum of the used Al_2O_3 spheres also allowed us to confirm the presence of carbonaceous deposits on the solid surface, as indicated by the strong C1s peak observed (Figure 7). Its asymmetric shape, with decreasing intensity towards higher binding energies indicate that there is a contribution from oxygenated functional groups [23]. The C1s spectrum was fitted to three peaks: a peak for polyaromatic carbons (284.8 eV, C–C and

C–H), for carbon singly bonded to one oxygen atom (285.8 eV, C–O) and for carboxylate groups (288.5, O–C=O) [24].

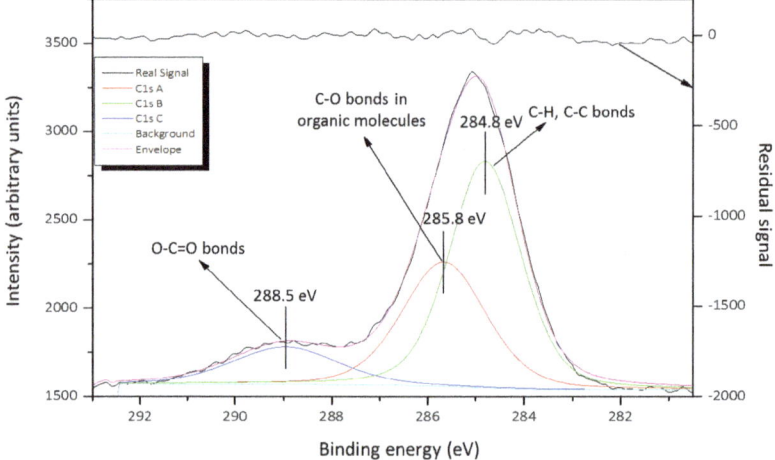

Figure 7. Deconvolution spectrum of C1s for the Al_2O_3 spheres after being used in the phenol WPO-O_2.

2.3. Kinetic Modeling

The kinetic model developed in Part I of this work [16] for phenol WPO-O_2 in the presence of quartz beads explained that the condensation by-products catalyzed the decomposition of H_2O_2 into oxidizing radicals able to degrade the organic pollutant, according to the following autocatalytic rate equations:

$$(-r_{H_2O_2})\left(\frac{mol}{L\cdot min}\right) = k'_{H_2O_2} \cdot C_{H_2O_2} \cdot (C_{H_2O_2,0} - C_{H_2O_2}) \quad (1)$$

$$(-r_{Phenol})\left(\frac{mol}{L\cdot min}\right) = k'_{Phenol} \cdot C_{Phenol} \cdot (C_{H_2O_2,0} - C_{H_2O_2}) \quad (2)$$

$$(-r_{TOC})\left(\frac{mol}{L\cdot min}\right) = k'_{TOC} \cdot C_{TOC} \cdot (C_{H_2O_2,0} - C_{H_2O_2}) \quad (3)$$

The concentration profiles of phenol, TOC and H_2O_2 vs. residence time (t_r) obtained in the presence of Al_2O_3 at different temperatures are shown in Figure 8 (in symbols). As observed, the disappearance of phenol is not significantly affected by the temperature, the same for TOC and H_2O_2 decomposition at temperatures above 127 °C.

The numerical integration of the rate equations of Equations (1–3) assuming plug-flow reactor, with the initial conditions [Phenol] = [Phenol]$_0$, [H_2O_2] = [H_2O_2]$_0$ at t = 0 was solved by using the Microsoft Excel Solver (Microsoft Office, 2010, Microsoft Corp., Redmond, WA, USA) based on the Generalized Reduced Gradient (GRG) algorithm for least squares minimization. The equations were solved at all temperatures simultaneously considering the Arrhenius equation to determine activation energy and the pre-exponential factor.

The resulting differential equations of this autocatalytic model are:

$$(-r_{H_2O_2})\left(\frac{mol}{L\cdot min}\right) = 17.8\cdot 10^3 \cdot \exp\left(\frac{-1095}{T}\right)\cdot C_{H_2O_2}\cdot (0.074 - C_{H_2O_2}) \quad (R^2 = 0.99) \quad (4)$$

$$(-r_{Phenol})\left(\frac{mol}{L\cdot min}\right) = 267 \cdot \exp\left(\frac{-20.5}{T}\right)\cdot C_{Phenol}\cdot (0.074 - C_{H_2O_2}) \quad (R^2 = 0.96) \quad (5)$$

$$(-r_{TOC})\left(\frac{mol}{L \cdot min}\right) = 21.1 \cdot 10^3 \cdot \exp\left(\frac{-1949}{T}\right) \cdot C_{TOC} \cdot (0.074 - C_{H_2O_2}) \qquad (R^2 = 0.94) \qquad (6)$$

The good agreement found between the experimental (symbols) and calculated values (lines) validates this model, as can be seen in Figure 8. Nevertheless, it must be noted that the values obtained for the pre-exponential factor (L mol^{-1} min^{-1}) and activation energy (E$_a$/R) are actually a lumping of different reactions including H$_2$O$_2$ decomposition, CP production and phenol or TOC evolution rate. The existence of these various phenomena occurring simultaneously but with opposite effects makes it difficult to assign a physical meaning to those values.

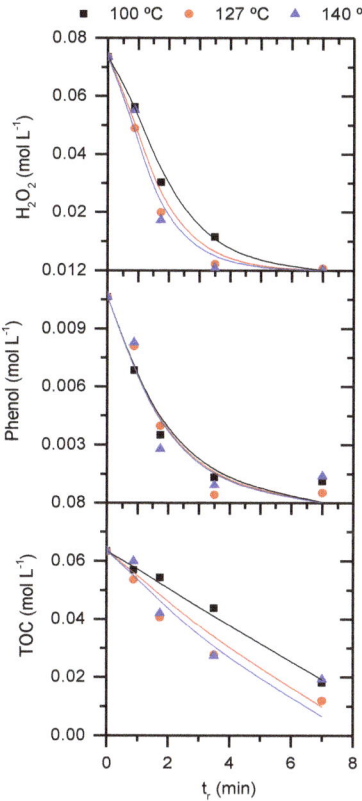

Figure 8. Experimental (symbols) and predicted (lines) concentrations of H$_2$O$_2$, phenol and TOC by the autocatalytic kinetic model of the phenol WPO-O$_2$ over Al$_2$O$_3$ at different temperatures. Operating conditions: [Phenol]$_0$ = 1000 mg L^{-1}, [H$_2$O$_2$]$_0$ = 2500 mg L^{-1}, P$_{O2}$ = 8 bar, natural pH$_0$ and W$_{Al2O3}$ = 2 g.

2.4. Stability of the Condensation By-Product Deposits

The condensation by-products generated upon reaction can also be consumed, depending on the operating conditions. It is then clear that reaching a stable concentration of these species is a key issue to ensure the efficiency and reproducibility of the results. The main aspects that are expected to determine the generation and consumption of condensation by-products during reactions are the initial target pollutant concentration and the dose of H$_2$O$_2$, respectively [16]. Both process variables are analyzed below.

Figure 9 shows the initial oxidation rate values obtained for phenol disappearance, TOC removal and H$_2$O$_2$ consumption when the initial concentration of phenol was evaluated within 100–5000 mg L^{-1}, maintaining the dose of H$_2$O$_2$ at 50% of the stoichiometric amount for its complete mineralization.

As observed, the highest initial oxidation rate was achieved with the highest phenol concentration tested (5000 mg L^{-1}). The difference was quite significant, around one and even two orders of magnitude higher, when comparing 5000 mg L^{-1} to 1000 and 100 mg L^{-1}, respectively. This is indicative of a higher amount of "catalytic" carbonaceous deposits formed by increasing the initial concentration of phenol. On the other hand, it should be mentioned that the efficiency of the consumption of H_2O_2 (calculated as the ratio between the conversion of TOC and that of H_2O_2) also increased with the amount of phenol, from 15 to 60%, when the concentration was varied from 100 to 5000 mg L^{-1}. Therefore, the amount of condensation by-products deposited on the Al_2O_3 spheres not only affected the kinetics but also the extension of the reaction. In fact, the intermediate distribution was strongly dependent on the initial concentration of phenol tested (see Figure S3 in the Supplementary Material for experimental data). Notably, the amount of dissolved condensation by-products increased with the initial concentration of phenol (from 1 to 13% of the initial TOC was attributed to dissolved condensation by-products with 100 and 5000 mg L^{-1} initial phenol concentration, respectively). This is consistent with the fact that those species are generated by condensation of phenolic radicals, which would be in significantly higher amounts at higher initial phenol concentrations.

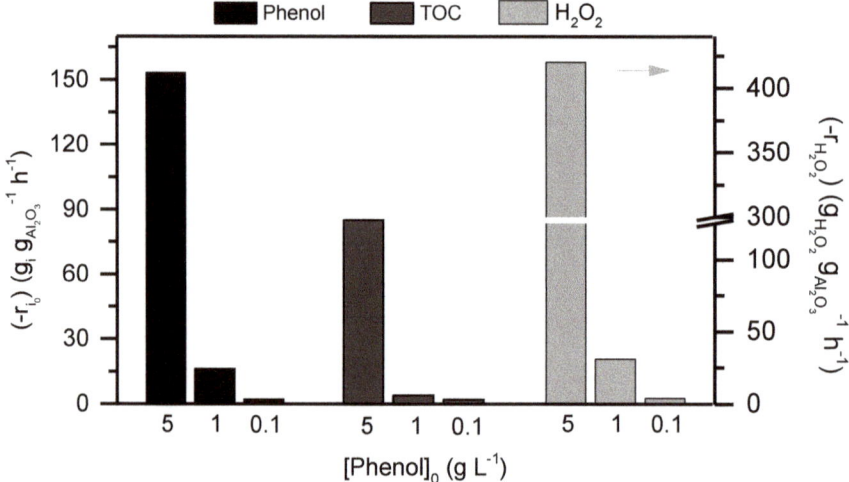

Figure 9. Effect of the initial phenol concentration on the initial rate of phenol disappearance, TOC removal and H_2O_2 consumption upon the phenol WPO-O_2 over Al_2O_3 spheres. Operating conditions: [H_2O_2]$_0$ equivalent to the 50% stoichiometric dose for phenol mineralization, P_{O2} = 8 bar (92 N mL$_{O2}$ min^{-1}), T = 140 °C, natural pH$_0$ and W$_{Al2O3}$ = 2g.

The results obtained in terms of TOC conversion at different H_2O_2 doses (from 50–300% of the stoichiometric amount for complete phenol oxidation), maintaining the initial phenol concentration at 1000 mg L^{-1}, are summarized in Figure 10. It should be noted that at the residence time established for all experiments (7 min), complete phenol degradation and total H_2O_2 consumption was reached. As observed, the increase in H_2O_2 concentration resulted in higher mineralization yields, even above 90% when two-times the stoichiometric amount of H_2O_2 was used. Nevertheless, a further increase in the amount of H_2O_2 did not lead to a higher degree of mineralization, which could be due to the inefficient consumption of this reagent by the radical auto-scavenging reactions, and also to the fact that the amount of carbonaceous deposits present to catalyze the reaction would be significantly lower as they would be oxidized due to the harsher operating conditions. The images of the Al_2O_3 spheres at the exit of the reactor with the lowest and highest H_2O_2 doses tested confirmed this hypothesis (Figure 10). While the used solids showed a brown color when H_2O_2 doses below the stoichiometric amount were used, Al_2O_3 spheres recovered their original white color at or above that threshold value, which proved that the condensation by-products deposits were completely removed. It should also be noted that the efficiency of H_2O_2 consumption (calculated as the ratio between the conversion of TOC and that of H_2O_2) was progressively decreased with increasing the H_2O_2 dose. Values of the efficiency of 43, 31, 26, 14 and 9.3% were obtained with H_2O_2 doses of 50%, 75%, 100%, 200% and 300% of the stoichiometric amount, respectively.

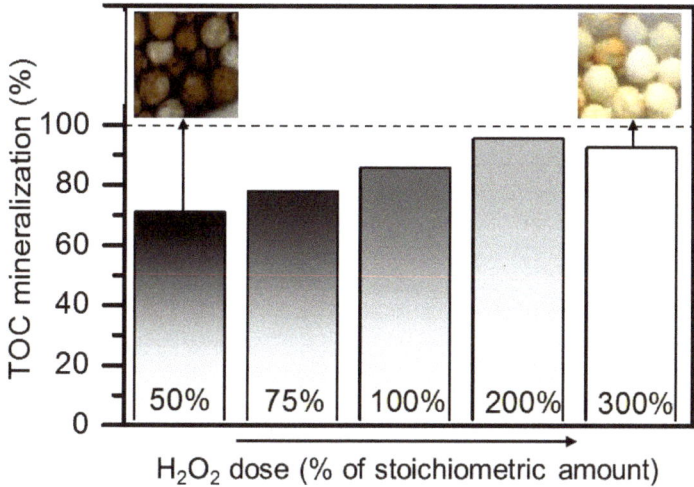

Figure 10. Effect of the H_2O_2 dose on the TOC conversion achieved upon the phenol WPO-O_2 over Al_2O_3 at 7 min of reaction time. Operating conditions: [Phenol]$_0$ = 1000 mg L^{-1}, P_{O2} = 8 bar (92 N mL$_{O2}$ min^{-1}), T = 140 °C, natural pH$_0$ and W$_{Al2O3}$ = 2 g. The inset images show the Al_2O_3 spheres at the exit of the reactor.

3. Experimental Methods

3.1. Al_2O_3 Spheres and Characterization

Commercial Al_2O_3 spheres (ϕ_p = 24 mm) supplied by Sigma Aldrich (ref.: 414069, Kyoto, Japan) were used as inert material to arrange a bed to adsorb the condensation by-products formed during the WPO-O_2 process.

Al_2O_3 spheres before and after the WPO-O_2 experiments (fresh and used Al_2O_3) were characterized by different techniques. The porous structure of the catalysts was characterized from the 77 K N_2 adsorption/desorption using a Micromeritics Tristar apparatus. The samples were first outgassed overnight at 150 °C to a reduced pressure < 10^{-3} Torr in order to ensure a dry clean surface. The morphology of the Al_2O_3 spheres was observed by scanning electron microscopy (SEM) using a Hitachi S-3000N apparatus at an accelerating voltage of 20 kV. Evidence for the presence of carbonaceous deposits on the spent Al_2O_3 spheres were obtained by thermogravimetric analysis (TGA/SDTA851e thermobalance, Mettler-Toledo) coupled to a Thermostat (model GSD 301 TC) from Pfeiffer Vacuum MS. The samples were heated in helium from room temperature to 900 °C at 10 °C min^{-1}. Besides, the chemical composition of the condensation by-products was investigated by EA and ICP-MS. EA for carbon and hydrogen were carried out with a LECO Model CHNS-932 analyzer. Aluminum quantification by ICP-MS in a PerkinElmer's NexION 2000. The samples were prepared by closed microwave digestion in acid solution, containing H_3PO_4 and HF, at 270 °C for 2 h. XRD was performed with a PANalytical X'Pert Pro Theta/2Theta diffractometer with CuK radiation on Al_2O_3 spheres previously crushed into powder. Data were collected from 10° to 100° (2θ), with steps of 0.0167°. Information about the functionalization of the condensation by-products was obtained by the FTIR (Bruker IFS66v spectrophotometer, 4 cm^{-1} resolution, 250 scans in normal conditions, 7000–550 cm^{-1}) and XPS analyses (Physical Electronics, mod. ESCA 5701, equipped with a monochromatic Mg Ka X-ray excitation source, $h\upsilon$ = 1253.6 eV). The deconvolution of the C 1s profile at 284.6 eV was carried out with the provided software 'Multipak v8.2b', which obtains the relative amounts of the different surface groups. The procedure involves smoothing, background subtraction and mixed Gaussian–Lorentzian peak shape by a least-squared method curve fitting.

3.2. Oxidation Experiments

Oxidation experiments were conducted in a fixed-bed reactor consisting of a titanium tube with a 0.91 cm internal diameter and 30 cm long (reactor volume, V_R = 4 cm^3), loaded with 2 g of glass beads (ϕ_{beads} = 2.4 mm and ε_L = 0.45). The temperature was measured by a thermocouple located in the bed. The liquid and gas phases were passed through the bed in concurrent up-flow. Detailed information about the components and operation procedure of this setup has been reported elsewhere [25]. An aqueous solution of H_2O_2 with or without phenol at natural pH_0 was continuously fed to the reactor along with a 92 NmL min^{-1} pure oxygen stream. H_2O_2 decomposition reactions were performed with fresh and used Al_2O_3. The same results were obtained in the presence and absence of Al_2O_3, which demonstrated that it is an inert material for the decomposition of H_2O_2. The runs were conducted at $[H_2O_2]_0$ = 2500 mg L^{-1}, P_{O2} = 8 bar (92 N mL$_{O2}$ min^{-1}), T = 150 °C, natural pH_0 and at different flow rates (Q_L) = 10 and 5 mL min^{-1} equivalent to residence times (t_r) of c.a. 5.5 and 11 s. This t_r is calculated as the ratio of the liquid volume (V_L) and flow rate (Q_L). V_L was equal to $\varepsilon_L \cdot V_R$. The used Al_2O_3 was obtained after 145 h on stream upon phenol WPO-O_2 at the following operating conditions: $[phenol]_0$ = 1000 mg L^{-1}, $[H_2O_2]_0$ = 2500 mg L^{-1}, P_{O2} = 8 bar (92 N mL$_{O2}$ min^{-1}), T = 140 °C, natural pH_0 and t_r = 4.5 min.

The phenol WPO-O_2 runs were conducted in a wide range of operating conditions: $[phenol]_0$ = 100–5000 mg L^{-1}, $[H_2O_2]_0$ = 250–12500 mg L^{-1} (from 50 to 300% stoichiometric dose of H_2O_2, depending on the initial phenol concentration), P_{O2} = 8 bar (92 N mL$_{O2}$ min^{-1}), T = 100–140 °C and Q_L = 1–0.125 mL min^{-1} to cover the experimental range of t_r = 0.9–7 min.

Liquid samples were periodically withdrawn from the reactors and immediately injected in a vial (submerged in crushed ice) containing a known volume of cold distilled water. The diluted samples were filtered (0.45 µm Nylon filter) and subsequently analyzed by different techniques.

3.3. Analytical Methods

High performance liquid chromatography (Ultimate 3000, Thermo Scientific, Waltham, MA, USA) was used to follow the evolution of phenol and aromatic intermediates during WPO-O_2 reactions. A C18 5 µm column (Kinetex from Phenomenex, 4.6 mm diameter, 15 cm long) and a 4 mM H_2SO_4 aqueous solution were employed as stationary and mobile phases, respectively. Wavelengths of 210 and 246 nm were used for the quantification. Short-chain organic acids were analyzed by means of chromatography (Personal IC mod. 790, Metrohm). A mixture of Na_2CO_3 (3.2 mM) and $NaHCO_3$ (1 mM) aqueous solutions and a Metrosep A sup 5–250 column (4 mm diameter, 25 cm long) were used. A TOC analyzer (Shimadzu TOC VSCH, Kyoto, Japan) was used to quantify total organic carbon (TOC).

4. Conclusions

This work demonstrates that the condensation by-products formed during the WPO process promote H_2O_2 decomposition into radical species. This is achieved by the electron donor-acceptor groups in the condensation species (which are represented as aromatic carbon functionalized with quinone, phenolic, aldehyde, carboxylic and ketones groups) that make the electron-transfer to the H_2O_2 molecules and vice versa possible. Their beneficial effect on the oxidation process can be described through an autocatalytic kinetic model.

The oxidation rate and the extension of the reaction is determined by the amount of the condensation by-product deposits. In the presence of porous Al_2O_3 spheres in a packed bed, the oxidation rate values were one order of magnitude higher than the quartz beads, the mineralization yield was also significantly greater (X_{TOC} = 72% *vs.* 5%, respectively), and the intermediate distribution was very different, with a 90% less concentration of aromatic by-products in the media.

The amount of the deposits is a trade-off between the initial concentration of phenol (which favors the polycondensation reactions and therefore the production of condensation by-products) and the dose of H_2O_2 (which mineralizes the condensation by-products when used at high enough concentrations).

Supplementary Materials: The following are available online at http://www.mdpi.com/2073-4344/9/6/518/s1, Figure S1. EDX spectra of the Al_2O_3 spheres after being used in the $WPO-O_2$ of phenol; Figure S2. XRD spectra of the fresh Al_2O_3; Figure S3. Influence of the initial phenol concentration on the by-product distribution in the $WPO-O_2$ process.

Author Contributions: Conceptualization, J.A.C. and A.Q.; Data curation, J.L.D.T. and C.F.; Formal analysis, A.Q., J.L.D.T., C.F. and M.M.; Funding acquisition, J.A.C.; Investigation, A.Q., J.L.D.T., C.F. and M.M.; Methodology, A.Q., J.L.D.T. and C.F.; Project administration, J.A.C.; Supervision, J.A.C. and A.Q.; Validation, J.L.D.T. and C.F.; Writing—original draft, A.Q. and M.M.; Writing—review & editing, A.Q., M.M. and J.A.C.

Funding: This research was supported by the Spanish MINECO through the project CTM-2016-76454-R and by the CM through the project P2018/EMT-4341. M. Munoz thanks the Spanish MINECO for the Ramón y Cajal postdoctoral contract (RYC-2016-20648).

Conflicts of Interest: The authors declare no conflict of interest.

References

1. Delgado, J.J.; Chen, X.; Pérez-Omil, J.A.; Rodríguez-Izquierdo, J.M.; Cauqui, M.A. The effect of reaction conditions on the apparent deactivation of Ce–Zr mixed oxides for the catalytic wet oxidation of phenol. *Catal. Today* **2012**, *180*, 25–33. [CrossRef]
2. Vallejo, M.; Fernández-Castro, P.; San Román, M.F.; Ortiz, I. Assessment of PCDD/Fs formation in the Fenton oxidation of 2-chlorophenol: Influence of the iron dose applied. *Chemosphere* **2015**, *137*, 135–141. [CrossRef] [PubMed]
3. Domínguez, C.M.; Ocón, P.; Quintanilla, A.; Casas, J.A.; Rodriguez, J.J. Highly efficient application of activated carbon as catalyst for wet peroxide oxidation. *Appl. Catal. B* **2013**, *140–141*, 663–670. [CrossRef]
4. Poerschmann, J.; Trommler, U.; Górecki, T.; Kopinke, F.D. Formation of chlorinated biphenyls, diphenyl ethers and benzofurans as a result of Fenton-driven oxidation of 2-chlorophenol. *Chemosphere* **2009**, *75*, 772–780. [CrossRef] [PubMed]
5. Munoz, M.; de Pedro, Z.M.; Casas, J.A.; Rodriguez, J.J. Assessment of the generation of chlorinated byproducts upon Fenton-like oxidation of chlorophenols at different conditions. *J. Hazard. Mater.* **2011**, *190*, 993–1000. [CrossRef] [PubMed]
6. Diaz de Tuesta, J.L.; Quintanilla, A.; Casas, J.A.; Rodriguez, J.J. Kinetic modeling of wet peroxide oxidation with a carbon black catalyst. *Appl. Catal. B* **2017**, *209*, 701–710. [CrossRef]
7. Munoz, M.; de Pedro, Z.M.; Casas, J.A.; Rodriguez, J.J. Combining efficiently catalytic hydrodechlorination and wet peroxide oxidation (HDC–CWPO) for the abatement of organochlorinated water pollutants. *Appl. Catal. B* **2014**, *150 151*, 197 203. [CrossRef]
8. Hamoudi, S.; Larachi, F.; Adnot, A.; Sayari, A. Characterization of Spent MnO_2/CeO_2 Wet Oxidation Catalyst by TPO–MS, XPS, and S-SIMS. *J. Catal.* **1999**, *185*, 333–344. [CrossRef]
9. Kim, S.; Ihm, S. Nature of carbonaceous deposits on the alumina supported transition metal oxide catalysts in the wet air oxidation of phenol. *Top. Catal.* **2005**, *33*, 171–179. [CrossRef]
10. Lee, D.K.; Kim, D.S.; Kim, T.H.; Lee, Y.K.; Jeong, S.E.; Le, N.T.; Cho, M.J.; Henam, S.D. Deactivation of Pt catalysts during wet oxidation of phenol. *Catal. Today* **2010**, *154*, 244–249. [CrossRef]
11. Keav, S.; de los Monteros, A.E.; Barbier, J.; Duprez, D. Wet Air Oxidation of phenol over Pt and Ru catalysts supported on cerium-based oxides: Resistance to fouling and kinetic modelling. *Appl. Catal. B* **2014**, *150–151*, 402–410. [CrossRef]
12. Yadav, A.; Verma, N. Carbon bead-supported copper-dispersed carbon nanofibers: An efficient catalyst for wet air oxidation of industrial wastewater in a recycle flow reactor. *J. Ind. Eng. Chem.* **2018**, *67*, 448–460. [CrossRef]
13. Catrinescu, C.; Teodosiu, C.; Macoveanu, M.; Miehe-Brendlé, J.; Le Dred, R. Catalytic wet peroxide oxidation of phenol over Fe-exchanged pillared beidellite. *Water Res.* **2003**, *37*, 1154–1160. [CrossRef]

14. Collett, C.H.; McGregor, J. Things go better with coke: the beneficial role of carbonaceous deposits in heterogeneous catalysis. *Catal. Sci. Technol.* **2016**, *6*, 363–378. [CrossRef]
15. Catrinescu, C.; Arsene, D.; Teodosiu, C. Catalytic wet hydrogen peroxide oxidation of para-chlorophenol over Al/Fe pillared clays (AlFePILCs) prepared from different host clays. *Appl. Catal. B* **2011**, *101*, 451–460. [CrossRef]
16. Quintanilla, A.; Díaz de Tuesta, J.L.; Figueruelo, C.; Munoz, M.; Casas, J.A. Condensation by-products in wet peroxide oxidation: fouling or catalytic promotion? Part I. Evidences of an autocatalytic process. *Catalysts* **2019**, in press.
17. Becerra, M.E.; Arias, N.P.; Giraldo, O.H.; López-Suárez, F.E.; Illán-Gómez, M.J.; Bueno-López, A. Alumina-supported manganese catalysts for soot combustion prepared by thermal decomposition of KMnO4. *Catalysts* **2012**, *2*, 352–357. [CrossRef]
18. Lefävre, G.; Duc, M.; Lepeut, P.; Caplain, R.; Fadoroff, M. Hydration of $\^3$-alumina in water and its effects on surface reactivity. *Langmuir* **2002**, *18*, 7530–7537. [CrossRef]
19. El-Naggar, A.Y. Thermal analysis of the modified, coated and bonded alumina surfaces. *J. Emer. Trends Eng. Appl. Sci.* **2014**, *5*, 30–34.
20. Ram, S. Infrared spectral study of molecular vibrations in amorphous, nanocrystalline and AlO(OH)·αH$_2$O bulk crystals. *Infrared Phys. Technol.* **2001**, *42*, 547–560. [CrossRef]
21. Nyquist, R.A.; Lengers, M.A. Handbook of infrared and ramen spectra of inorganic compounds and organic salts. *Search PubMed* **1997**, *4*, 72.
22. Lambert, J.B.; Shurvell, H.F.; Lightner, D.A.; Cooks, R.G. *Introduction to Organic Spectroscopy*; Macmillan Publishing Company: London, UK, 1987.
23. Estrade-Szwarckopf, H. XPS photoemission in carbonaceous materials: A "defect" peak beside the graphitic asymmetric peak. *Carbon* **2004**, *42*, 1713–1721. [CrossRef]
24. Ishimaru, K.; Hata, T.; Bronsveld, P.; Meier, D.; Imamura, Y. Spectroscopic analysis of carbonization behavior of wood, cellulose and lignin. *J. Mater. Sci.* **2007**, *42*, 122–129. [CrossRef]
25. Quintanilla, A.; Casas, J.A.; Zazo, J.A.; Mohedano, A.F.; Rodríguez, J.J. Wet air oxidation of phenol at mild conditions with a Fe/activated carbon catalyst. *Appl. Catal. B* **2006**, *62*, 115–120. [CrossRef]

© 2019 by the authors. Licensee MDPI, Basel, Switzerland. This article is an open access article distributed under the terms and conditions of the Creative Commons Attribution (CC BY) license (http://creativecommons.org/licenses/by/4.0/).

Article

Kinetic and Mechanistic Study on Catalytic Decomposition of Hydrogen Peroxide on Carbon-Nanodots/Graphitic Carbon Nitride Composite

Zhongda Liu, Qiumiao Shen, Chunsun Zhou, Lijuan Fang, Miao Yang * and Tao Xia *

School of Chemistry, Chemical Engineering and Life Sciences, Wuhan University of Technology, Wuhan 430070, China; liuzhongda@whut.edu.cn (Z.L.); shen_qiumiao@163.com (Q.S.); zhouchunsun@whut.edu.cn (C.Z.); fanglijuan@whut.edu.cn (L.F.)
* Correspondence: yangmiao@whut.edu.cn (M.Y.); hsiatao@hotmail.com (T.X.);
  Tel.: +188-7225-6465 (M.Y.); +134-0714-2371 (T.X.)

Received: 12 September 2018; Accepted: 8 October 2018; Published: 11 October 2018

Abstract: The metal-free CDots/g-C_3N_4 composite, normally used as the photocatalyst in H_2 generation and organic degradation, can also be applied as an environmental catalyst by in-situ production of strong oxidant hydroxyl radical (HO·) via catalytic decomposition of hydrogen peroxide (H_2O_2) without light irradiation. In this work, CDots/g-C_3N_4 composite was synthesized via an electrochemical method preparing CDots followed by the thermal polymerization of urea. Transmission electron microscopy (TEM), X-Ray diffraction (XRD), Fourier Transform Infrared (FTIR), N_2 adsorption/desorption isotherm and pore width distribution were carried out for characterization. The intrinsic catalytic performance, including kinetics and thermodynamic, was studied in terms of catalytic decomposition of H_2O_2 without light irradiation. The second-order rate constant of the reaction was calculated to be $(1.42 \pm 0.07) \times 10^{-9}$ m·s^{-1} and the activation energy was calculated to be (29.05 ± 0.80) kJ·mol^{-1}. Tris(hydroxymethyl) aminomethane (Tris) was selected to probe the produced HO· during the decomposing of H_2O_2 as well as to buffer the pH of the solution. The composite was shown to be base-catalyzed and the optimal performance was achieved at pH 8.0. A detailed mechanism involving the *adsorb-catalyze* double reaction site was proposed. Overall, CDots/g-C_3N_4 composite can be further applied in advanced oxidation technology in the presence of H_2O_2 and the instinct dynamics and the mechanism can be referred to further applications in related fields.

Keywords: CDots/g-C_3N_4; H_2O_2; hydroxyl radical; Tris; advanced oxidation technology

---

1. Introduction

Advanced oxidation technology (AOT) is one of the most effective and economical approaches dealing with non-biodegradable organic pollutants (NBDOPs) in water, such as dyestuffs, pesticides, pharmaceutical and personal care products (PPCPs), synthetic chemicals and leachate of landfills [1–5]. In typical AOTs, different strategies like chemical, photochemical, sonochemical and electrochemical pathways, are employed to produce intermediate active oxidant radicals [1,6–8]. With an oxidation potential of 2.7 eV and nanosecond-level life time, hydroxyl radical (HO·) is one of the most typical radicals, which can decompose NBDOPs non-selectively, forming CO_2, H_2O, inorganic ions or other biodegradable molecules [9–11]. It is worthy to note that the degradation of NBDOPs and the generation of HO· take place simultaneously [12]. Thus, the core process of various AOTs is to improve the yield of HO·, which mainly leads to the decomposition of NBDOPs.

The concentration of instantaneous HO· can be hardly determined directly but can be determined indirectly by probes like Rhodamine B [13], terephthalic acid [14], dimethyl sulfoxide [15], phenylalanine [16] and Tris(hydroxymethyl)aminomethane (Tris) [17]. Among the various probes, Tris can be applied in both homogeneous and heterogeneous systems as HO· scavenger and pH buffer at the same time [18]. Hydroxyl radical captures hydrogen atom from Tris, producing formaldehyde (CH_2O) and other compounds. Since the produced CH_2O can be quantified by the modified Hantzsch method [19], the concentration of HO· can be indirectly quantified. The detailed mechanism of the reaction between HO· and Tris was reported, involving the effects of O_2 and pH [17].

H_2O_2 is one of the most common sources of HO· in the presence of metal salt solution, carbon-based species, metal or metal oxide via Fenton/Fenton-like reaction, electron-transfer mechanism or catalytic decomposition on the solid-liquid interface [18,20–24]. The well-known Fenton/Fenton-like reaction may occur in both homogeneous and heterogeneous system according to many works [18,20–23]. Similar to the Fenton reaction, HO· and HO_2· can be formed on the surface of carbon-based catalysts via the electron-transfer mechanism due to the donor-acceptor properties of the carbon surface. The redox cycle is necessary to keep the production of HO· and HO_2· species [24]. The catalytic decomposition of H_2O_2 on the surface of metal or metal oxides has also been studied to some extent in recent years, including Fe, W, Cu, UO_2, ZrO_2, CuO, CuO_2 and so on [18,22,25–27]. It is known that HO· and HO_2· will be formed as intermediates during the decomposition of H_2O_2 while the disproportion reaction of HO_2· ends up with H_2O_2 and O_2. The disproportionation may also occur in the Fenton/Fenton-like reaction and the reaction between H_2O_2 and carbon-based catalyst. From previous work, it is known that the reaction between scavenger and HO· will affect the production of O_2 [28].

Despite its high efficiency and effectiveness, the application of classic Fenton reaction faces the disadvantages of strict pH restrictions, iron precipitation and the cost for catalyst recycling [29–31]. The formation rate of HO· is strongly dependent on the pH value while the oxidation potential of HO· declines as the pH increases [31,32]. Furthermore, the generation of HO· is directly limited by the formation of iron sludge in alkaline condition [30]. Since iron precipitation remains the bottleneck of classic iron-based Fenton reaction, non-ferrous heterogeneous catalysts with multiple oxidation states and redox stability (Ce, Cu, Mn and Ru) [11] and transition metal substituted iron oxide (Cr, Co and Ti) [32], have been developed for the replacement. Nevertheless, the abovementioned metal materials still face the drawbacks of high cost, high toxicity and/or environmental unfriendliness. Hence, a number of metal-free catalysts have been developed for the generation and/or decomposition of H_2O_2 regarding to their high earth abundance, good biocompatibility and environment-friendly properties, including graphene [6], carbon nanotubes [33], activated carbon fibers [34], graphitic carbon nitride (g-C_3N_4) [35–38] and carbon nanodots (CDots) [39].

As a metal-free polymer semiconductor material with suitable band gap and band position, g-C_3N_4 has embodied its research value in the field of H_2 production, CO_2 reduction, selective oxidation of alcohols and pollutant degradation [40–43]. The combination of CDots and g-C_3N_4 was firstly introduced in 2015 by J. Liu and her co-workers for water splitting, solving the chock point that g-C_3N_4 is poisoned by in-situ generated H_2O_2 in hydrogen evolution [44]. H_2O can be catalytically split into H_2O_2 and H_2 by g-C_3N_4 in the presence of photo irradiation. However, with the two-dimensional structure and large accessible area on the surface of g-C_3N_4, the in-situ generated hydrogen peroxides are strongly bonded and difficult to remove, which leads to the poisoning of catalyst, thereby limiting the yield of H_2 [45–47]. CDots was introduced to solve this problem by decomposing the bonded H_2O_2 on the surface of g-C_3N_4 into H_2O and O_2, thereby remitting the poisoning of g-C_3N_4. It is known that intermediate HO· will be formed via electron-transfer on the surface of carbon-based catalyst [24]. Inspired from these, it can be hypothesized that the CDots/g-C_3N_4 composite can be used as a catalyst providing promising yield of HO· via decomposing adsorbed H_2O_2 on the surface of g-C_3N_4 by embedded CDots. To the best of our knowledge, the kinetics and mechanism of catalytic decomposition of H_2O_2 on CDots/g-C_3N_4 composite has been rarely studied.

In this work, CDots/g-C_3N_4 composite was synthesized via an electrochemical method followed by a thermal polymerization process. The obtained composites were characterized by TEM, FTIR,

Brunauer-Emmett-Teller (B.E.T) and XRD. The catalytic performance of CDots/g-C$_3$N$_4$ composite for H$_2$O$_2$ decomposition was also investigated. The second-order reaction rate constant of H$_2$O$_2$ decomposition and reaction activation energy were obtained by varying the dosage of composite and temperature. Furthermore, a detailed mechanism involving the *adsorb-catalyze* double reaction sites was proposed.

2. Results and Discussion

2.1. Morphology of the Catalyst

The obtained CDots/g-C$_3$N$_4$ composite was prepared via an electrochemical method followed by a thermal polymerization process. To confirm the modification of CDots on g-C$_3$N$_4$, FTIR spectra and XRD patterns of pure g-C$_3$N$_4$ and CDots/g-C$_3$N$_4$ composite were obtained and exhibited in Figure 1A,B. The influence of CDots modification on the specific surface area was investigated by the B.E.T. method with isothermal adsorption and desorption of high purity nitrogen. The N$_2$ adsorption-desorption isotherms and pore size distributions of g-C$_3$N$_4$ and CDots/g-C$_3$N$_4$ composite are shown in Figure 1C. The TEM images of CDots/g-C$_3$N$_4$ are shown in Figure 1D,E.

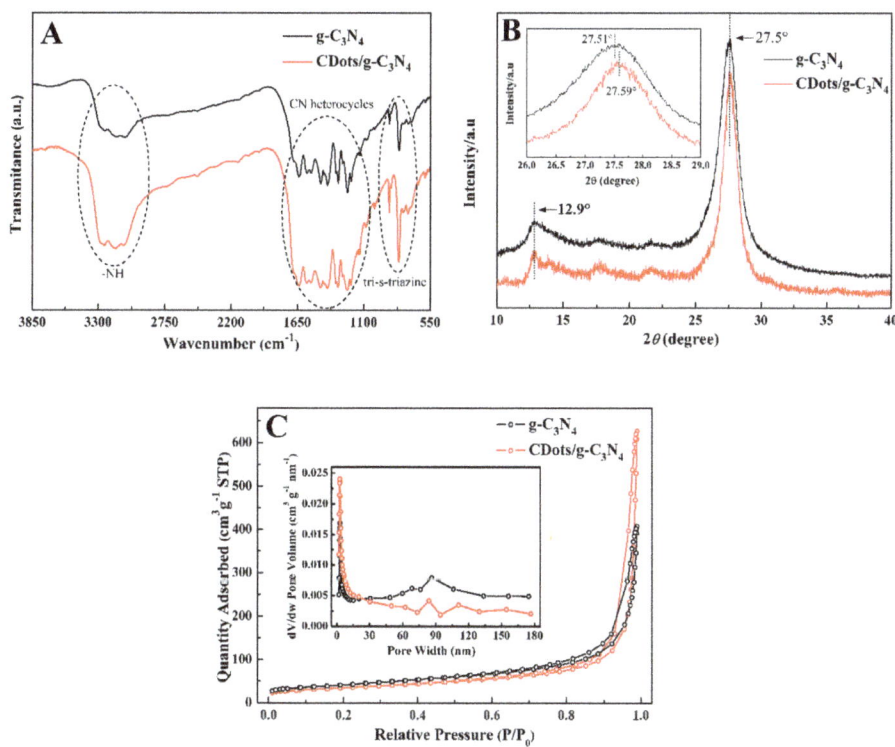

Figure 1. Cont.

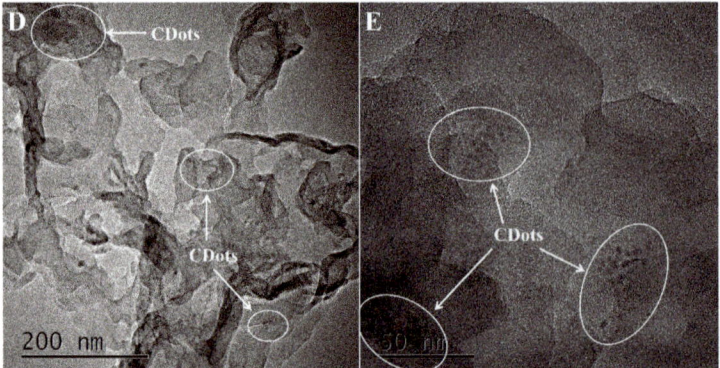

Figure 1. (**A**) Fourier Transform Infrared (FTIR) spectra; (**B**) X-Ray diffraction (XRD) patterns; (**C**) N_2 adsorption/desorption isotherm and pore width distribution of pure g-C_3N_4 and CDots/g-C_3N_4 composite; (**D,E**) Transmission electron microscopy (TEM) images of CDots/g-C_3N_4 composite.

As can be seen in Figure 1A, the sharp peak for g-C_3N_4 at 810 cm^{-1} is attributed to stretching vibration bond of tri-s-triazine [48]. Vibration peaks between 1200–1650 cm^{-1} corresponds to the typical stretching modes of CN heterocycles [49]. A wider band can be seen at 3100–3300 cm^{-1}, which belongs to the stretching vibration modes for the unreacted –NH [50]. The same characteristic peaks are observed in CDots/g-C_3N_4 composite and the peak at 1405 cm^{-1} can be seen as the indicator for the coupling of CDots and g-C_3N_4 [48].

XRD patterns of g-C_3N_4 and CDots/g-C_3N_4 are displayed in Figure 1B. The main diffraction peaks observed at 12.9° and 27.5° in both g-C_3N_4 and CDots/g-C_3N_4 composite are indexed to the (100) peak of the in-plane structure of tri-s-triszine unit and (002) crystal facets of the inter-layer stacking of aromatic segments [51,52]. The two patterns fit well with graphitic carbon nitride (JCPDS 87-1526) and no significant difference is observed, implying the low content of CDots in CDots/g-C_3N_4 composite. However, it is remarkable that the difference in relative intensity, together with the shift observed in the (002) peak location from 27.51° for g-C_3N_4 to 27.59° for CDots/g-C_3N_4, can be seen as the evidence of CDots introduction [45]. As can be seen in Figure 1C, the introduction of CDots in CDots/g-C_3N_4 leads to a 20% increase (from 120.92 to 145.24 m^2/g) in specific surface area, which favors the decomposition of H_2O_2.

Figure 1D clearly shows the two-dimensional structure of g-C_3N_4 together with the embedding of CDots (the white circles). The close look of the CDots embedded in g-C_3N_4 matrix is given in Figure 1E. The CDots are non-uniformly distributed, ranging from 2 to 10 nm, which is in line with previous studies [45,51,53].

From the results and analysis above, it can be confirmed that CDots have been successfully decorated in g-C_3N_4 and the inlay of CDots brings an improvement in specific surface area of the composite.

2.2. Kinetic Study

The effect of CDots content in the catalyst has been investigated in several previous works proving that a certain amount of CDots can enhance the catalytic properties of the catalysts while an excessive loading may work opposite [45,51,53]. Thus, in this work the CDots/g-C_3N_4 composite was fabricated with a fixed fraction of CDots (1.26 wt.%) selected by preliminary experiments. It is known that the surface reaction is dominating in the present heterogeneous system, therefore surface area to solution volume ratio (SA/V) is used to represent the dosage of the composites other than the mass concentration, which has been applied in many reported works [17,18,22,27,28,54–56]. The SA/V value is obtained by combining the mass concentration (g/L) with specific surface area (m^2/g) and can be normalized to m^{-1}.

To verify the synergistic effect of CDots and g-C$_3$N$_4$ in H$_2$O$_2$ decomposition, five samples were prepared according to the proportion of CDots and g-C$_3$N$_4$ in the composite. Sample 5 is the CDots/g-C$_3$N$_4$ composite with SA/V of 4×10^5 m^{-1}. The pure g-C$_3$N$_4$ with the same SA/V value was identified as Sample 2. The single-component CDots solution containing the equivalent amount of CDots as Sample 5 (13.4 mg/L) was named as Sample 1 and it was obtained by directly diluting the originally prepared CDots solution. The physical mixture of Sample 1 and 2 is identified as Sample 3. Additionally, the originally prepared CDots solution with a high concentration (133 mg/L) was named as Sample 4.

Detailed descriptions of the samples are listed in Table 1. It should be noted that sample 1, 3 and 5 has equivalent amount of CDots and sample 3 is a simple mixture of CDots and g-C$_3$N$_4$ while sample 5 is the composite. The catalytic decomposition of H$_2$O$_2$ by each sample was investigated under the same experimental condition where initial concentration of H$_2$O$_2$ ([H$_2$O$_2$]$_0$) is 0.5 mM and the temperature is 298K. Normalized concentration of H$_2$O$_2$ ([H$_2$O$_2$]/[H$_2$O$_2$]$_0$) of each case is plotted against reaction time respectively (shown in Figure 2).

Table 1. Samples prepared for the investigation of synergistic effect of CDots and g-C$_3$N$_4$.

Sample	Description
1	13.4 mg/L CDots
2	4×10^5 m^{-1} g-C$_3$N$_4$
3	physical mixture of Sample 1 and Sample 2
4	133 mg/L CDots
5	4×10^5 m^{-1} CDots/g-C$_3$N$_4$ composite

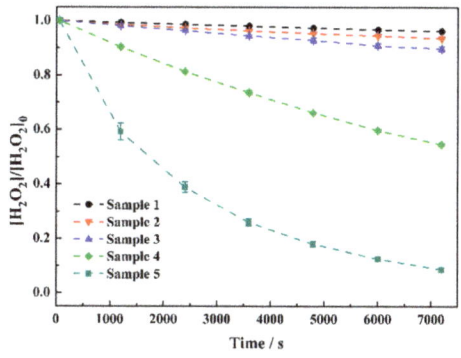

Figure 2. [H$_2$O$_2$]/[H$_2$O$_2$]$_0$ ([H$_2$O$_2$]$_0$ = 0.5 mM) as a function of reaction time in the presence of sample 1–5 at 298 K.

As shown in Figure 2, Sample 4 (the originally prepared CDots, 133 mg/L) incurs a gradely decline in H$_2$O$_2$ concentration in darkness, demonstrating the inherent catalytic property of CDots for H$_2$O$_2$ decomposition. However, the catalytic performance becomes faint after diluting CDots to a 10.2% concentration when comparing sample 1 and 4, indicating that such catalysis process is strongly dependent on the applied CDots concentration, which is in accordance with reported work [57]. Sample 2 (pure g-C$_3$N$_4$) also incurs slight decline in H$_2$O$_2$, which can be attributed to the adsorption of g-C$_3$N$_4$ and catalytic decomposition by carbon-based material through the delocalization of electrons on the surface [24]. However, the decomposition of H$_2$O$_2$ catalyzed by the pure g-C$_3$N$_4$ should not be considered as the main process in the system with CDots/g-C$_3$N$_4$ composite. It is remarkable that the consumption rate of H$_2$O$_2$ for sample 5 is much larger than sample 3, implying the thermal polymerization process gives rise to the remarkable synergy and the proximity between the CDots and the adsorption sites of H$_2$O$_2$ in g-C$_3$N$_4$ is necessary for the high efficiency of H$_2$O$_2$ decomposition [44].

It has been previously reported that [18,54], the catalytic decomposition of H_2O_2 in the heterogeneous system follows pseudo first-order kinetics with respect to H_2O_2 when solid is excess to H_2O_2 and the reaction rate equation can be described as $\frac{-d[H_2O_2]}{dt} = k_1 \times [H_2O_2]$, which can be integrated as

$$\ln \frac{[H_2O_2]}{[H_2O_2]_0} = -k_1 t \quad (1)$$

where k_1 is the pseudo first-order rate constant at a given temperature and dosage of the solid, t is the reaction time, $[H_2O_2]$ is the concentration of H_2O_2 at a time and $[H_2O_2]_0$ is the concentration of H_2O_2 at t = 0. When the solid catalyst is excess to H_2O_2, the second-order rate constant in the system can be determined by studying the pseudo first-order rate constant (k_1) as a function of SA/V (surface area of solid to volume of solution). The second-order rate expression is given as

$$\frac{-d[H_2O_2]}{dt} = k_2 \times \frac{SA}{V} \times [H_2O_2] \quad (2)$$

where k_2 denotes the second-order reaction rate constant, SA denotes the surface area of the CDots/g-C_3N_4 and V is the volume of the reaction solution. The term SA/V has been applied to denote the catalyst concentration in a number of studies regarding the heterogeneous catalysis system [18,22,27,54].

According to the preliminary experiments, the lower limit of SA/V is 3.2×10^5 m^{-1} after which it is excess to the fixed initial H_2O_2 concentration (0.5 mM). A series of experiments were carried out by varying the dosage of catalyst (SA/V) from 3.2 to 6.4×10^5 m^{-1} under the same condition at 298 K to explore the kinetics of the present system. The logarithm of normalized H_2O_2 concentration is plotted as a function of reaction time (Figure 3A) and the slope of the linearly fitted curve of these plots (k_1) is plotted against SA/V accordingly (Figure 3B).

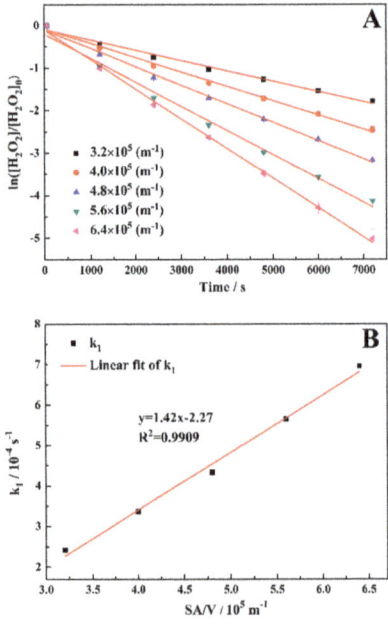

Figure 3. (A) $\ln([H_2O_2]/[H_2O_2]_0)$ ($[H_2O_2]_0$ = 0.5 mM) as a function of reaction time with different SA/V values of catalyst (3.2–6.4 × 10^5 m^{-1}) at 298 K; (B) pseudo first-order reaction rate constant as a function of SA/V at 298 K. The k_1 values were obtained from Figure 3A.

From Figure 3A, it can be seen that all the $\ln([H_2O_2]/[H_2O_2]_0)$ plots are linearly fitted with reaction time which indicates it follows pseudo first-order kinetics at given dosage of the composite and the slopes of the fitted curves are denoted as k_1. In addition, it is clearly that the observed decomposition rate of H_2O_2 increases with increasing the dosage of the composite. The key parameters of the fitted curves, including SA/V, slopes (k_1), standard deviation and R^2, are listed in Table 2.

Table 2. The key parameters of the fitted curves with different SA/V values.

SA/V (10^5 m^{-1})	k_1 (10^{-4} s^{-1})	Standard Deviation (10^{-4} s^{-1})	R^2 (%)
3.2	2.61	0.078	99.38
4.0	3.52	0.068	99.74
4.8	4.53	0.084	99.76
5.6	6.05	1.063	99.49
6.4	7.16	0.094	99.88

The obtained k_1 values from Table 2 were plotted in Figure 3B against SA/V. As can be seen from Figure 3B, k_1 is linearly correlated to SA/V in the range of 3.2–6.4 × 10^5 m^{-1} and the slope of the fitted curve is calculated as $(1.42 \pm 0.07) \times 10^{-9}$ m·s^{-1} which can be denoted as the overall second-order rate constant. This value is far from the rate constant of a diffusion controlled reaction in the order of 10^{-5} m·s^{-1} but still higher than some metal oxide catalysts like ZrO_2 ($(2.39 \pm 0.09) \times 10^{-10}$ m·s^{-1}), CuO ($(1.23 \pm 0.06) \times 10^{-9}$ m·s^{-1}) and Gd_2O_3 ($(9.4 \pm 1.0) \times 10^{-10}$ m·s^{-1}) [18,54].

Generally, the first-order rate constant k_1 is strongly related to the reaction temperature according to the Arrhenius equation:

$$k_1 = Ae^{-E_a/RT} \qquad (3)$$

where E_a denotes the activation energy for the reaction, R is the gas constant, T is the absolute temperature and A is the pre-exponential factor. The logarithm of k_1 obtained by plotting $\ln([H_2O_2]/[H_2O_2]_0)$ against T (shown in Figure 4A) is plotted as a function of 1/T in Figure 4B so as to calculate E_a.

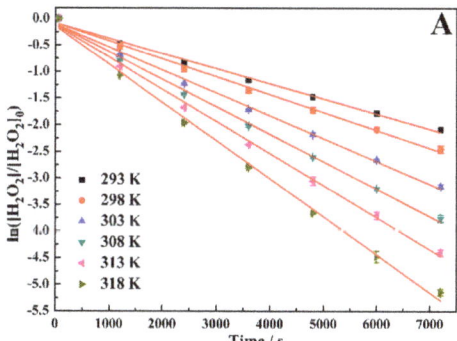

Figure 4. Cont.

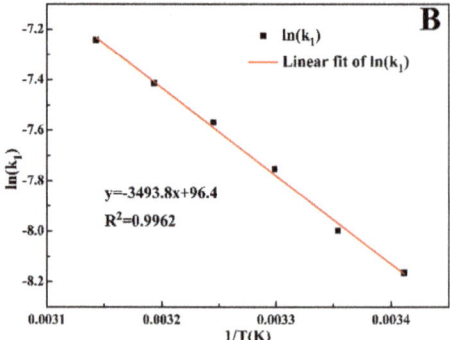

Figure 4. (**A**) $\ln([H_2O_2]/[H_2O_2]_0)$ ($[H_2O_2]_0 = 0.5$ mM) as a function of reaction time at different reaction temperatures (293–318 K) with the catalyst dosage of 4.0×10^5 m^{-1}; (**B**) $\ln(k_1)$ as a function of $(1/T)$ where k_1 were obtained from Figure 4A.

As shown in Figure 4A, $\ln([H_2O_2]/[H_2O_2]_0)$ declines with reaction time and the slope of the fitted curves increases as temperature increases. In addition, $\ln([H_2O_2]/[H_2O_2]_0)$ drops sharply in the initial period (~20 min) indicating the adsorption dominates this period, after which adsorbed H_2O_2 on the surface of the CDots/g-C_3N_4 composite reaches an equilibrium state and the decomposition of H_2O_2 catalyzed by embedded CDots turns the dominant role.

It can be seen from Figure 4B, $\ln(k_1)$ is linearly dependent on $1/T$ and the slope of the fitted curve is obtained. Based on the slope, the activation energy of the reaction is calculated to be (29.05 ± 0.80) kJ·mol^{-1}, which is to some extent lower than a series of metal oxides developed before, including ZrO_2 ((33 ± 1) kJ·mol^{-1}), TiO_2 ((37 ± 1) kJ·mol^{-1}), Y_2O_3 ((47 ± 5) kJ·mol^{-1}), Fe_2O_3 ((47 ± 1) kJ·mol^{-1}), CuO ((76 ± 1) kJ·mol^{-1}), CeO_2 ((40 ± 1) kJ·mol^{-1}), Gd_2O_3 ((63 ± 1) kJ·mol^{-1}) and HfO_2 ((60 ± 1) kJ·mol^{-1}) [18,54,58].

The key parameters of the fitted curves, including T, pseudo first-order reaction rate constants (k_1), standard deviation and R^2, are listed in Table 3.

Table 3. The key parameters of the fitted curves with different temperatures.

T(K)	k_1 (10^{-4} s^{-1})	Standard Deviation (10^{-4} s^{-1})	R^2 (%)
293	3.02	0.071	99.61
298	3.52	0.068	99.74
303	4.50	0.088	99.74
308	5.37	0.090	99.80
313	6.27	0.104	99.81
318	7.44	0.129	99.80

2.3. The Effect of pH

To investigate the mechanism of the present system containing H_2O_2 and CDots/g-C_3N_4 composite, it is significant to study the pH effect as well as quantify the in-situ produced hydroxyl radicals. Due to its scavenging capacity against HO· and pH buffering ability, Tris is chosen to carry out the mechanistic study. The pKa and the buffering range of Tris are 8.07 and 7.0–9.0 respectively, so the pH values were selected within this range. The decline of H_2O_2 together with the production of CH_2O against the reaction time with different pH are exhibited in Figure 5.

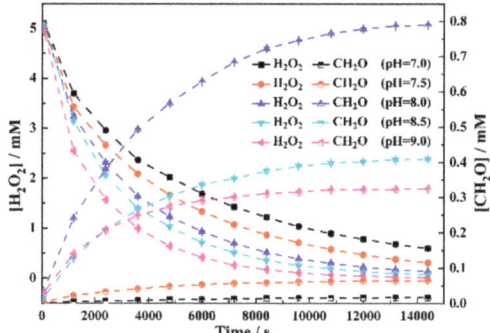

Figure 5. The concentration of H_2O_2 ($[H_2O_2]_0$ = 5 mM) and CH_2O as a function of reaction time in the presence of Tris ($[Tris]_0$ = 100 mM) with the catalyst dosage of 4.0×10^5 m^{-1} at 298.15K in pH from 7.0 to 9.0.

It can be clearly seen in Figure 5 that the decomposition rate of H_2O_2 increases with pH increases in the whole range. However, the formation of CH_2O shows a different trend. The evolution of CH_2O remains relatively low in neutral condition (pH 7.0), then it starts to accelerate until pH 8.0, after which the formation rate of CH_2O declines as pH increases. This indicates that the formation of CH_2O and probably the production of HO· are alkaline favored while the production is to some degree related to the pKa of Tris [54].

To figure out whether the production of HO· is also pH dependent, it is necessary to introduce the yield (Y) of CH_2O formed by HO· and Tris. Y is defined by the equation:

$$Y = [CH_2O]/[HO·] \qquad (4)$$

where [HO·] is the production of HO· and [CH_2O] is the accumulated CH_2O in H_2O_2 decomposition experiment. According to a previous study using γ-radiation in homogeneous system [17], the yield (Y) of CH_2O increases from 25% to 51% as increasing pH from 7.0 to 9.0. Provided that the yield (Y) in heterogeneous system is consistent with that in homogeneous system, the production of HO· in H_2O_2 decomposition on CDots/g-C_3N_4 composite can be estimated by this value together with the final concentration of CH_2O. The results are shown in Figure 6.

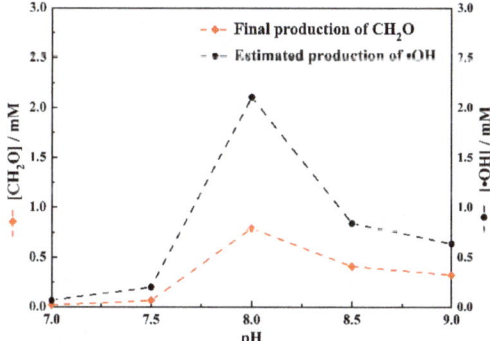

Figure 6. Final production of CH_2O and estimated production of HO· as a function of pH. The final productions of CH_2O were extracted from Figure 5. The estimated [HO·] = [CH_2O]/Y and the yields (Y) were derived from a previous work [17].

As appeared in Figure 6, the plots of estimated production of HO· exhibit the similar tendency as that of final production of CH_2O and it still peaks at pH 8.0 with the maximum concentration of 2.10 mM ([HO·] = [CH_2O]/Y where Y = 37.5% at pH 8.0). This means the stoichiometry between H_2O_2 and CH_2O is approximately 1:0.158 and 42.1% of H_2O_2 ([H_2O_2]$_0$ = 5 mM, Tris is excess to H_2O_2) end up with HO· at the optimal pH. The efficiency of H_2O_2 consumption towards HO· is much higher as compared to that of the H_2O_2/ZrO_2/Tris system (13.4% at pH 8.0) [17]. Therefore, it can be concluded that the production of HO· is also pH-dependent and there is an optimal pH which may have something to do with the Tris [22].

Based on the results in present work, the intrinsic chemical catalytic properties of the synthesized CDots/g-C_3N_4 composite, other than the photocatalytic properties, have been revealed to some extent. The hypothesis in the introduction section could be confirmed as following: firstly, as demonstrated in Figure 2, the pure CDots synthesized via an electrochemical pathway showed excellent catalytic ability, in line with the literature [59]; secondly, similar as the reported property of g-C_3N_4 [60], Figure 2 also shows that the concentration of H_2O_2 in the solution decreases slightly in the presence of the pure g-C_3N_4 which indicates g-C_3N_4 does provide sufficient reaction site for H_2O_2 to adsorb; thirdly, by analyzing the results in Figures 2, 3, 5 and 6, it is known that hydroxyl radical can be formed during the decomposition of H_2O_2 catalyzed by the CDots/g-C_3N_4 composite in the heterogeneous system as hypothesized in the former section. Besides the strong affinity of g-C_3N_4 towards H_2O_2 and the catalytic property of CDots against the adsorbed H_2O_2 [47], the delocalization of the electrons on the surface of g-C_3N_4 may also leads to the decomposition of H_2O_2 via electron-transfer mechanism [24]. In conclusion, the synthesized CDots/g-C_3N_4 composite exhibits the synergy of adsorption of H_2O_2 and delocalization of electrons on g-C_3N_4 and catalytic decomposition of H_2O_2 by g-C_3N_4 and CDots producing hydroxyl radicals.

Based on the results and discussion above, the mechanism of catalytic decomposition of H_2O_2 in the heterogeneous system with CDots/g-C_3N_4 composite is proposed and illustrated in Figure 7.

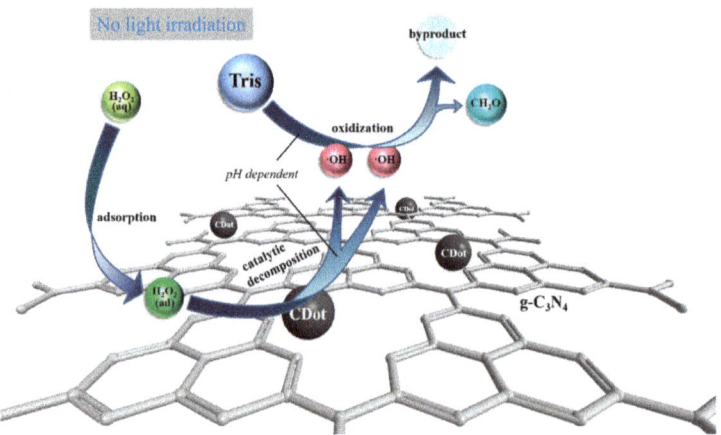

Figure 7. The mechanism of catalytic decomposition of H_2O_2 on CDots/g-C_3N_4 composite.

Key of the mechanism is the so-called *adsorb-catalyze* double reaction sites. With plenty of accessible adsorption sites on the surface of g-C_3N_4, CDots/g-C_3N_4 composite shows high selective adsorption ability towards aqueous H_2O_2. From previous works studying similar heterogeneous system with H_2O_2 and solid catalyst [18,22,54], it is known that H_2O_2 concentration exhibits an initial drop indicating the adsorption on the surface of catalyst, after which the H_2O_2 decomposition obeys pseudo first-order kinetic when the surface reaches equilibrium state. As can be seen in Figure 3A, the H_2O_2 concentration follows the similar trend and kinetics. Hence, it can be demonstrated that the adsorption of H_2O_2 is also dominating in the initial short period. Tris was introduced in the present work as a probe of hydroxyl radical and pH buffer. It is known that Tris can be partially oxidized to CH_2O and other byproducts and the ratio between the concentration of hydroxyl radicals and formed CH_2O is relatively fixed under given condition (pH and dissolved oxygen concentration) [17,27]. Therefore, the formation of CH_2O can be used to probe the formed hydroxyl radicals in the heterogeneous system containing H_2O_2 and CDots/g-C_3N_4 composite. It is in line with previous works that [17,18,22,27], CH_2O formation is reflected by the decomposition of H_2O_2 and is pH-dependent (Figures 5 and 6). Hence, it can be deduced that after the initial period, the adsorbed H_2O_2 on the surface of the CDots/g-C_3N_4 composite reaches an equilibrium state and the decomposition of H_2O_2 catalyzed by embedded CDots on the surface sites turns the dominant role. During this procedure, large quantities of HO· was produced, exhibiting strong oxidation ability towards scavengers like Tris. It should be noted that the production of HO· is strongly pH dependent. To sum up, the CDots/g-C_3N_4 composite shows synergetic effect on the decomposition of H_2O_2 via *adsorb-catalyze* double reaction sites and more importantly, is proved to be a promising catalyst for the degradation of NBDOPs since it is a metal-free pathway of producing HO· efficiently.

3. Experimental Section

3.1. Instrumentation

The morphology and microstructure of samples were observed by JET-2100F (JEOL, Wuhan, China) transmission electron microscope (TEM). The Fourier transform infrared spectroscopy (FTIR) of the samples were recorded by Nicolet iS5 (Thermo Fisher Scientific, Wuhan, China) FTIR spectrometer with KBr pellets. The specific surface area of CDots/g-C_3N_4 composite and pure g-C_3N_4 were determined by the Brunauer-Emmett-Teller (B.E.T) method via isothermal adsorption and desorption of high purity nitrogen using a TriStar II 3020 (Micromeritics, Wuhan, China) instrument. X-ray diffraction (XRD) patterns were recorded with D8 advance (Bruker, Wuhan, China) diffractometer using Bragg-Brentano geometry in the 2θ angle from 10° to 40° and Cu Kα irradiation (λ = 1.54 Å). The samples were weighted to ± 10^{-4} g in a ME104E (Mettler Toledo, Wuhan, China) microbalance. UV/Vis spectra were collected by V-5600 (METASH, Wuhan, China) and UV-5500PC (METASH, Wuhan, China) spectrophotometer. The pH of reaction solution was measured by PHS-3C (YOKE, Wuhan, China) pH meter with an accuracy of ± 0.01 pH units.

3.2. Reagents and Experiments

All the solutions used in this study were prepared using deionized water.

Preparation of the catalyst: CDots were synthesized via an electrochemical method based on previous reported work [59]. In a typical preparation process, two graphite rods were insert parallel into 300 mL ultrapure water as electrodes with a separation of 7.5 cm and 4 cm depth under water. 60 V static potentials were applied to the rods by a direct-current (DC). After electrolyzing for 120 h, the anode graphite rod corroded and a dark brown solution was formed. The solution was filtered with slow-speed quantitative filter paper and then centrifuged at 10,000 rpm for 10 min. Finally, the soluble CDots was obtained and the concentration can be quantified by drying and weighting.

A thermal polymerization method was applied for the synthesis of pure g-C_3N_4 [61]. Typically, 40 g urea (CAS[57-13-6], 99%, Sinopharm, Wuhan, China) was dissolved in 40 mL ultrapure water in a quartz crucible, then heated to 550 °C with the rate of 7 °C/min in a muffle furnace and kept at 550 °C for 2 h. After naturally cooling down to room temperature, the resultant yellow product was collected and ground into powder to obtain pure g-C_3N_4. CDots/g-C_3N_4 composite was synthesized via in-situ thermal polymerization [51]. Following the same procedure, 40 g urea was dissolved in 40 mL CDots solution and calcined in muffle furnace. The dark gray product CDots/g-C_3N_4 was collected and ground for further use.

Kinetic studies: Hydrogen peroxide H_2O_2 (CAS[7722-84-1], 30 wt.%, Sinopharm, Wuhan, China), glacial acetic acid HAc (CAS[64-19-7], 99.5%, Sinopharm, Wuhan, China), sodium acetate NaAc

(CAS[127-09-3], 99%, Sinopharm, Wuhan, China), ammoniumdimolybdate ADM (CAS[27546-07-2], Mo 56.5%, Macklin, Wuhan, China) and potassium iodide KI (CAS[7681-11-0], 99%, Sinopharm, Wuhan, China) were used in kinetic studies. H_2O_2 decomposition experiments were performed in lucifugal bottles with different dosages of CDots/g-C_3N_4 composite and under variable temperatures. The suspension was dispersed with ultrasonic sound for 1 min before the reaction. Afterwards, H_2O_2 was added to the suspension to trigger the reaction. Samples were extracted with a syringe and a filter (220 nm, cellulose membrane) at fixed time intervals. The concentration of H_2O_2 as a function of time was determined by Ghormley triiodide method, where I^- can be oxidized to I_3^- by H_2O_2 in faintly acid conditions catalyzed by ADM [62,63]. In detail, 0.2 mL sample was added to 1.6 mL water, followed by the addition of 0.1 mL 1M KI and 0.1 mL 1M HAc/NaAc containing 0.03% ADM. The absorbance of produced I_3^- was measured at 350 nm by UV-vis spectrophotometer and the calibration curve of the absorbance of I_3^- as a function of H_2O_2 concentration was obtained with a linear correlation between absorbance and concentration in the range of 0.1–1 mM H_2O_2. The experimental error in the determination of H_2O_2 was less than 2%.

Mechanistic studies: Tris(hydroxymethyl) aminomethane Tris (CAS[77-86-1], 99%, Aldrich, Wuhan, China), acetoacetanilide AAA (CAS[102-01-2], 98%, Macklin, Wuhan, China), ammonium acetate NH_4Ac (CAS[631-61-8], 98%, Sinopharm, Wuhan, China) and formaldehyde CH_2O (CAS[50-00-0], Sinopharm, 37% wt.%, Wuhan, China) were used in mechanistic studies. The experiments were carried out by using Tris as HO· scavenger (forming CH_2O) to quantify the produced HO· indirectly. It is known that the amount of formed CH_2O is quantitatively correlated with that of HO· [17]. The decomposition experiments were carried out in 100 mM Tris solution with fixed quantities of CDots/g-C_3N_4 (SA/V = 4 × 10^5 m^{-1}, SA and V stands for the surface area of solid and the volume of solution) and H_2O_2 (5 mM). The pH values of solution were selected within the valid buffering range of Tris, namely pH 7.0–9.0 for investigating the effect of pH. The produced CH_2O was quantitatively determined by a modified Hantzsch method, where CH_2O reacts with AAA in the presence of NH_4Ac to form a dihydropyridine derivative with a maximum absorbance wavelength at 368 nm [19]. The calibration curve where the absorbance of produced dihydropyridine derivative was plotted as a function of CH_2O concentration with linear correlation was obtained at 368 nm in the range of 0.04–1.3 mM CH_2O for the conversion of absorbance to CH_2O concentration. The experimental error in the determination of CH_2O was less than 2%.

4. Conclusions

In this work, a promising catalyst CDots/g-C_3N_4 composite for degradation of non-biodegradable organic pollutants was synthesized via a two-step pathway including electrochemical exfoliation of graphite rod preparing CDots and thermal polymerization of CDots mixed urea. Through different characterization methods and kinetic experiments, it has been confirmed that CDots embed in g-C_3N_4 matrix and such structure accounts for the synergetic catalytic performance of the composite. Kinetics of catalytic decomposition of H_2O_2 on CDots/g-C_3N_4 composite were researched. The second-order rate constant (k_2) was measured to be (1.42 ± 0.07) × 10^{-9} m·s^{-1} and the activation energy of the reaction was measured to be (29.05 ± 0.80) kJ·mol^{-1} under the applied conditions. The effect of pH (pH 7.0–9.0) on the production of HO· was also investigated by using Tris as a probe. It has been shown that the production of HO· is strongly alkaline dependent and the maximum reaches at pH 8.0 which is close to the pKa of Tris. A mechanism based on the *adsorb-catalyze* double reaction site theory has been proposed. This work implies that the photocatalyst (CDots/g-C_3N_4 composite) for water splitting or H_2 evolution may also be applied as an alternative catalyst in degradation of non-biodegradable organic pollutants. The instinct kinetics and the mechanism can be referred to for further applications in related fields.

Author Contributions: M.Y. conceived and designed the experiments; Z.L. and Q.S. performed the experiments; C.Z. and L.F. analyzed the data; T.X. contributed reagents/materials/analysis tools; M.Y. and Z.L. wrote the paper.

Funding: This work is supported by National Natural Science Foundation of China (21707108) and Independent Innovation Foundation of Wuhan University of Technology (20411057 and 20410962).

Conflicts of Interest: The authors declare no conflict of interest.

References

1. He, J.; Yang, X.F.; Men, B.; Wang, D.S. Interfacial mechanisms of heterogeneous Fenton reactions catalyzed by iron-based materials: A review. *J. Environ. Sci.* **2016**, *39*, 97–109. [CrossRef] [PubMed]
2. Nidheesh, P.V.; Gandhimathi, R.; Ramesh, S.T. Degradation of dyes from aqueous solution by Fenton processes: A review. *Environ. Sci. Pollut. Res.* **2013**, *20*, 2099–2132. [CrossRef] [PubMed]
3. Aval, A.E.; Hasani, A.H.; Omrani, G.A.; Karbassi, A. Removal of Landfill Leachate's Organic load by modified Electro-Fenton process. *Int. J. Electrochem. Sci.* **2017**, *12*, 9348–9363. [CrossRef]
4. Huang, J.; Shi, Q.; Feng, J.; Chen, M.; Li, W.; Li, L. Facile pyrolysis preparation of rosin-derived biochar for supporting silver nanoparticles with antibacterial activity. *Compos. Sci. Technol.* **2017**, *145*, 89–95. [CrossRef]
5. Freyria, F.S.; Armandi, M.; Compagnoni, M.; Ramis, G.; Rossetti, I.; Bonelli, B. Catalytic and Photocatalytic Processes for the Abatement of N-Containing Pollutants from Wastewater. Part 2: Organic Pollutants. *J. Nanosci. Nanotechnol.* **2017**, *17*, 3654–3672. [CrossRef]
6. Chen, C.Y.; Tang, C.; Wang, H.F.; Chen, C.M.; Zhang, X.Y.; Huang, X.; Zhang, Q. Oxygen Reduction Reaction on Graphene in an Electro-Fenton System: In Situ Generation of H_2O_2 for the Oxidation of Organic Compounds. *Chemsuschem* **2016**, *9*, 1194–1199. [CrossRef] [PubMed]
7. Santos, L.V.D.; Meireles, A.M.; Lange, L.C. Degradation of antibiotics norfloxacin by Fenton, UV and UV/H_2O_2. *J. Environ. Manag.* **2015**, *154*, 8–12. [CrossRef] [PubMed]
8. Li, Y.G.; Hsieh, W.P.; Mahmudov, R.; Wei, X.M.; Huang, C.P. Combined ultrasound and Fenton (US-Fenton) process for the treatment of ammunition wastewater. *J. Hazard. Mater.* **2013**, *244*, 403–411. [CrossRef] [PubMed]
9. Nidheesh, P.V. Heterogeneous Fenton catalysts for the abatement of organic pollutants from aqueous solution: A review. *RSC Adv.* **2015**, *5*, 40552–40577. [CrossRef]
10. Oller, I.; Malato, S.; Sanchez-Perez, J.A. Combination of Advanced Oxidation Processes and biological treatments for wastewater decontamination-A review. *Sci. Total Environ.* **2011**, *409*, 4141–4166. [CrossRef] [PubMed]
11. Bokare, A.D.; Choi, W. Review of iron-free Fenton-like systems for activating H_2O_2 in advanced oxidation processes. *J. Hazard. Mater.* **2014**, *275*, 121–135. [CrossRef] [PubMed]
12. Li, P.; Xie, T.; Duan, X.; Yu, F.B.; Wang, X.; Tang, B. A New Highly Selective and Sensitive Assay for Fluorescence Imaging of ·OH in Living Cells: Effectively Avoiding the Interference of Peroxynitrite. *Chem. Eur. J.* **2010**, *16*, 1834–1840. [CrossRef] [PubMed]
13. Cao, Y.Q.; Sui, D.D.; Zhou, W.J.; Lu, C. Highly selective chemiluminescence detection of hydroxyl radical via increased pi-electron densities of rhodamine B on montmorillonite matrix. *Sens. Actuators B Chem.* **2016**, *225*, 600–606. [CrossRef]
14. Jing, Y.; Chaplin, B.P. Mechanistic Study of the Validity of Using Hydroxyl Radical Probes to Characterize Electrochemical Advanced Oxidation Processes. *Environ. Sci. Technol.* **2017**, *51*, 2355–2365. [CrossRef] [PubMed]
15. Yang, X.F.; Guo, X.Q. A novel fluorescence probe for the determination of hydroxyl radicals. *Chem. J. Chin. Univ.* **2001**, *22*, 396–398.
16. Fisher, S.C.; Schoonen, M.A.A.; Brownawell, B.J. Phenylalanine as a hydroxyl radical-specific probe in pyrite slurries. *Geochem. Trans.* **2012**, *13*, 1–18. [CrossRef] [PubMed]
17. Yang, M.; Jonsson, M. Evaluation of the O_2 and pH Effects on Probes for Surface Bound Hydroxyl Radicals. *J. Phys. Chem. C* **2014**, *118*, 7971–7979. [CrossRef]
18. Lousada, C.M.; Jonsson, M. Kinetics, Mechanism, and Activation Energy of H_2O_2 Decomposition on the Surface of ZrO_2. *J. Phys. Chem. C* **2010**, *114*, 11202–11208. [CrossRef]
19. Nash, T. The colorimetric estimation of formaldehyde by means of the Hantzsch reaction. *Biochem. J.* **1953**, *55*, 416–421. [CrossRef] [PubMed]

20. Zheng, P.; Pan, Z.; Zhang, J. Synergistic Enhancement in Catalytic Performance of Superparamagnetic Fe3O4@Bacilus subtilis as Recyclable Fenton-Like Catalyst. *Catalysts* **2017**, *7*, 349. [CrossRef]
21. Yang, H.; Shi, B.; Wang, S. Fe Oxides Loaded on Carbon Cloth by Hydrothermal Process as an Effective and Reusable Heterogenous Fenton Catalyst. *Catalysts* **2018**, *8*, 207. [CrossRef]
22. Yang, M.; Zhang, X.; Grosjean, A.; Soroka, I.; Jonsson, M. Kinetics and Mechanism of the Reaction between H_2O_2 and Tungsten Powder in Water. *J. Phys. Chem. C* **2015**, *119*, 22560–22569. [CrossRef]
23. Hiroki, A.; LaVerne, J.A. Decomposition of Hydrogen Peroxide at Water−Ceramic Oxide Interfaces. *J. Phys. Chem. B* **2005**, *109*, 3364–3370. [CrossRef] [PubMed]
24. Domínguez, C.M.; Quintanilla, A.; Ocón, P.; Casas, J.A.; Rodriguez, J.J. The use of cyclic voltammetry to assess the activity of carbon materials for hydrogen peroxide decomposition. *Carbon* **2013**, *60*, 76–83. [CrossRef]
25. Lousada, C.M.; Trummer, M.; Jonsson, M. Reactivity of H_2O_2 towards different UO2-based materials: The relative impact of radiolysis products revisited. *J. Nuclear Mater.* **2013**, *434*, 434–439. [CrossRef]
26. Vilardi, G.; Di Palma, L.; Verdone, N. On the critical use of zero valent iron nanoparticles and Fenton processes for the treatment of tannery wastewater. *J. Water Process Eng.* **2018**, *22*, 109–122. [CrossRef]
27. Björkbacka, Å.; Yang, M.; Gasparrini, C.; Leygraf, C.; Jonsson, M. Kinetics and mechanisms of reactions between H_2O_2 and copper and copper oxides. *Dalton Trans.* **2015**, *44*, 16045–16051. [CrossRef] [PubMed]
28. Lousada, C.M.; LaVerne, J.A.; Jonsson, M. Enhanced hydrogen formation during the catalytic decomposition of H_2O_2 on metal oxide surfaces in the presence of HO radical scavengers. *Phys. Chem. Chem. Phys.* **2013**, *15*, 12674–12679. [CrossRef] [PubMed]
29. Pignatello, J.J.; Oliveros, E.; MacKay, A. Advanced oxidation processes for organic contaminant destruction based on the Fenton reaction and related chemistry. *Crit. Rev. Environ. Sci. Technol.* **2006**, *36*, 1–84. [CrossRef]
30. Diya'uddeen, B.H.; Aziz, A.R.A.; Daud, W.M.A.W. On the Limitation of Fenton Oxidation Operational Parameters: A Review. *Int. J. Chem. React. Eng.* **2012**, *10*, 1498–1502. [CrossRef]
31. Garcia-Segura, S.; Bellotindos, L.M.; Huang, Y.H.; Brillas, E.; Lu, M.C. Fluidized-bed Fenton process as alternative wastewater treatment technology-A review. *J. Taiwan Inst. Chem. Eng.* **2016**, *67*, 211–225. [CrossRef]
32. Pouran, S.R.; Raman, A.A.A.; Daud, W.M.A.W. Review on the application of modified iron oxides as heterogeneous catalysts in Fenton reactions. *J. Clean. Prod.* **2014**, *64*, 24–35. [CrossRef]
33. Yao, Y.J.; Chen, H.; Lian, C.; Wei, F.Y.; Zhang, D.W.; Wu, G.D.; Chen, B.J.; Wang, S.B. Fe, Co, Ni nanocrystals encapsulated in nitrogen-doped carbon nanotubes as Fenton-like catalysts for organic pollutant removal. *J. Hazard. Mater.* **2016**, *314*, 129–139. [CrossRef] [PubMed]
34. Zhou, F.Y.; Lu, C.; Yao, Y.Y.; Sun, L.J.; Gong, F.; Li, D.W.; Pei, K.M.; Lu, W.Y.; Chen, W.X. Activated carbon fibers as an effective metal-free catalyst for peracetic acid activation: Implications for the removal of organic pollutants. *Chem. Eng. J.* **2015**, *281*, 953–960. [CrossRef]
35. Zhu, Z.D.; Pan, H.H.; Murugananthan, M.; Gong, J.Y.; Zhang, Y.R. Visible light-driven photocatalytically active g-C_3N_4 material for enhanced generation of H_2O_2. *Appl. Catal. B Environ.* **2018**, *232*, 19–25. [CrossRef]
36. Jiang, G.D.; Yang, X.X.; Wu, Y.; Li, Z.W.; Han, Y.H.; Shen, X.D. A study of spherical TiO_2/g-C_3N_4 photocatalyst: Morphology, chemical composition and photocatalytic performance in visible light. *Mol. Catal.* **2017**, *432*, 232–241. [CrossRef]
37. Yan, J.; Fan, Y.M.; Lian, J.B.; Zhao, Y.; Xu, Y.G.; Gu, J.M.; Song, Y.H.; Xu, H.; Li, H.M. Kinetics and mechanism of enhanced photocatalytic activity employing ZnS nanospheres/graphene-like C_3N_4. *Mol. Catal.* **2017**, *438*, 103–112. [CrossRef]
38. Bicalho, H.A.; Lopez, J.L.; Binatti, I.; Batista, P.F.R.; Ardisson, J.D.; Resende, R.R.; Lorencon, E. Facile synthesis of highly dispersed Fe(II)-doped g-C_3N_4 and its application in Fenton-like catalysis. *Mol. Catal.* **2017**, *435*, 156–165. [CrossRef]
39. Zhang, M.L.; Yao, Q.F.; Guan, W.J.; Lu, C.; Lin, J.M. Layered Double Hydroxide-Supported Carbon Dots as an Efficient Heterogeneous Fenton-Like Catalyst for Generation of Hydroxyl Radicals. *J. Phys. Chem. C* **2014**, *118*, 10441–10447. [CrossRef]
40. Wang, K.; Li, Q.; Liu, B.S.; Cheng, B.; Ho, W.K.; Yu, J.G. Sulfur-doped g-C_3N_4 with enhanced photocatalytic CO_2-reduction performance. *Appl. Catal. B Environ.* **2015**, *176*, 44–52. [CrossRef]
41. Zhao, D.M.; Chen, J.; Dong, C.L.; Zhou, W.; Huang, Y.C.; Mao, S.S.; Guo, L.J.; Shen, S.H. Interlayer interaction in ultrathin nanosheets of graphitic carbon nitride for efficient photocatalytic hydrogen evolution. *J. Catal.* **2017**, *352*, 491–497. [CrossRef]

42. Zhou, M.; Yang, P.J.; Yuan, R.S.; Asiri, A.M.; Wakeel, M.; Wang, X.C. Modulating Crystallinity of Graphitic Carbon Nitride for Photocatalytic Oxidation of Alcohols. *Chemsuschem* **2017**, *10*, 4451–4456. [CrossRef] [PubMed]
43. Lin, K.Y.A.; Lin, J.T. Ferrocene-functionalized graphitic carbon nitride as an enhanced heterogeneous catalyst of Fenton reaction for degradation of Rhodamine B under visible light irradiation. *Chemosphere* **2017**, *182*, 54–64. [CrossRef] [PubMed]
44. Liu, J.; Liu, Y.; Liu, N.Y.; Han, Y.Z.; Zhang, X.; Huang, H.; Lifshitz, Y.; Lee, S.T.; Zhong, J.; Kang, Z.H. Metal-free efficient photocatalyst for stable visible water splitting via a two-electron pathway. *Science* **2015**, *347*, 970–974. [CrossRef] [PubMed]
45. Liu, Q.; Chen, T.; Guo, Y.; Zhang, Z.; Fang, X. Ultrathin g-C_3N_4 nanosheets coupled with carbon nanodots as 2D/0D composites for efficient photocatalytic H_2 evolution. *Appl. Catal. B Environ.* **2016**, *193*, 248–258. [CrossRef]
46. Oh, J.; Lee, J.M.; Yoo, Y.; Kim, J.; Hwang, S.J.; Park, S. New insight of the photocatalytic behaviors of graphitic carbon nitrides for hydrogen evolution and their associations with grain size, porosity, and photophysical properties. *Appl. Catal. B Environ.* **2017**, *218*, 349–358. [CrossRef]
47. Liu, J.H.; Zhang, Y.W.; Lu, L.H.; Wu, G.; Chen, W. Self-regenerated solar-driven photocatalytic water-splitting by urea derived graphitic carbon nitride with platinum nanoparticles. *Chem. Commun.* **2012**, *48*, 8826–8828. [CrossRef] [PubMed]
48. Su, Y.H.; Chen, P.; Wang, F.L.; Zhang, Q.X.; Chen, T.S.; Wang, Y.F.; Yao, K.; Lv, W.Y.; Liu, G.G. Decoration of TiO_2/g-C_3N_4 Z-scheme by carbon dots as a novel photocatalyst with improved visible-light photocatalytic performance for the degradation of enrofloxacin. *RSC Adv.* **2017**, *7*, 34096–34103. [CrossRef]
49. Fang, S.; Xia, Y.; Lv, K.L.; Li, Q.; Sun, J.; Li, M. Effect of carbon-dots modification on the structure and photocatalytic activity of g-C_3N_4. *Appl. Catal. B Environ.* **2016**, *185*, 225–232. [CrossRef]
50. Wang, X.F.; Cheng, J.J.; Yu, H.G.; Yu, J.G. A facile hydrothermal synthesis of carbon dots modified g-C_3N_4 for enhanced photocatalytic H_2-evolution performance. *Dalton Trans.* **2017**, *46*, 6417–6424. [CrossRef] [PubMed]
51. Wang, F.L.; Chen, P.; Feng, Y.P.; Xie, Z.J.; Liu, Y.; Su, Y.H.; Zhang, Q.X.; Wang, Y.F.; Yao, K.; Lv, W.Y.; et al. Facile synthesis of N-doped carbon dots/g-C_3N_4 photocatalyst with enhanced visible-light photocatalytic activity for the degradation of indomethacin. *Appl. Catal. B Environ.* **2017**, *207*, 103–113. [CrossRef]
52. Dadigala, R.; Bandi, R.; Gangapuram, B.R.; Guttena, V. Carbon dots and Ag nanoparticles decorated g-C_3N_4 nanosheets for enhanced organic pollutants degradation under sunlight irradiation. *J. Photochem. Photobiol. A* **2017**, *342*, 42–52. [CrossRef]
53. Zhang, H.; Zhao, L.X.; Geng, F.L.; Guo, L.H.; Wan, B.; Yang, Y. Carbon dots decorated graphitic carbon nitride as an efficient metal-free photocatalyst for phenol degradation. *Appl. Catal. B Environ.* **2016**, *180*, 656–662. [CrossRef]
54. Lousada, C.M.; Yang, M.; Nilsson, K.; Jonsson, M. Catalytic decomposition of hydrogen peroxide on transition metal and lanthanide oxides. *J. Mol. Catal. A-Chem.* **2013**, *379*, 178–184. [CrossRef]
55. Yang, M.; Jonsson, M. Surface reactivity of hydroxyl radicals formed upon catalytic decomposition of H_2O_2 on ZrO_2. *J. Mol. Catal. A Chem.* **2015**, *400*, 49–55. [CrossRef]
56. Fidalgo, A.B.; Dahlgren, B.; Brinck, T.; Jonsson, M. Surface Reactions of H_2O_2, H_2, and O_2 in Aqueous Systems Containing ZrO_2. *J. Phys. Chem. C* **2016**, *120*, 1609–1614. [CrossRef]
57. Pirsaheb, M.; Moradi, S.; Shahlaei, M.; Farhadian, N. Application of carbon dots as efficient catalyst for the green oxidation of phenol: Kinetic study of the degradation and optimization using response surface methodology. *J. Hazard. Mater.* **2018**, *353*, 444–453. [CrossRef] [PubMed]
58. Lousada, C.M.; Johansson, A.J.; Brinck, T.; Jonsson, M. Mechanism of H_2O_2 Decomposition on Transition Metal Oxide Surfaces. *J. Phys. Chem. C* **2012**, *116*, 9533–9543. [CrossRef]
59. Ming, H.; Ma, Z.; Liu, Y.; Pan, K.; Yu, H.; Wang, F.; Kang, Z. Large scale electrochemical synthesis of high quality carbon nanodots and their photocatalytic property. *Dalton Trans.* **2012**, *41*, 9526–9531. [CrossRef] [PubMed]
60. Wei, Q.Y.; Yan, X.Q.; Kang, Z.; Zhang, Z.; Cao, S.Y.; Liu, Y.C.; Zhang, Y. Carbon Quantum Dots Decorated C_3N_4/TiO_2 Heterostructure Nanorod Arrays for Enhanced Photoelectrochemical Performance. *J. Electrochem. Soc.* **2017**, *164*, H515–H520. [CrossRef]

61. Guo, F.; Shi, W.L.; Guan, W.S.; Huang, H.; Liu, Y. Carbon dots/g-C$_3$N$_4$/ZnO nanocomposite as efficient visible-light driven photocatalyst for tetracycline total degradation. *Sep. Purif. Technol.* **2017**, *173*, 295–303. [CrossRef]
62. Hochanadel, C.J. Effects of Cobalt γ-Radiation on Water and Aqueous Solutions. *J. Phys. Chem.* **1952**, *21*, 587–594. [CrossRef]
63. Ghormley, J.A.; Stewart, A.C. Effects of γ-Radiation on Ice[1]. *J. Am. Chem. Soc.* **1956**, *78*, 2934–2939. [CrossRef]

© 2018 by the authors. Licensee MDPI, Basel, Switzerland. This article is an open access article distributed under the terms and conditions of the Creative Commons Attribution (CC BY) license (http://creativecommons.org/licenses/by/4.0/).

Article

Catalytic Degradation of Textile Wastewater Effluent by Peroxide Oxidation Assisted by UV Light Irradiation

Sarto Sarto *, Paesal Paesal, Irine Bellina Tanyong, William Teja Laksmana, Agus Prasetya and Teguh Ariyanto

Department of Chemical Engineering, Faculty of Engineering, Universitas Gadjah Mada, Jl Grafika No 2, Yogyakarta 55281, Indonesia; ariya.chemeng@gmail.com (P.P.); ariya_chemeng@yahoo.co.id (I.B.T.); william.teja.l13@gmail.com (W.T.L.); aguspras@ugm.ac.id (A.P.); teguh.ariyanto@ugm.ac.id (T.A.)
* Correspondence: sarto@ugm.ac.id; Tel.: +62-274-649-2171

Received: 30 April 2019; Accepted: 30 May 2019; Published: 4 June 2019

Abstract: Textile industries produce a complex wastewater which is difficult to be treated. In this work, a catalytic degradation of wastewater effluent composed of sulphur black coloring agent discharged by industry was studied. UV lamp power, peroxide concentration, pH, and iron oxide catalyst were varied to determine the best conditions for oxidative treatment. Kinetic parameters were evaluated based on the reaction model proposed. In the absence of iron oxide catalyst, chemical oxygen demand (COD) and biological oxygen demand (BOD) degradation of up to 80% and 75%, respectively, were observed as resulting from using an H_2O_2 concentration of 0.61 moles/L, UV lamp power of 30 watts, and pH of 6. When using an iron oxide catalyst combined with UV light irradiation, the degradation rate could be increased significantly, while similar final COD and BOD degradation percentages resulted. It is found that the reaction rate order was shifted from first order to second order when using an $H_2O_2/UV/Fe_2O_3$ system. The results could be an alternative for treating textile industry wastewater, and the parameters obtained can be used for equipment scale-up.

Keywords: biochemical oxygen demand; chemical oxygen demand; Fenton reaction; wastewater effluent

---

1. Introduction

Textile industries produce wastewater which is considered dangerous for the environment. The combined waste from every process and stage can produce solutions with high chemical oxygen demand (COD), biological oxygen demand (BOD), thick color, and a total suspended solids (TSS) parameter which is alkaline [1]. It is reported that ~20% of dyes, which are the main component in wastewater effluent, are released to natural water environments [2–4]. Solutions with high COD and BOD can cause a rapid reduction in dissolved oxygen and are toxic to biological life if they are directly discharged into the environment. Furthermore, due to aesthetic concerns, the textile industry could face a serious challenge from the public to treat its dye effluent before releasing it to water bodies.

Traditional physical treatment can be used to treat wastewater effluent, with methods such as adsorption using activated carbon [5], ultrafiltration [6], ion exchange [7], and coagulation [8]. However, these treatments just transfer the pollutant to another phase, hence a further handling is needed. Destructive methods are an alternative approach, i.e., biological treatment using microorganisms and advanced oxidation processes (AOPs). Degradation by microorganism typically requires a large area and volume due to slow rates. Also, it is possible that dyes are adsorbed in the sludge, hence not degraded [9]. AOPs have the potential to be effective and efficient methods, since they can quickly destruct the contaminants, forming other acceptable constituents [10–12]. However, when the colorants are complexes, a modification of this process is necessary, like using photo assisted oxidation [13–15].

Fenton and photo-Fenton reactions have significant advantages compared to other advanced oxidation processes. Energy utilization is reduced due to the dark usage of cheap reagent hydrogen peroxide (H_2O_2) in a Fenton reaction. By adding iron, there will be more hydroxyl radical formation when H_2O_2 is mixed with wastewater following the reaction of Equation (1) [16].

$$Fe^{2+} + H_2O_2 + H^+ \rightarrow Fe^{3+} + {}^*OH + H_2O \qquad (1)$$

UV radiation can decompose H_2O_2 more efficiently and initiate photo-reduction of ferric to ferrous iron. Free solar radiation can be utilized to carry out the photochemical stages of the photo-Fenton reaction so that complete mineralization of organic compounds is possible in a short period of time [17]. The photo-Fenton reaction has been proven to be efficient intreating water contaminated with a variety of organic and biological pollutants and in organic synthesis [18].

In this work, a comprehensive study was performed to obtain the optimal conditions to treat a complex and real textile wastewater effluent based on sulphur black coloring agent using Fenton reaction assisted by UV light irradiation. UV lamp power, peroxide concentration, pH and iron oxide catalyst were adjusted and their effects on the degradation of sulphur black wastewater were studied. Because the photo-Fenton reaction is very complex, reaction models for the reaction rate prediction are based on empirical correlations. The reaction behaviors using only UV and combination of UV and iron oxide catalyst is presented. We demonstrate that with a good combination of different factors, it is possible to treat wastewater quickly and effectively.

2. Results and Discussion

2.1. Influence of UV Lamp Power on the Dye Effluent Degradation

UV lamp power is one of the factors that influences textile wastewater degradation. The lamp power was varied to 10, 20, and 30 watts. Figure 1 shows that with the usage of stronger UV lamp power, the degradation of textile wastewater will be greater, as can be seen from COD (Figure 1a) and BOD (Figure 1b) values. That is because the stronger the UV lamp power is, the faster the decomposition rate of H_2O_2 to *OH radical will be. However, the increase of degradation seems not to be linear with increasing lamp power. There is only a slight change of degradation rates of COD and BOD with UV lamp powers of 10 and 20 watts. There is a deviation, that the BOD degradation rate at 30 min at 20 watts power was lower than the rate measured at 10 watts power. This is likely due to the limitations of measurement accuracy. When the UV lamp power was 30 watts, the degradation rates of COD and BOD were more powerful than those of 10 and 20 watts as shown by a strong decrease of COD and BOD values.

The influence of the research variable with the performance of oxidation reaction can be depicted from the reaction rate constant. The constant value of the reaction rate is the slope of the graph between $-\ln(1 - x)$ and reaction time. The relation between $-\ln(1 - x)$ and reaction time is shown in Figure 2a. The values of the constants used are summarized in Table 1. The higher the power of the UV lamp, the greater the value of the reaction rate constant. This can be seen from the UV lamp power of 30 watts, the reaction rate constant is 0.025 min^{-1}. UV radiation has photonic energy of sufficient strength to break the chemical bonds. Absorption of a photon provides the energy required to force electrons into higher energy states. Upon absorbing sufficient energy, electrons may be freed from an atom. Some of this released energy becomes heat, fluorescence, or activating light by chemicals that can lead to a reaction, which does not exist in lightless media.

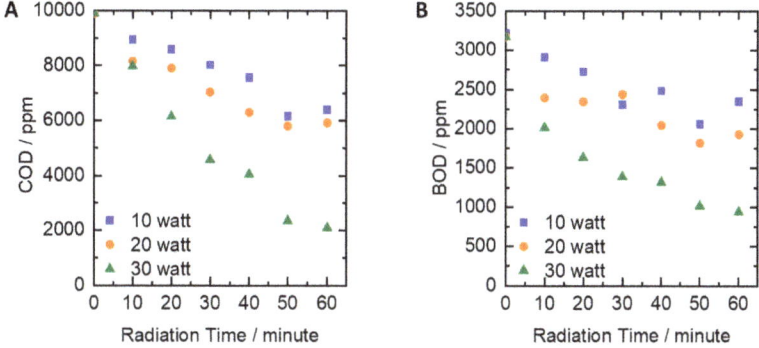

Figure 1. (**A**) Chemical oxygen demand (COD) and (**B**) biological oxygen demand (BOD) degradations of textile wastewater as functions of UV lamp power and irradiation time.

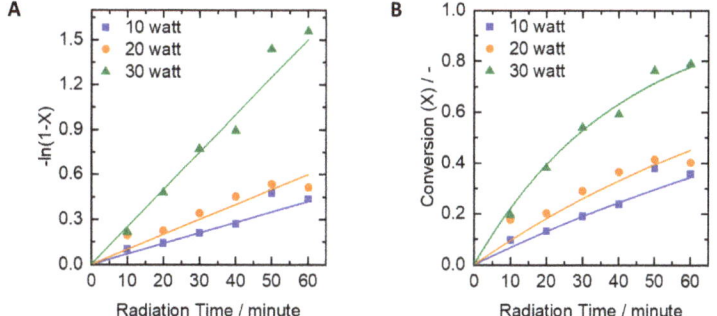

Figure 2. (**A**) Correlation of −ln(1 − x) and reaction time in UV lamp power variation. (**B**) Data fitted with a reaction model in the case of UV lamp power variation. Concentration of peroxide: 0.61 moles/L solution, pH 9.

Table 1. Kinetic parameters as a function of UV lamp power variation.

UV Lamp Power, watt	k', min^{-1}
10	0.0070
20	0.0100
30	0.0250

The correlation between lamp power with k' value reviewed from COD degradation can be seen in Table 1. The greater the UV lamp power, the faster the reaction rate. This is because the UV lamp power of 30 watts can decompose and change hydrogen peroxide to hydroxyl radical optimally. This is marked by a 78.15% decrease of COD and a 70.45% decrease of BOD, with the k' value of 0.025 min^{-1}.

2.2. Influence of Peroxide Concentration

Figure 3a,b depicts the correlation between radiation time and the degradation of COD and BOD. It is meant to find the optimum concentration of H_2O_2 needed to degrade textile wastewater with the COD parameter. Without H_2O_2, it is shown that there is no degradation because there is no hydroxyl radical formed without addition of hydrogen peroxide. But when H_2O_2 is added with a concentration of 0.20 moles/L and 0.41 moles/L, there are some changes in the degradation, by about 19.97% and 46.85% in 60 min, respectively. When the concentration of H_2O_2 added was increased to 0.61 moles/L, there was a remarkable change of degradation that reached 78.91% in 60 min. Addition of higher concentrations of H_2O_2 (0.82 moles/L and 1.02 moles/L) resulted in the degradation efficiency becoming less than before, (66.35% and 56.83%, respectively).

When the addition of H_2O_2 exceeded a concentration of 0.61 moles/L, it did not give more degradation efficiency. On the contrary, it resulted in lower efficiency, which made the degradation percentage lower too. This is because an excessive amount of *OH can react with H_2O_2 to produce *HO_2 which has a weaker ability to degrade organic compounds compared to hydroxyl radicals. Formation of *HO_2 reduces the amount of hydroxyl radical that is produced from photolysis reaction of H_2O_2 and UV light, where the *OH radical formed will react with H_2O_2 according to Equation (2) [14].

$$^*OH + H_2O_2 \rightarrow {^*HO_2} + H_2O \qquad (2)$$

By the occurrence of this reaction, the hydroxyl radical produced is not proportional with the addition of H_2O_2 concentration. Therefore, it is understandable that the addition of higher H_2O_2 concentrations does not give a significant change of degradation percentage.

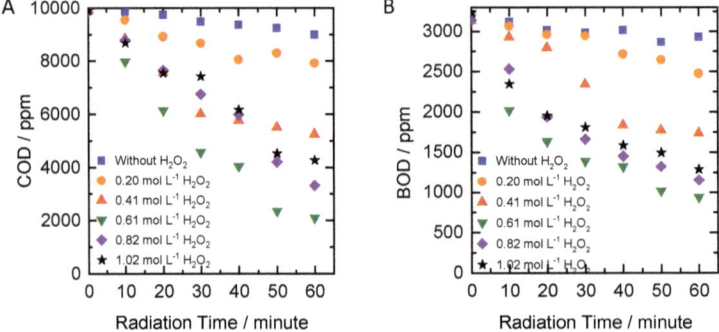

Figure 3. (**A**) COD and (**B**) BOD degradation of textile wastewater with concentration of hydrogen peroxide variation.

From Figure 4a, the reaction rate constant value increased with the increase of H_2O_2 concentration, which is because the amount of hydroxyl radical formed also increased as an oxidation agent. However, when the concentration of H_2O_2 exceeded 0.61 moles/L, the reaction rate constant decreased. As mentioned before, this is because an excessive amount of H_2O_2 concentration will form hydroperoxyl radicals (*HO_2) which are less reactive compared to hydroxyl radicals and result in a decrease of its reaction rate or its oxidation rate. The values of the reaction rate constant can be seen in Table 2 and with the reaction constraints, the data are well fitted with reaction model (Figure 4b). It can be concluded from the degradation of COD and BOD that the optimum reaction rate constant is 0.0250 min^{-1}.

Table 2. Kinetic parameters of hydrogen peroxide concentration variation.

Hydrogen Peroxide Concentration, moles/L	k', min^{-1}
0	0.0010
0.2	0.0040
0.41	0.0120
0.61	0.0250
0.82	0.0160
1.02	0.0130

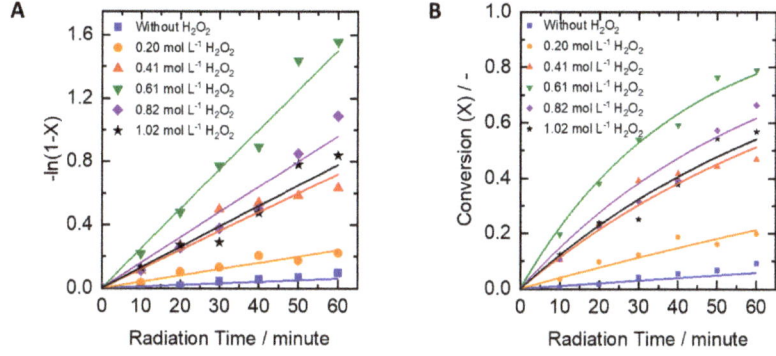

Figure 4. (**A**) Correlation of −ln(1 − x) and reaction time in concentration of hydrogen peroxide variation. (**B**) Data fitted with reaction model in the case of peroxide variation. UV lamp power: 30 watts.

2.3. Influence of pH

To examine the effect of pH, reaction experiments were carried out in different pH levels but at a fixed concentration of H_2O_2 (0.61 moles/L) and lamp power (30 watts). Figure 5a shows degradation of COD over the reaction time. The results display that acidic pH is more favorable for dye degradation. As described in the literature [19], at higher pH (more OH^-) it is possible that hydroxyl reacts with peroxide to produce water and oxygen, hence finally, it will reduce the production of hydroxyl radicals. The reaction rate constant was evaluated based on the experimental data (see Figure 5b). In the acidic and basic regions, the values of k' were ~0.0250 min^{-1} and ~0.020 min^{-1}, respectively.

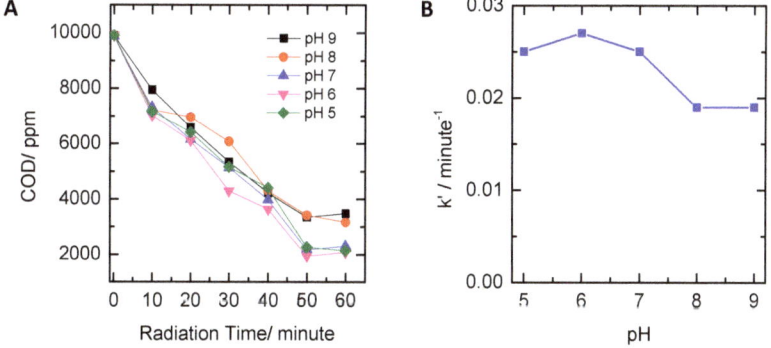

Figure 5. (**A**) Effects of pH in the degradation of COD and (**B**) reaction rate constant evaluated from experiment when varying pH of solutions.

2.4. Influence of Iron Oxide Catalyst

Finally, the oxide catalyst was added to the reaction system to further promote the degradation of wastewater. The UV lamp power, H_2O_2 concentration, and pH were set to the optimal values of 30 watts, 0.61 moles/L and pH 6, respectively. As shown in Figure 6a,b the addition of Fe_2O_3 made the degradation of textile wastewater faster. The greater the concentration of Fe_2O_3 added, the greater was the degradation of COD and BOD. This is because Fe_2O_3 is functioning as a catalyst which able to decompose H_2O_2 to hydroxyl radicals, and it can further react with the organic compounds in the textile wastewater [20]. According to Koprivana and Kusic, the minimal concentration of

Fe^{2+} or Fe^{3+} required for the Fenton reaction is in the range 3–15 mg/L (0.05–0.3 mM) [21]. Hence, the concentration of catalyst in this work is enough to decompose pollutant in wastewater.

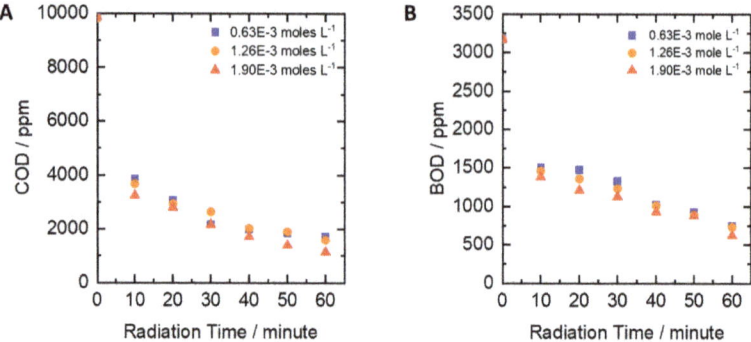

Figure 6. (**A**) COD and (**B**) BOD degradation of textile wastewater with concentration of iron (III) oxide catalyst variation. Conditions: pH 6, 0.61 moles/L concentration of H_2O_2, and 30-watt lamp power.

From the data observed, it is likely that, the reaction order changes with the addition of Fe_2O_3 catalyst. The data were fitted with reaction order one (Figure 7a) and order two (Figure 7b). It can be seen that the order two reaction is more precise to be used in this catalytic reaction. The change of the reaction order also showed that the reaction speed was increased by adding Fe_2O_3 catalyst to decompose H_2O_2 to hydroxyl radicals. The comparison of the values of the reaction rate constants between order one and order two can be seen in Table 3.

Table 3. Kinetic parameter of iron (III) oxide concentration variation.

Iron (III) Oxide Concentration, moles/L	First Order k', min^{-1}	Second Order k', L.(mg·min)$^{-1}$
6×10^{-4}	0.0254	1.12×10^{-5}
1×10^{-3}	0.0255	9.71×10^{-6}
2×10^{-3}	0.0410	1.28×10^{-5}

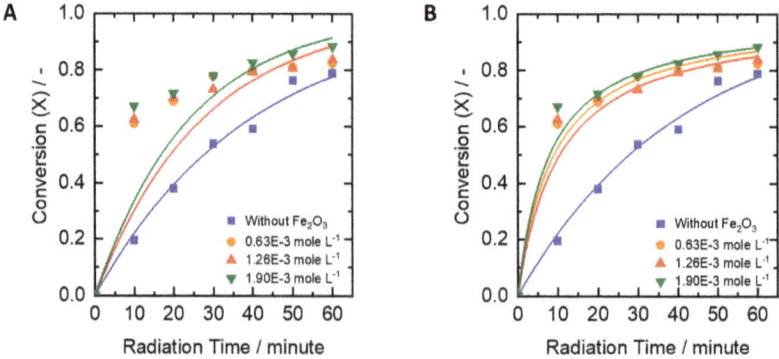

Figure 7. Fitting conversion data with reaction model order one (**A**) and order two (**B**). Conditions: pH 6, 0.61 moles/L concentration of H_2O_2, and 30-watt lamp power.

3. Materials and Methods

3.1. Wastewater-Effluent Characteristics

Wastewater was obtained from textile industry located in Bandung, Indonesia. The sulphuric coloring agent was the main component. Before given treatment, the initial conditions of the waste were measured including COD, BOD, color, and pH parameters. The parameter measurement results can be seen in Table 4.

Table 4. Initial characteristics of sulphuric coloring agent of the textile industry.

No.	Parameter	Value
1	Color (TCU)	47,000
2	COD (mg/L)	9906
3	BOD_5 (mg/L)	3175
4	pH	9

3.2. Fenton Reactions

Up to 1000 mL of wastewater was placed in the reactor (shown in Figure 8) and then peroxide was added. The reaction was started when the UV lamp (254 nm) and stirrer were turned on. Samples were taken (as much as 50 mL) with radiation time 0, 10, 20, 30, 40, 50, and 60 min to analyze the COD and BOD. The experimental set-up and reaction conditions were based on reports of [22,23]. Variations of conditions were performed to evaluate the effects of reaction condition to degradation of dyes as follows.

(i) Hydrogen peroxide (H_2O_2) concentration variation: concentrations of hydrogen peroxide were set to 0.20; 0.41; 0.61; 0.82; and 1.02 moles/L while UV lamp was set to 30 watts;
(ii) Variation of UV lamp power: the UV lamp power was set to 10, 20, and 30 watts while peroxide concentration employed was 0.61 moles/L;
(iii) Variation of pH: pH was changed in the range of 5–9, while peroxide concentration employed was 0.61 moles/L and UV lamp power was 30 watts. pH was measured using a pH meter (Hanna Instruments, USA);
(iv) The effect of the reaction with Fe_2O_3 catalyst (phase: α-Fe_2O_3, particle size <5 µm, pro-analysis, from Merck, Germany): pH, peroxide concentration, and UV lamp power were set to 6, 0.61 moles/L, and 30 watts, respectively. Iron (III) oxide was added at the concentrations of 0.00063 moles/L, 0.00126 moles/L, and 0.00190 moles/L.

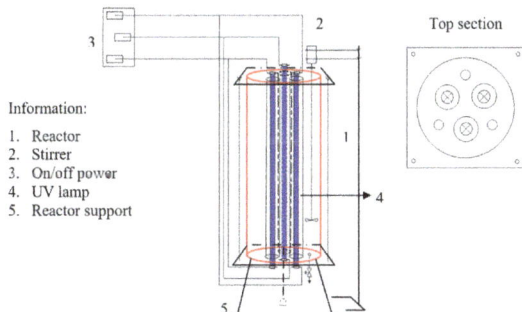

Information:
1. Reactor
2. Stirrer
3. On/off power
4. UV lamp
5. Reactor support

Figure 8. Scheme of reactor to perform textile wastewater degradation by peroxide oxidation assisted by UV light irradiation. Reactor dimensions: 32 cm length, 9 cm diameter, and 1.2 L capacity. Intensity of UV light: 35 watt/m².

3.3. Sample Analyses

3.3.1. Determination of COD Values

The H_2SO_4 pro-COD was prepared by dissolving 1 g of $AgSO_4$ with 100 mL of H_2SO_4 98%. Ferrous ammonium sulfate (FAS) 0.1 N was made by solving 39 g of $(NH_4)_2SO_4 \cdot 6H_2O$ with 20 mL of H_2SO_4 98% and then distilled water was added until the volume was 1 L. The ferroin indicator was prepared by dissolving 1.485 g of phenanthroline monohydrate and 0.695 g $FeSO_4 \cdot 7H_2O$ with distilled water until the volume was 100 mL. Two screw cap tubes were provided, one was filled with 2 mL of distilled water and the other was filled with the sample. After adding 40 mg of $HgSO_4$, 3 mL of H_2SO_4 pro COD, and 1 mL of $K_2Cr_2O_7$ 0.25 N to the sample, it was then heated in the COD reactor for 2 h. After that, the sample was cooled down until room temperature, moved to a 100 mL Erlenmeyer flask, and three drops of ferroin indicator were added. The sample was titrated with FAS 0.1 N until an equivalent point or until the color changed to reddish brown. The volume of FAS needed was recorded.

The COD value was calculated by Equations (3) and (4).

$$Normality = \frac{mL_{K_2Cr_2O_7} \times 0.1N}{mL_{FAS\ needed}} \tag{3}$$

$$COD = \frac{(V_{blank\ titration} - V_{sample\ titration}) \times (N_{FAS}) \times (8000) \times (Dilution\ Factor)}{2} \tag{4}$$

3.3.2. Determination of BOD Values

As much as 25 mL of sample which had been incubated with Sanyo MIR-153 incubator was tested to get the initial BOD values and the pH was measured (about 6.5–7.5). If the pH was higher or lower, neutralization was performed. The temperature was set to be about $20 \pm 1\ °C$, after the temperature was reached, as much as 40 mL of samples were put into BOD bottles with a magnetic stirrer and added with 100 mL of distilled water. Then the bottle was placed in the BOD unit, a seal was installed that had been given two drops of KOH 40%, and a cap was used to close the bottle. The BOD unit was connected to an electric current, and the motor was turned on. The screw cap of the pressure meter and the cap of the sample bottle were loosened up until about 30 min to get the temperature to 20 °C. After the temperature was reached, the screw cap of the temperature meter and the sample bottle's cap were fastened. The mercury column was adjusted at zero points and the measurement time was started. The BOD values were determined on day five and entered in the datasheet.

The BOD was calculated by Equation (5).

$$BOD_b = BOD_a \times n \tag{5}$$

BOD_a = readable value, mg/L
BOD_b = actual BOD, mg/L
n = dilution factor.

3.4. Kinetic Models

The oxidation reaction with hydrogen peroxide was performed through three steps: (i) degradation reaction of hydrogen peroxide to hydroxyl free radicals with UV ray; (ii) reaction between COD and hydroxyl free radical; and (iii) degradation of COD* to other molecules as described in Equations (6)–(8) [11].

$$H_2O_2 \xrightarrow{UV} {}^{\bullet}OH \tag{6}$$

$${}^{\bullet}OH + COD \rightarrow COD^{*} + H_2O \tag{7}$$

$$COD^* + O_2/H^+ \rightarrow CO_2 + H_2O \tag{8}$$

The degradation reaction of hydrogen peroxide with UV ray is fast, so the determining reaction is the reaction between COD with hydroxyl radicals (Equation (7)). The reaction between COD and *OH is a one-way reaction. The reaction model is shown in Equation (9).

$$(-r_{COD}) = k(C_{COD})^\alpha (C_{*OH})^\beta \tag{9}$$

With:

$-r_{COD}$ = Oxidation of COD reaction rate, (mg/L.minute)
K = Reaction rate constant, mg/(L.minute)/(mg/L)$^{\alpha+\beta}$
C_{COD} = COD concentration, (mg/L)
C_{*OH} = Hydroxy radical concentration, (mg/L)
α = COD concentration order reaction
β = Hydroxy ion concentration order reaction

By making excessive hydroxyl radicals relatively unchanged during the reaction, Equation (9) becomes Equation (10).

$$(-r_{COD}) = k\prime(C_{COD})^\alpha \tag{10}$$

where $k\prime$ is defined as Equation (11).

$$k\prime = k(C_{*OH})^\beta \approx k(C_{*OH_0})^\beta \tag{11}$$

For order one reaction, or $\alpha = 1$, Equation (10) becomes Equation (12).

$$(-r_{COD}) = k\prime(C_{COD}) \tag{12}$$

where, $-r_{COD}$ = Oxidation of COD reaction rate, (mg/L.minute);
$k\prime$ = Reaction rate constant with C*OH relatively constant, (min^{-1});
C_{COD} = COD concentration, (mg/L).
Using mass balance and integration, we obtain Equation (13):

$$-\ln\left(\frac{C_{CODi}}{C_{COD0}}\right) = k\prime t \tag{13}$$

By defining conversion, x, as Equation (14):

$$x = \frac{C_{COD0} - C_{CODi}}{C_{COD0}}, \tag{14}$$

Equation (14) is inserted to Equation (13), hence:

$$-\ln(1-x) = k\prime t \tag{15}$$

For order one reaction or $\alpha = 2$, Equation (10) becomes Equation (16):

$$(-r_{COD}) = k\prime(C_{COD})^2 \tag{16}$$

Using mass balance and integration, Equation (16) becomes Equation (17):

$$C_{CODi} = \frac{C_{COD0}}{1 + k\prime t C_{COD0}} \tag{17}$$

where, t = oxidation time, (minute) and $k\prime$ = kinetics constant, (L.mg^{-1} min^{-1}).

4. Conclusions

An oxidative method assisted by UV light irradiation was carried out to treat textile wastewater based sulphur coloring agent. It was found that the optimal conditions without iron oxide addition were an H_2O_2 concentration of 0.61 moles/L, UV lamp power of 30 watts, and pH 6; from which COD and BOD degradations of up to 80% and 75% (respectively) resulted. In this condition, the reaction rate followed a first order reaction with k' of 0.0250 min^{-1}. The addition of Fe_2O_3 could increase degradation rate significantly, while similar COD and BOD degradation percentages were obtained. When Fe_2O_3 was present, the reaction followed a second order reaction model.

Author Contributions: Conceptualization, S.S., A.P.; methodology, S.S., A.P, P.P., I.B.T.; investigation, P.P., I.B.T.; writing—original draft preparation, S.S.; writing—review and editing, T.A.; visualization, W.T.L.; supervision, S.S.

Funding: The APC was funded by Publisher and Publication Board of Universitas Gadjah Mada.

Acknowledgments: In this section you can acknowledge any support given which is not covered by the author contribution or funding sections. This may include administrative and technical support, or donations in kind (e.g., materials used for experiments).

Conflicts of Interest: The authors declare no conflict of interest.

References

1. Doble, M.; Kumar, A. *Biotreatment of Industrial Effluents*; Elsevier Science: Amsterdam, Netherlands, 2005; ISBN 9780080456218.
2. Zollinger, H. *Color Chemistry: Syntheses, Properties, and Applications of Organic Dyes and Pigments*; Wiley: Hoboken, NJ, USA, 2003; ISBN 9783906390239.
3. Weber, E.J.; Stickney, V.C. Hydrolysis kinetics of Reactive Blue 19-Vinyl Sulfone. *Water Res.* **1993**, *27*, 63–67. [CrossRef]
4. Babaei, A.A.; Kakavandi, B.; Rafiee, M.; Kalantarhormizi, F.; Purkaram, I.; Ahmadi, E.; Esmaeili, S. Comparative treatment of textile wastewater by adsorption, Fenton, UV-Fenton and US-Fenton using magnetic nanoparticles-functionalized carbon (MNPs@C). *J. Ind. Eng. Chem.* **2017**, *56*, 163–174. [CrossRef]
5. Ariyanto, T.; Kurniasari, M.; Laksmana, W.T.; Prasetyo, I. Pore size control of polymer-derived carbon adsorbent and its application for dye removal. *Int. J. Environ. Sci. Technol.* **2018**. [CrossRef]
6. Lin, J.; Ye, W.; Baltaru, M.C.; Tang, Y.P.; Bernstein, N.J.; Gao, P.; Balta, S.; Vlad, M.; Volodin, A.; Sotto, A.; et al. Tight ultrafiltration membranes for enhanced separation of dyes and Na2SO4 during textile wastewater treatment. *J. Membr. Sci.* **2016**, *514*, 217–228. [CrossRef]
7. Hassan, M.M.; Carr, C.M. A critical review on recent advancements of the removal of reactive dyes from dyehouse effluent by ion-exchange adsorbents. *Chemosphere* **2018**, *209*, 201–219. [CrossRef] [PubMed]
8. Merzouk, B.; Gourich, B.; Sekki, A.; Madani, K.; Vial, C.; Barkaoui, M. Studies on the decolorization of textile dye wastewater by continuous electrocoagulation process. *Chem. Eng. J.* **2009**, *149*, 207–214. [CrossRef]
9. Hu, T.-L. Removal of reactive dyes from aqueous solution by different bacterial genera. *Water Sci. Technol.* **1996**, *34*, 89–95.
10. Martins, A.F. Advanced oxidation processes applied to effluent streams from an agrochemical industry. *Pure Appl. Chem.* **1998**, *70*, 2271–2279. [CrossRef]
11. Munter, R. Advanced oxidation processes current status and prospect. *Proc. Est. Acad. Sci. Chem* **2001**, *50*, 59–80.
12. Munoz, M.; Mora, F.J.; de Pedro, Z.M.; Alvarez-Torrellas, S.; Casas, J.A.; Rodriguez, J.J. Application of CWPO to the treatment of pharmaceutical emerging pollutants in different water matrices with a ferromagnetic catalyst. *J. Hazard. Mater.* **2017**, *331*, 45–54. [CrossRef] [PubMed]
13. Ajmal, A.; Majeed, I.; Malik, R.N.; Idriss, H.; Nadeem, M.A. Principles and mechanisms of photocatalytic dye degradation on TiO2 based photocatalysts: A comparative overview. *RSC Adv.* **2014**, *4*, 37003–37026. [CrossRef]
14. Schrank, S.G.; dos Santos, J.N.R.; Souza, D.S.; Souza, E.E.S. Decolourisation effects of Vat Green 01 textile dye and textile wastewater using H_2O_2/UV process. *J. Photochem. Photobiol. A Chem.* **2007**, *186*, 125–129. [CrossRef]

15. Wang, Q.; Pang, W.; Mao, Y.; Sun, Q.; Zhang, P.; Ke, Q.; Yu, H.; Dai, C.; Zhao, M. Study of the degradation of trimethoprim using photo-Fenton oxidation technology. *Water (Switzerland)* **2019**, *11*, 207. [CrossRef]
16. Vorontsov, A.V. Advancing Fenton and photo-Fenton water treatment through the catalyst design. *J. Hazard. Mater.* **2019**, *372*, 103–112. [CrossRef] [PubMed]
17. Wang, J.L.; Xu, L.J. Advanced oxidation processes for wastewater treatment: Formation of hydroxyl radical and application. *Crit. Rev. Environ. Sci. Technol.* **2012**, *42*, 251–325. [CrossRef]
18. Babuponnusami, A.; Muthukumar, K. A review on Fenton and improvements to the Fenton process for wastewater treatment. *J. Environ. Chem. Eng.* **2014**, *3*, 557–572. [CrossRef]
19. Shu, H.Y.; Hunag, C.R.; Chang, M.C. Decolorization of mono-azo dyes in wastewater by advanced oxidation process: a case study of acid red 1 and acid yellow 23. *Chemosphere* **1994**, *29*, 2597–2607. [CrossRef]
20. Amelia, S.; Sediawan, W.B.; Prasetyo, I.; Munoz, M.; Ariyanto, T. Role of the pore structure of Fe/C catalysts on heterogeneous Fenton oxidation. *J. Environ. Chem. Eng.* **2019**, 102921. (In Press) [CrossRef]
21. Natalija Koprivance, H.K. AOP as an effective tool for the mineralization of hazardous organic pollutants in colored wastewater; chemical and photochemical processes. In *Hazardous Materials and Wastewater: Treatment, Removal and Analysis*; Lewinsky, A.A., Ed.; Nova Science Publishers Inc.: New York, NY, USA, 2007; pp. 149–199. ISBN 1-60021-257-3.
22. Tanyong, I.B. Penurunan kadar chemical oxygen demand limbah rhodamin B. Bachelor Thesis, Universitas Gadjah Mada, Yogyakarya, Indonesia, 31 January 2012.
23. Paesal, P. Studi degradasi limbah cair industri tekstil dengan menggunakan metode proses oksidasi lanjut. Master's Thesis, Universitas Gadjah Mada, Yogyakarta, Indonesia, 22 August 2008.

© 2019 by the authors. Licensee MDPI, Basel, Switzerland. This article is an open access article distributed under the terms and conditions of the Creative Commons Attribution (CC BY) license (http://creativecommons.org/licenses/by/4.0/).

Article

Solar-Driven Removal of 1,4-Dioxane Using WO$_3$/nγ-Al$_2$O$_3$ Nano-Catalyst in Water

Xiyan Xu [1], Shuming Liu [1,*], Yong Cui [2], Xiaoting Wang [1], Kate Smith [1] and Yujue Wang [1]

1. School of Environment, Tsinghua University, Beijing 100084, China; xiyanxu@tsinghua.edu.cn (X.X.); wang-xt15@mails.tsinghua.edu.cn (X.W.); skt15@mails.tsinghua.edu.cn (K.S.); wangyujue@tsinghua.edu.cn (Y.W.)
2. Boda Water Co., Ltd., Beijing 100176, China; cuiyong@bdawater.com
* Correspondence: shumingliu@tsinghua.edu.cn; Tel./Fax: +86-10-6278-7964

Received: 5 April 2019; Accepted: 24 April 2019; Published: 25 April 2019

Abstract: Increasing demand for fresh water in extreme drought regions necessitates potable water reuse. However, current membrane-based water reclamation approaches cannot effectively remove carcinogenic 1,4-dioxane. The current study reports on the solar-driven removal of 1,4-dioxane (50 mg L^{-1}) using a homemade WO$_3$/nγ-Al$_2$O$_3$ nano-catalyst. Characterization methods including scanning electron microscope (SEM), X-ray photoelectron spectroscopy (XPS) and X-ray fluorescence (XRF) analyses are used to investigate the surface features of the catalyst. The 1,4-dioxane mineralization performance of this catalyst under various reaction conditions is studied. The effect of the catalyst dosage is tested. The mean oxidation state carbon (MOSC) values of the 1,4-dioxane solution are followed during the reaction. The short chain organic acids after treatment are measured. The results showed that over 75% total organic carbon (TOC) removal was achieved in the presence of 300 mg L^{-1} of the catalyst with a simulated solar irradiation intensity of 40 mW cm^{-2}. Increasing the dose of the catalyst from 100 to 700 mg L^{-1} can improve the treatment efficiency to some extent. The TOC reduction curve fits well with an apparent zero-order kinetic model and the corresponding constant rates are within 0.0927 and 0.1059 mg L^{-1} s^{-1}, respectively. The MOSC values of the 1,4-dioxane solution increase from 1.3 to 3 along the reaction, which is associated with the formation of some short chain acids. The catalyst can be effectively reused 7 times. This work provides an oxidant-free and energy saving approach to achieve efficient removal of 1,4-dioxane and thus shows promising potential for potable reuse applications.

Keywords: 1,4-dioxane; photocatalysis; WO$_3$/nγ-Al$_2$O$_3$; solar radiation; water treatment; potable reuse

1. Introduction

Rapid population expansion of cities results in increasing water consumption and requires exploitation of alternative water resources for potable purposes, especially in extremely water-scarce regions [1–4]. Membrane-based water purification techniques including ultrafiltration (UF) and reverse osmosis (RO) can effectively remove the major proportion of salts and organic contaminants in municipal wastewater secondary effluent (MWSE) [5–9]. However, some micropollutants including 1,4-dioxane (1,4-D) present in the MWSE in trace concentrations cannot be easily removed [10]. 1,4-D is a reagent stabilizer in industrial chlorinated solvents, which commonly exists in cosmetics, toiletries, and food addictive [11–13]. Uncontrolled exposure to 1,4-D can cause failures of human organs including kidney and liver [14,15]. It can even cause cancer when it exists in drinking water. Thus, it has been classified as a Group 2B human carcinogen [1].

Therefore, 1,4-D should be eliminated if the reclamation of MWSE for potable reuse is required [10]. Advanced oxidation processes (AOPs) have been extensively applied to decompose or even mineralize organic pollutants in water [8,16–21]. Reactive free radicals, or more specifically reactive oxygen species (ROS) like HO$^\bullet$ and HO$_2^\bullet$/O$^{\bullet-}$, can be formed from the activation of oxidants by reagents or energy in these systems. These ROS tend to attack organic pollutants and result in their degradation [22–26]. Photolytic AOPs can take advantage of the photons' energy, especially ultraviolet (UV), to activate the oxidants [27]. In order to eliminate 1,4-D from water for potable reuse, recent studies provided several possible solutions based on photolytic AOP approaches [1,28,29]. Patton et al. investigated the performance of UV/H$_2$O$_2$-based system for 1,4-D removal in the presence of mono- and dichloramines [1]. On that basis, Li et al. used another oxidant (persulfate, S$_2$O$_8^{2-}$) to establish a UV/S$_2$O$_8^{2-}$ system to eliminate 1,4-D [5]. Although these works revealed that mono- or dichloramines can promote treatment efficiency to some extent by participating in the reaction processes, these systems inevitably introduce external reagents (the oxidants) into the target solution. This compromises the practical application of those methods for potable water reuse since the effect of the residual reagents on human health needs to be further considered.

Photocatalytic oxidation systems are possible candidates to avoid that drawback. In photocatalytic systems, oxidants are not indispensable since ROS can be formed from the activation of dissolved oxygen or even water molecules by the photocatalysts under photoirradiation [30–32]. Electrons on the valence band (VB) of the catalyst can be excited to transit to the conduction band (CB) when the energy of photoirradiation reaches the band gap [33]. Electron/hole (h$^+$/e$^-$) pairs can be formed on the catalysts during this process and lead to formation of ROS from dissolved oxygen and water molecules [30,34]. Photocatalytic approaches have been used to treat 1,4-D contaminated water [35–37]. However, these systems were mainly activated by UV light rather than visible light, since the band gap of traditional photocatalysts (like TiO$_2$) is so broad that photons with lower energy (visible light) are not capable of inducing the formation of ROS [38,39].

Tungsten-based catalysts have been developed in recent years since they are sensitive to visible light and thus can improve the photocatalytic efficiency [40–43]. Meanwhile, γ-Al$_2$O$_3$ supporter showed more desirable stability in AOP systems than conventional supporters like activated carbon due to its high active phase-supporter interaction [16]. Moreover, using nano-size supporters may provide a stronger active phase-supporter link due to their unique features like extremely large specific surface area. Thus, the nano-size supporters have the potential to improve the reusability of the catalyst [44]. However, to the best of our knowledge, no previous study applied nano tungsten-based catalysts with γ-Al$_2$O$_3$ as the supporter for the photocatalytic breakdown of 1,4-D.

The current study aimed at the removal of 1,4-D using an oxidant-free photocatalytic system with a homemade nano-size tungsten-based catalyst (WO$_3$/nγ-Al$_2$O$_3$) under solar light irradiation. The characteristics of the catalyst were studied using scanning electron microscope (SEM), X-ray photoelectron spectroscopy (XPS) and X-ray fluorescence (XRF). 1,4-D mineralization efficiency under various conditions was considered and the reaction kinetic rates were calculated. The effect of the catalyst dosage was tested. The oxidation state of the effluent solution during the reaction was followed and the formation of short chain organic acids after the reaction was measured.

2. Results and Discussion

2.1. Characterization of the Catalyst

To have an insight into the features of the WO$_3$/nγ-Al$_2$O$_3$ catalyst, characterization was conducted. The SEM image (Figure 1a) shows the appearance of the prepared catalyst. It can be seen that the shape of the catalyst is roughly sphere-like.

The diameter of the catalyst is around 50 nm (Figure 1a). XPS analyses of the WO$_3$/nγ-Al$_2$O$_3$ catalyst were also conducted. Figure 1b depicts the W 4f core level XPS spectra. There are two symmetric peaks at binding energies 35.9 and 37.4 eV, which are associated with the W-Al band and WO$_3$, respectively [45,46]. The O 1s XPS profile (Figure 1c) shows two symmetric peaks at 530.8 and 532.7 eV. These two peaks correspond to WO$_3$ and Al$_2$O$_3$, respectively [47,48]. Al 2p XPS spectra profile (Figure 1d) shows two peaks at 74.5 and 75.9 eV, which are associated to the Al-W band and Al_2O_3 [45,49], respectively. The XPS spectra profiles support the fact that a certain amount of active phase (WO_3) has been loaded on the carrier (nγ-Al_2O_3). It can be inferred from the results that the three elements (W, O, and Al) are interconnected in terms of WO_3 and Al_2O_3 and the former two compounds are probably connected by the W-Al band.

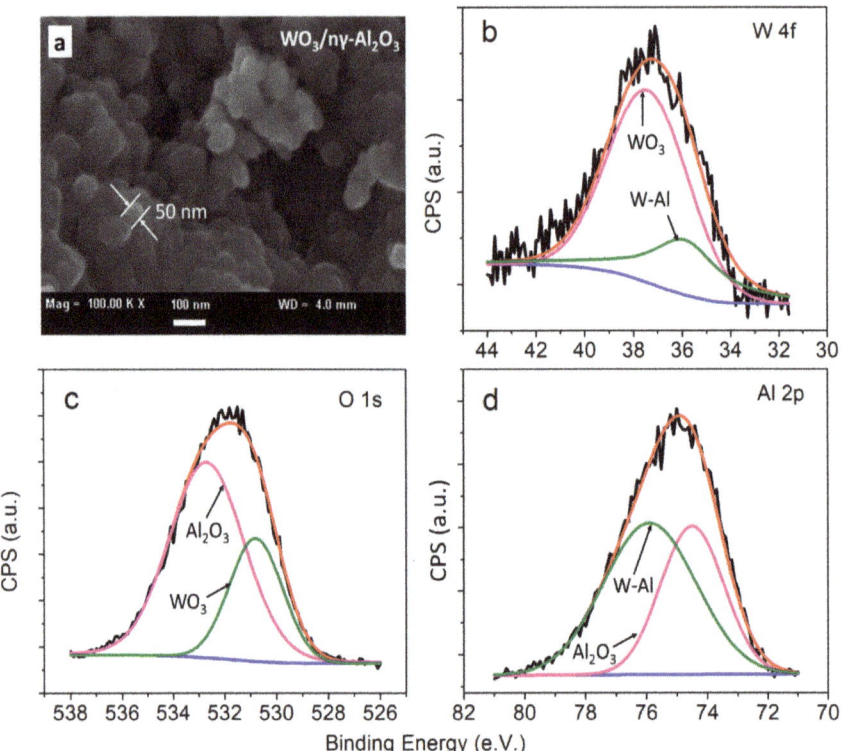

Figure 1. (a) SEM image and (b) W 4f, (c) O 1s and (d) Al 2p core level X-ray photoelectron spectroscopy (XPS) spectra of the WO_3/nγ-Al_2O_3 catalyst.

XRF profile provides quantitative proportions of the surface compounds on the catalyst (Table 1) where the weight percentage of WO_3 is 4.73% on the catalyst surface, confirming that WO_3 was formed and fixed on the surface of the catalysts.

Table 1. Percentages of surface compounds on the WO_3/nγ-Al_2O_3 catalyst from X-ray fluorescence (XRF).

Compound	Wt (%) *	Error
Al_2O_3	94.15	0.12
WO_3	4.73	0.11
Others	1.12	–

* Wt (%), the weight percent.

2.2. Solar-Driven Removal of 1,4-D

Solar-driven oxidation of 1,4-D was firstly conducted without the catalyst. As can be seen in Figure 2a, less than 17% of mineralization can be achieved after 4 h. This limited removal is due to the non-photosensitive structure of 1,4-D [15,36]. A comparative experiment using the catalyst in dark conditions was also carried out to check the contribution of adsorption for 1,4-D removal. Figure 2a shows that almost no total organic carbon (TOC) removal can be observed under these conditions, indicating the negligible effect of adsorption. Then, 1,4-D was exposed under solar radiation in the presence of the $WO_3/n\gamma$-Al_2O_3 nano-catalyst. Under this condition, a significant TOC reduction (over 75%) was obtained following a reaction time of 4 h Figure 2a. The above results indicate that the current $WO_3/n\gamma$-Al_2O_3 nano-catalyst can effectively mineralize 1,4-D and the effect is mainly due to photocatalytic decomposition.

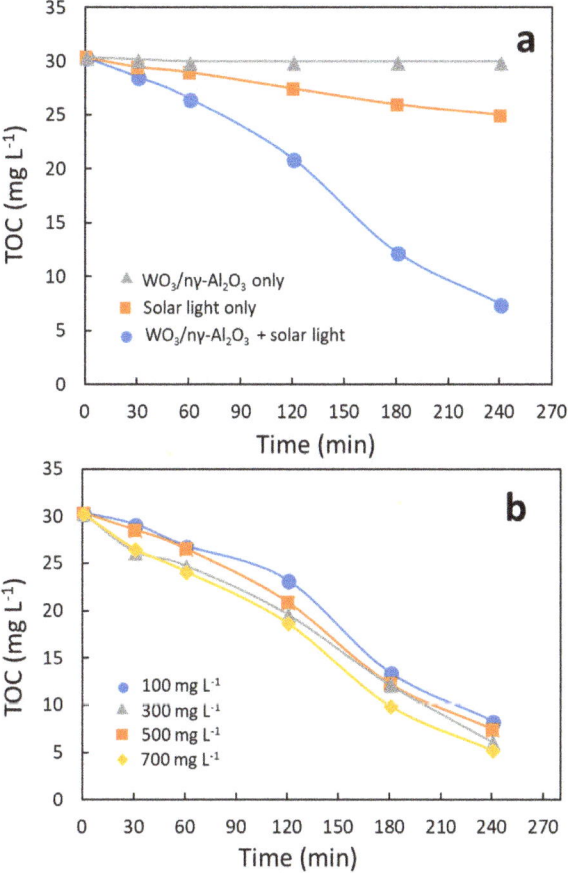

Figure 2. (a) Total organic carbon (TOC) reduction of 1,4-D at different reaction conditions and (b) the effect of catalyst dose on TOC reduction. $[1,4\text{-}D]_0 = 50$ mg L^{-1}. In (a) for both runs in the presence and absence of solar light, [catalyst] = 300 mg L^{-1}.

To check the effect of catalyst dose on mineralization of 1,4-D, various amounts from 100 to 700 mg L^{-1} were tested (Figure 2b). It can be observed that increasing the catalyst dose can improve the treatment efficiency with the final mineralization extent within 67% and 85%. The TOC evolution was fitted to a zero-order kinetic model which has been used in previous photo-oxidation systems [26].

Table 2 collects the corresponding mineralization rate constants with the correlation coefficients (≥ 0.95). It also includes the rate constants obtained using 300 mg L^{-1} of the catalyst alone and using solar light alone. It can be seen that the rate constant of the run using 700 mg L^{-1} of the catalyst reaches up to 0.1059 mg L^{-1} s^{-1}, which is over 4 folds and 81 folds the results achieved when using 300 mg L^{-1} catalyst alone and using solar light alone, respectively. The results confirm the high efficiency of the current $WO_3/n\gamma\text{-}Al_2O_3$ nano-catalyst for improving the solar-driven photocatalytic systems for the breakdown of 1,4-D.

Table 2. Apparent zero-order kinetic constant rates at different conditions.

Catalysts Dosage (mg L^{-1})	k_{TOC} (mg L^{-1} s^{-1})	R^2
700	0.1059	0.99
500	0.0990	0.99
300	0.0971	0.98
100	0.0927	0.97
300 mg L^{-1} catalyst only	0.0227	0.99
Solar light only	0.0013	0.95

2.3. Oxidation State and Formation of Short-Chain Acids

The mean oxidation state carbon (MOSC) of a solution provides an overall oxidation state of all the compounds in the solution in terms of their averaged MOSC values [50]. MOSC value can be calculated considering both chemical oxidation demand (COD) and TOC via the following equation (Equation (1)).

$$MOSC = 4 - 1.5\frac{COD}{TOC} \qquad (1)$$

A higher MOSC value indicates a higher oxidation state, whereas negative ones stand for a higher potential for further oxidation of the compounds in the solution. A previous study reported that the MOSC values of a chemical industrial wastewater containing 1,4-D were within the interval of [4,−4] and 1,4-D had an MOSC value of around 1 [36].

In the current study, to better reveal the oxidation state of the effluent during the course of the reaction, the evolution of solution MOSC was followed and the corresponding results are included in Figure 3a. As can be observed in the figure, the MOSC value of the solution increases from around 1.3 to approximately 3 after 4 h of reaction. This trend is in agreement with a previous work [36]. Extension of reaction time can hardly increase the MOSC value, indicating that the solution reaches a relatively stable oxidation state in the current photocatalytic system.

Some short-chain organic acids were formed after 4 h, including acetic, formic, and fumaric ones (Figure 3b). Among them, acetic acid shows the highest concentration. These short-chain acids have been frequently reported as the final oxidation byproducts in advanced oxidation systems with low ecotoxicity [16,18,51]. It can be concluded that the high MOSC value after reaction must be associated with the formation of these reaction byproducts.

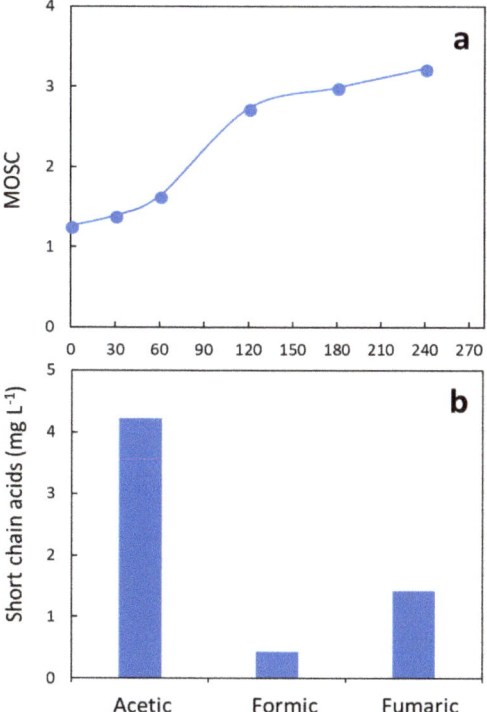

Figure 3. (a) Evolution of the mean oxidation state carbon (MOSC) values of 1,4-D aqueous solution during the solar-driven catalytic oxidation and (b) the amounts of short-chain acids formed after 4 h of reaction. $[1,4-D]_0 = 50$ mg L^{-1}; [catalyst] = 300 mg L^{-1}.

2.4. Reusability of the Catalyst

Reusability is an important factor impacting on the potential for practical application of a photocatalyst. In that respect, 300 mg L^{-1} catalyst was added to the system to lead the solar-driven decomposition of 1,4-D for 4 h. Then, the catalyst was separated and dried at 60 °C after reaction, and the same amount of 1,4-D was added again to repeat the reaction. As is shown in Figure 4, a TOC removal of 72.1% is achieved after 7 times of reuse of the catalyst, showing fairly desirable reusability. The slight decline of mineralization extent may be due to the inactivation of the catalyst by adsorption of some reaction byproducts on the active sites.

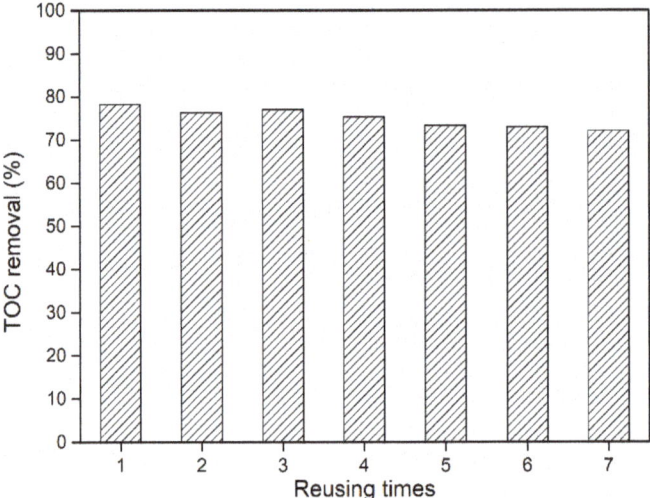

Figure 4. TOC removal of 1,4-D after reuse. $[1,4\text{-D}]_0 = 50$ mg L^{-1}; [catalyst] = 300 mg L^{-1}; reaction time = 4 h.

2.5. Postulation of Photocatalytic Mechanisms

The results so far prove that the current WO$_3$/nγ-Al$_2$O$_3$-based photocatalytic system can effectively remove 1,4-D and the byproducts include some short-chain organic acids. HO$^{\bullet}$ and O$_2^{\bullet-}$/HO$_2^{\bullet}$ radicals must be responsible for the high efficiency of the current system. These reactive free radicals can be generated through the following scheme: the photon from solar light with certain energy excites the electrons, causing them to transit from the valence band (VB) to the conduction band (CB) and thus, holes are formed on VB. The electron-hole pairs can further excite the dissolved oxygen and water to form ROS (Figure 5). The electrons have a strong reducing ability and can promote the formation of O$_2^{\bullet-}$/HO$_2^{\bullet}$ radicals from dissolved oxygen. In the meantime, the holes act as positions with high oxidation ability and thus tend to take electrons from water molecules to generate HO$^{\bullet}$ radicals. In addition, the combination of excessive HO$^{\bullet}$ radicals may occur to further form hydrogen peroxide (H$_2$O$_2$) [34,52–54]. The above mechanisms enable the current system for the efficient mineralization of the target pollutant 1,4-D in the absence of oxidant (Figure 5).

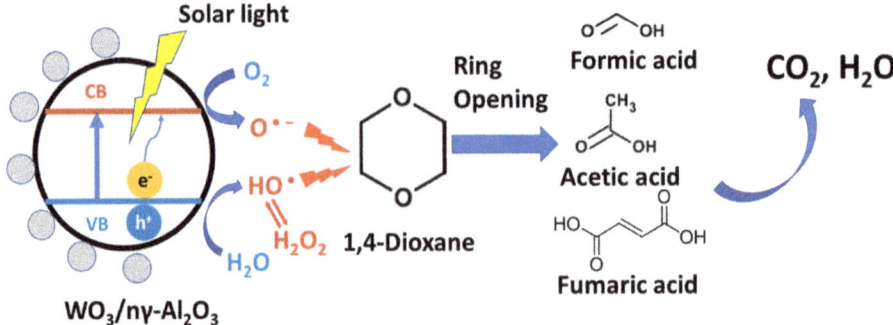

Figure 5. Mechanisms of the degradation of 1,4-D in the current WO$_3$/nγ-Al$_2$O$_3$-based photocatalytic system.

3. Materials and Methods

3.1. Materials

$Na_2WO_4·2H_2O$ was purchased from Sigma-Aldrich company, St. Louis, MO, USA. 1,4-D, nano-size γ-Al_2O_3 (nγ-Al_2O_3, diameter < 20 nm) and other reagents were supplied by Aladdin Reagent company, Shanghai, China. All the reagents were of AR grade and used without further purification. Ultrapure water was used throughout the experiments.

The WO_3/nγ-Al_2O_3 catalyst was prepared by incipient wetness impregnation. $Na_2WO_4·2H_2O$ (0.08 M) was added drop by drop onto nγ-Al_2O_3 to adjust W load to achieve nominal 5% (w/w). The impregnated nγ-Al_2O_3 was shaken at room temperature at 60 rpm for 2 h and then dried at 60 °C for 12 h. After that, the sample was calcinated at 300 °C with a heating rate of 4.5 °C min^{-1} from room temperature. The total calcination time was 8 h.

3.2. Characterization of the Catalyst

Surface morphology of the catalyst was observed using SEM (Zeiss Merlin VP Compact, Oberkochen, Germany).

XPS (PHI Quantera SXM, produced by ULVAC-PHI, Chigasaki, Japan) was used to analyze the components on the surface of the catalyst. An Al-Kα X-radiation source was used with a vacuum in the analysis chamber lower than 1.0×10^{-7} Pa. The X-ray beam spot size was 200 μm at an incident angle of 45° with pass energy of 55 eV and step length of 0.1 eV. XPS profiles of W 4f, O 1s and Al 2p were recorded and fitted using XPS Peak 4.1 software.

Quantitative analysis of the compounds on the catalyst surface was conducted using XRF (Shimadzu XRF-1800 SEQUENTIAL, Kyoto, Japan). Based on the protocol, the weight percentage of each compound was presented.

3.3. Solar-Driven Photocatalytic Processes

The solar-driven photocatalytic experiments were conducted in cylinder quartz reactors with a diameter of 6 cm and a height of 7 cm, which were placed in a solar simulator. For a typical run, 100 mL 1,4-D (50 mg L^{-1}) and certain amount of catalyst (100 to 700 mg L^{-1}) were added in the reactor with magnetic stirring at 200 rpm. Then, the lamp was turned on to start the reaction. The light source was a Xe lamp with an irradiation intensity of 40 mW cm^{-2}. The spectrum of the light source is in the range of 190–1100 nm with peaks at 436 nm and 546 nm. The initial pH value of the 1,4-D solution was around 6.8 and not adjusted during the reaction. Solution temperature was controlled by water circulation (< 40 °C after 4 h of reaction), during which no obvious evaporation was observed.

3.4. Analytical Methods

TOC of solution was measured using a TOC analyzer (Shimadzu, model 5000A, Kyoto, Japan). COD was determined by a Hach COD reactor (DRB200, Hach, Loveland, CO, USA) equipped with a spectrophotometer. Organic short-chain acids were measured by an ionic chromatography (IC) equipped with a conductivity detector (Thermo ICS5000+, Waltham, MA, USA). 3.2 mM Na_2CO_3 and 1mM of $NaHCO_3$ were used as mobile phases at pumping rate of 0.7 mL min^{-1}.

4. Conclusions

The current work reports an oxidant-free solar-driven photocatalytic system for the breakdown of 1,4-D based on a homemade WO_3/nγ-Al_2O_3 nano-catalyst. More than 75% mineralization extent can be achieved using 300 mg L^{-1} catalyst at solar light intensity of 40 mW m^{-2} after 4 h of irradiation. Increasing the dose of catalyst to 700 mg L^{-1} can improve the TOC removal to 85%. The MOSC values of the solution were followed and it was found that the oxidation state was greatly raised after reaction. The short-chain acids formed during the reaction—including acetic, formic and fumaric acids—were believed to be associated with the high solution oxidation state. A reusability study indicates that the current catalyst can still be efficient after being reused seven times.

The current work provides a promising approach to eliminate 1,4-D from water for potable reuse. It is not only oxidant-free but can also make use of solar light radiation to lead 1,4-D mineralization. However, it should be noticed that around 20% of organic carbons still remain in the solution after treatment. In that sense, future studies should be conducted to either analyze the composition of these organic carbons or evaluate their ecotoxicity.

Author Contributions: Conceptualization, X.X., S.L. and K.S.; methodology, X.X., S.L., Y.C., X.W., K.S. and Y.W.; software, X.X. and X.W.; validation, X.X.; formal analysis, X.X.; investigation, X.X., S.L., Y.C. and X.W.; resources, Y.W.; data curation, X.X., Y.C. and X.W.; writing—original draft preparation, X.X.; writing—review and editing, X.X., S.L., Y.C., X.W., K.S. and Y.W.; visualization, X.X. and K.S.; supervision, S.L. and Y.W.; project administration, S.L. and Y.W.; funding acquisition, S.L.

Acknowledgments: We are grateful for the support of the National Natural Science Foundation of China (51808312, 51879139), China Postdoctoral Science Foundation (No. 2018M631495), and the National Key Research and Development Program of China for International Science & Innovation Cooperation Major Project between Governments (2016YFE0118800).

Conflicts of Interest: The authors declare no conflict of interest.

References

1. Patton, S.; Romano, M.; Naddeo, V.; Ishida, K.P.; Liu, H. Photolysis of mono- and dichloramines in UV/Hydrogen peroxide: Effects on 1,4-Dioxane removal and relevance in water reuse. *Environ. Sci. Technol.* **2018**, *52*, 11720–11727. [CrossRef] [PubMed]
2. Du, Y.; Wu, Q.Y.; Lv, X.T.; Ye, B.; Zhan, X.M.; Lu, Y.; Hu, H.Y. Electron donating capacity reduction of dissolved organic matter by solar irradiation reduces the cytotoxicity formation potential during wastewater chlorination. *Water Res.* **2018**, *145*, 94–102. [CrossRef] [PubMed]
3. Liu, S.; Che, H.; Smith, K.; Lei, M.; Li, R. Performance evaluation for three pollution detection methods using data from a real contamination accident. *J. Environ. Manage.* **2015**, *161*, 385–391. [CrossRef] [PubMed]
4. Xu, X.; Liu, Y.; Liu, S.; Li, J.; Guo, G.; Smith, K. Real-time detection of potable-reclaimed water pipe cross-connection events by conventional water quality sensors using machine learning methods. *J. Environ. Manag.* **2019**, *238*, 201–209. [CrossRef]
5. Li, W.; Patton, S.; Gleason, J.M.; Mezyk, S.P.; Ishida, K.P.; Liu, H. UV Photolysis of chloramine and persulfate for 1,4-Dioxane removal in reverse-osmosis permeate for potable water reuse. *Environ. Sci. Technol.* **2018**, *52*, 6417–6425. [CrossRef] [PubMed]
6. Cheng, M.; Zeng, G.; Huang, D.; Lai, C.; Liu, Y.; Zhang, C.; Wan, J.; Hu, L.; Zhou, C.; Xiong, W. Efficient degradation of sulfamethazine in simulated and real wastewater at slightly basic pH values using Co-SAM-SCS/H_2O_2 Fenton-like system. *Water Res.* **2018**, *138*, 7–18. [CrossRef] [PubMed]
7. Zhang, T.Y.; Hu, H.Y.; Wu, Y.H.; Zhuang, L.L.; Xu, X.Q.; Wang, X.X.; Dao, G.H. Promising solutions to solve the bottlenecks in the large-scale cultivation of microalgae for biomass/bioenergy production. *Renew. Sust. Energ. Rev.* **2016**, *60*, 1602–1614. [CrossRef]
8. Cheng, M.; Lai, C.; Liu, Y.; Zeng, G.; Huang, D.; Zhang, C.; Qin, L.; Hu, L.; Zhou, C.; Xiong, W. Metal-organic frameworks for highly efficient heterogeneous Fenton-like catalysis. *Coordin. Chem. Rev.* **2018**, *368*, 80–92.
9. Chen, A.; Shang, C.; Shao, J.; Lin, Y.; Luo, S.; Zhang, J.; Huang, H.; Lei, M.; Zeng, Q. Carbon disulfide-modified magnetic ion-imprinted chitosan-Fe(III): A novel adsorbent for simultaneous removal of tetracycline and cadmium. *Carbohyd. Polym.* **2017**, *155*, 19–27. [CrossRef]
10. Warsinger, D.M.; Chakraborty, S.; Tow, E.W.; Plumlee, M.H.; Bellona, C.; Loutatidou, S.; Karimi, L.; Mikelonis, A.M.; Achilli, A.; Ghassemi, A.; et al. A review of polymeric membranes and processes for potable water reuse. *Prog. Poly. Sci.* **2018**, *81*, 209–237. [CrossRef]
11. Tahara, M.; Obama, T.; Ikarashi, Y. Development of analytical method for determination of 1,4-dioxane in cleansing products. *Int. J. Cosmetic Sci.* **2013**, *35*, 575–580. [CrossRef] [PubMed]
12. Guo, P.-y. Determination of dioxane in toiletries by headspace gas chromatography-mass spectrometry. *Deter. Cosm.* **2013**, *36*, 20–22.
13. Guo, W.Q.; Brodowsky, H. Determination of the trace 1,4-dioxane. *Microchem. J.* **2000**, *64*, 173–179. [CrossRef]

14. Nannelli, A.; De Rubertis, A.; Longo, V.; Gervasi, P. Effects of dioxane on cytochrome P450 enzymes in liver, kidney, lung and nasal mucosa of rat. *Arch. Toxicol.* **2005**, *79*, 74–82. [CrossRef] [PubMed]
15. Stickney, J.A.; Sager, S.L.; Clarkson, J.R.; Smith, L.A.; Locey, B.J.; Bock, M.J.; Hartung, R.; Olp, S.F. An updated evaluation of the carcinogenic potential of 1,4-dioxane. *Regul. Toxicol. Pharm.* **2003**, *38*, 183–195. [CrossRef]
16. Xu, X.; Pliego, G.; Garcia-Costa, A.L.; Zazo, J.A.; Liu, S.; Casas, J.A.; Rodriguez, J.J. Cyclohexanoic acid breakdown by two-step persulfate and heterogeneous Fenton-like oxidation. *Appl. Catal. B-Environ.* **2018**, *232*, 429–435. [CrossRef]
17. Xu, X.; Pliego, G.; Zazo, J.A.; Sun, S.; García-Muñoz, P.; He, L.; Casas, J.A.; Rodriguez, J.J. An overview on the application of advanced oxidation processes for the removal of naphthenic acids from water. *Crit. Rev. Environ. Sci. Technol.* **2017**, *47*, 1337–1370. [CrossRef]
18. Munoz, M.; Mora, F.J.; Pedro, Z.M.D.; Alvarez-Torrellas, S.; Casas, J.A.; Rodriguez, J.J. Application of CWPO to the treatment of pharmaceutical emerging pollutants in different water matrices with a ferromagnetic catalyst. *J. Hazard. Mater.* **2017**, *331*, 45–54. [CrossRef] [PubMed]
19. Munoz, M.; de Pedro, Z.M.; Casas, J.A.; Rodriguez, J.J. Preparation of magnetite-based catalysts and their application in heterogeneous Fenton oxidation–a review. *Appl. Catal. B-Environ.* **2015**, *176*, 249–265. [CrossRef]
20. Quintanilla, A.; Casas, J.; Rodriguez, J. Hydrogen peroxide-promoted-CWAO of phenol with activated carbon. *Appl. Catal. B-Environ.* **2010**, *93*, 339–345. [CrossRef]
21. Quintanilla, A.; Fraile, A.; Casas, J.; Rodríguez, J. Phenol oxidation by a sequential CWPO–CWAO treatment with a Fe/AC catalyst. *J. Hazard. Mater.* **2007**, *146*, 582–588. [CrossRef] [PubMed]
22. Anipsitakis, G.P.; Dionysiou, D.D.; Gonzalez, M.A. Cobalt-mediated activation of peroxymonosulfate and sulfate radical attack on phenolic compounds. Implications of chloride ions. *Environ. Sci. Technol.* **2006**, *40*, 1000–1007. [CrossRef] [PubMed]
23. Anipsitakis, G.P.; Dionysiou, D.D. Radical generation by the interaction of transition metals with common oxidants. *Environ. Sci. Technol.* **2004**, *38*, 3705–3712. [CrossRef]
24. Xu, X.; Pliego, G.; Zazo, J.A.; Casas, J.A.; Rodriguez, J.J. Mineralization of naphtenic acids with thermally-activated persulfate: The important role of oxygen. *J. Hazard. Mater.* **2016**, *318*, 355–362. [CrossRef] [PubMed]
25. Pliego, G.; Zazo, J.A.; Garcia-Muñoz, P.; Munoz, M.; Casas, J.A.; Rodriguez, J.J. Trends in the intensification of the Fenton process for wastewater treatment: An overview. *Crit. Rev. Environ. Sci. Technol.* **2015**, *45*, 2611–2692. [CrossRef]
26. García-Muñoz, P.; Pliego, G.; Zazo, J.; Barbero, B.; Bahamonde, A.; Casas, J. Modified ilmenite as catalyst for CWPO-Photoassisted process under LED Light. *Chem. Eng. J.* **2017**, *318*, 89–94.
27. Wols, B.A.; Hofman-Caris, C.H.M. Review of photochemical reaction constants of organic micropollutants required for UV advanced oxidation processes in water. *Water Res.* **2012**, *46*, 2815–2827. [CrossRef] [PubMed]
28. Barndok, H.; Hermosilla, D.; Negro, C.; Blanco, A. Comparison and Predesign Cost Assessment of different advanced oxidation processes for the treatment of 1,4-Dioxane-containing wastewater from the chemical industry. *ACS Sustain. Chem. Eng.* **2018**, *6*, 5888–5894. [CrossRef]
29. Barndok, H.; Merayo, N.; Blanco, L.; Hermosilla, D.; Blanco, A. Application of on-line FTIR methodology to study the mechanisms of heterogeneous advanced oxidation processes. *Appl. Catal. B-Environ.* **2016**, *185*, 344–352. [CrossRef]
30. Gomez-Aviles, A.; Penas-Garzon, M.; Bedia, J.; Rodriguez, J.J.; Belver, C. C-modified TiO2 using lignin as carbon precursor for the solar photocatalytic degradation of acetaminophen. *Chem. Eng. J.* **2019**, *358*, 1574–1582. [CrossRef]
31. Belver, C.; Bedia, J.; Rodriguez, J.J. Titania-clay heterostructures with solar photocatalytic applications. *Appl. Catal. B-Environ.* **2015**, *176–177*, 278–287. [CrossRef]
32. Belver, C.; Han, C.; Rodriguez, J.J.; Dionysiou, D.D. Innovative W-doped titanium dioxide anchored on clay for photocatalytic removal of atrazine. *Catal. Today* **2017**, *280*, 21–28. [CrossRef]
33. Zhang, C.; Li, Y.; Shuai, D.; Shen, Y.; Xiong, W.; Wang, L. Graphitic carbon nitride (g-C3N4)-based photocatalysts for water disinfection and microbial control: A review. *Chemosphere* **2019**, *214*, 462–479. [CrossRef] [PubMed]
34. Mazierski, P.; Mikolajczyk, A.; Bajorowicz, B.; Malankowska, A.; Zaleska-Medynska, A.; Nadolna, J. The role of lanthanides in TiO$_2$-based photocatalysis: A review. *Appl. Catal. B-Environ.* **2018**, *233*, 301–317. [CrossRef]

35. Hill, R.R.; Jeffs, G.E.; Roberts, D.R. Photocatalytic degradation of 1,4-Dioxane in aqueous solution. *J. Photoch. Photobio. A* **1997**, *108*, 55–58. [CrossRef]
36. Barndok, H.; Blanco, L.; Hermosilla, D.; Blanco, A. Heterogeneous photo-Fenton processes using zero valent iron microspheres for the treatment of wastewaters contaminated with 1,4-Dioxane. *Chem. Eng. J.* **2016**, *284*, 112–121. [CrossRef]
37. Lee, K.-C.; Beak, H.-J.; Choo, K.-H. Membrane photoreactor treatment of 1,4-Dioxane-containing textile wastewater effluent: Performance, modeling, and fouling control. *Water Res.* **2015**, *86*, 58–65. [CrossRef]
38. Tian, J.; Sang, Y.; Yu, G.; Jiang, H.; Mu, X.; Liu, H. A Bi_2WO_6-Based hybrid photocatalyst with broad spectrum photocatalytic properties under UV, visible, and near-infrared irradiation. *Adv. Mater.* **2013**, *25*, 5075–5080. [CrossRef]
39. Ribeiro, A.R.; Nunes, O.C.; Pereira, M.F.R.; Silva, A.M.T. An overview on the advanced oxidation processes applied for the treatment of water pollutants defined in the recently launched Directive 2013/39/EU. *Environ. Int.* **2015**, *75*, 33–51. [CrossRef]
40. Ravelli, D.; Protti, S.; Fagnoni, M. Decatungstate anion for photocatalyzed "window ledge" reactions. *Accounts Chem. Res.* **2016**, *49*, 2232–2242. [CrossRef]
41. El-Sheikh, S.M.; Rashad, M.M. Novel Synthesis of Cobalt Nickel Tungstate Nanopowders and its photocatalytic application. *J. Clust. Sci.* **2015**, *26*, 743–757. [CrossRef]
42. Huang, Z.F.; Song, J.; Pan, L.; Zhang, X.; Wang, L.; Zou, J.J. Tungsten oxides for photocatalysis, electrochemistry, and phototherapy. *Adv. Mater.* **2015**, *27*, 5309–5327. [CrossRef] [PubMed]
43. Visa, M.; Bogatu, C.; Duta, A. Tungsten oxide–fly ash oxide composites in adsorption and photocatalysis. *J. Hazard. Mater.* **2015**, *289*, 244–256. [CrossRef]
44. Xu, H.; Yu, T.; Liu, J. Photo-degradation of Acid Yellow 11 in aqueous on nano-ZnO/Bentonite under ultraviolet and visible light irradiation. *Mater. Lett.* **2014**, *117*, 263–265. [CrossRef]
45. Barrault, J.; Boulinguiez, M.; Forquy, C.; Maurel, R. Synthesis of methyl mercaptan from carbon oxides and H_2S with tungsten—alumina catalysts. *Appl. Catal.* **1987**, *33*, 309–330. [CrossRef]
46. Halada, G.; Clayton, C. Comparison of Mo–N and W–N synergism during passivation of stainless steel through X-ray photoelectron spectroscopy and electrochemical analysis. *J. Vac. Sci. Technol. A* **1993**, *11*, 2342–2347. [CrossRef]
47. Kerkhof, F.P.J.M.; Moulijn, J.A.; Heeres, A. The XPS spectra of the metathesis catalyst tungsten oxide on silica gel. *J. Electron. Spectrosc.* **1978**, *14*, 453–466. [CrossRef]
48. Carley, A.; Roberts, M. An X-ray photoelectron spectroscopic study of the interaction of oxygen and nitric oxide with aluminium. *P. Roy. Soc. Lond. A* **1978**, *363*, 403–424. [CrossRef]
49. Ealet, B.; Elyakhloufi, M.H.; Gillet, E.; Ricci, M. Electronic and crystallographic structure of γ-alumina thin films, Thin Solid Films, 1994, 250, 92–100. *Thin Solid Films* **1994**, *250*, 92–100. [CrossRef]
50. Vogel, F.; Harf, J.; Hug, A.; von Rohr, P.R. The mean oxidation number of carbon (MOC)—A useful concept for describing oxidation processes. *Water Res.* **2000**, *34*, 2689–2702. [CrossRef]
51. Peng, Y.; Fu, D.; Liu, R.; Zhang, F.; Xue, X.; Xu, Q.; Liang, X. $NaNO_2/FeCl_3$ dioxygen recyclable activator: An efficient approach to active oxygen species for degradation of a broad range of organic dye pollutants in water. *Appl. Catal. B-Environ.* **2008**, *79*, 163–170. [CrossRef]
52. Wang, C.; Ao, Y.; Wang, P.; Hou, J.; Qian, J. Preparation, characterization and photocatalytic activity of the neodymium-doped TiO_2 hollow spheres. *Appl. Sur. Sci.* **2010**, *257*, 227–231. [CrossRef]
53. Mehrvar, M.; Anderson, W.A.; Moo-Young, M. Photocatalytic degradation of aqueous tetrahydrofuran, 1,4-dioxane, and their mixture with TiO_2. *Inter. J. Photoenerg.* **2000**, *2*, 67–80. [CrossRef]
54. Asghar, A.; Abdul Raman, A.A.; Wan Daud, W.M.A. Advanced oxidation processes for in-situ production of hydrogen peroxide/hydroxyl radical for textile wastewater treatment: A review. *J. Clean. Prod.* **2015**, *87*, 826–838. [CrossRef]

© 2019 by the authors. Licensee MDPI, Basel, Switzerland. This article is an open access article distributed under the terms and conditions of the Creative Commons Attribution (CC BY) license (http://creativecommons.org/licenses/by/4.0/).

Article

Degradation of Crystal Violet by Catalytic Wet Peroxide Oxidation (CWPO) with Mixed Mn/Cu Oxides

Ana María Campos [1], Paula Fernanda Riaño [1], Diana Lorena Lugo [1], Jenny Alejandra Barriga [1], Crispín Astolfo Celis [1], Sonia Moreno [2] and Alejandro Pérez [1,*]

1. Línea de Investigación en Tecnología Ambiental y de Materiales (ITAM), Departamento de Química, Facultad de Ciencias, Pontificia Universidad Javeriana, Carrera 7 No. 43-82, Bogotá D.C., P.C. 110231, Colombia; amcamposr@yahoo.es (A.M.C.); paula.riano2@estudiantes.uamerica.edu.co (P.F.R.); d.moralesl@javeriana.edu.co (D.L.L.); barriga_j@javeriana.edu.co (J.A.B.); crispin.celis@javeriana.edu.co (C.A.C.)
2. Estado Sólido y Catálisis Ambiental (ESCA), Departamento de Química, Facultad de Ciencias, Universidad Nacional de Colombia, Carrera 30 45-03, Bogotá, P.C. 110231, Colombia; smorenog@unal.edu.co
* Correspondence: alejandroperez@javeriana.edu.co; Tel.: +571-3208-320 (ext. 4124)

Received: 30 April 2019; Accepted: 21 May 2019; Published: 13 June 2019

Abstract: The environment protection has been the starting point for the development of new technologies, which allow the control of highly toxic substances present in the effluents of various industries, whose removal is not feasible by conventional methods. In this research, mixed oxide catalysts Mn and Cu in different molar ratios were prepared from the autocombustion method and characterized by XRD, XRF, TPR-H_2, and N_2 adsorption–desorption isotherms. The solids were evaluated in the catalytic wet peroxide oxidation of crystal violet (CV) with mild conditions of reaction: 25 °C, normal pressure, airflow of 2 mL/min, and H_2O_2 0.1 M (2 mL/h). The experimental results indicated degradations of 100% of CV, conversion of the total organic carbon (TOC) of 74%, and elimination of chemical oxygen demand (COD) of 71% in 90 min of reaction. Additionally, the selectivity was monitored by CG-MS, finding that there was almost complete mineralization in a short reaction time, generating intermediate products such as carboxylic acids, alcohols, and amines that do not cause a serious risk to the environment. The Mn–Cu catalyst with molar ratios of 1:2 was the most promising catalyst, displaying a cooperative effect between the two metals, and demonstrating the importance of the redox properties for the elimination of CV dye in wastewater.

Keywords: catalytic wet peroxide oxidation; degradation; autocombustion; crystal violet

1. Introduction

The aim of wastewater treatment in industry is to reduce the load of pollutants from water and meet the discharge regulations. Wastewaters containing high concentrations of persistent, toxic, and nonbiodegradable organic pollutants (e.g., aromatics, pesticides, etc.) are hard to treat with conventional physical–chemical or biological methods.

Water treatment recalcitrant and toxic residuals have been studied because of their danger with various methods used in order to obtain efficiency and profitability at the industrial level [1]. Within the wastewater emanating from textile, leather, plastic, and pharmaceutical companies there are dissolved substances that generate an increase in the chemical oxygen demand (COD) due to organic chemical matter not being biodegradable. The synthetic dye crystal violet, used often in staining processes, histological staining, as a dermatological agent, and in forensic medicine, has been cataloged as a highly toxic cationic dye, even at very low concentrations, susceptible to oxidation reactions and hydrolysis, where toxic metabolites are produced in waters, which generate adverse effects in both animals and human health [2,3].

According to the multiple adverse effects of crystal violet (CV) as a carcinogenic and mutagenic agent, different methods, including adsorption techniques, bioremediation, photolysis, and photocatalysis, have been studied for removing water residuals, taking into account that many of these show limitations, such as a high cost for manipulation of conditions such as pressure and temperature, processes subsequent separation due to generated byproducts, and low efficiency [4]. On the other side, the advanced processes of oxidation (APOs) arises as one of the most promising alternatives, with high yields obtained; in which oxidation with hydrogen peroxide (CWPO: catalytic wet peroxide oxidation) is highlighted because it uses heterogeneous catalysts with low cost and easy handling [5]. The axis fundamental to the process is the reaction of Fenton, where hydrogen peroxide and Fe^{+2} salts are mainly used as reagents for the formation of the radical hydroxyl (HO·), being the active intermediate that facilitates the elimination of contaminants in the effluents [6,7]. The behavior of the CWPO reaction using salts of iron is very susceptible to the values of pH [8,9]. According to some authors, the maximum speed of elimination of this reaction is obtained in a range of pH values between 2.5 and 3.5, which coincides with the interval in which the decomposition of hydrogen peroxide is minimal [10]. Because of this, studies have been carried out that allow an extension of the pH range by implementing others metals such as Cu [11], presenting an advantage in wastewater treatment where pH ranges from around 6 to 9, which is characteristic for an alkaline medium.

CWPO has been employed as both a homogeneous and a heterogeneous catalyst, seeking to improve conditions of pressure, temperature, and residence time. However, an oxidation process that can employ a solid catalyst greatly facilitates the decontamination process [12]. The mixed oxides show large advantages as catalysts, due to the textural, morphological, structural, and catalytic properties conferred within the crystalline structure that characterizes them with high superficial areas, good stability, and activity [13]. Their synthesis is by numerous methods, such as hydrothermal, coprecipitation, sol-gel, autocombustion, and microemulsion impregnation, in which more efficient molecular structures are sought for the oxidation mechanisms involved [14]. Within these methods, the autocombustion method allows synthesis in short times, does not need intermediaries, and is simple and versatile to implement. The solids obtained have a wide range of particle sizes, crystallinity, and high purity [15,16].

Regarding the metal used for CWPO, the transition metal oxides Cu, Ni, Co, and Mn have advantages among other metals due to their easy availability and low cost. Concerning their oxides, Cu and Mn oxides are the most active potential candidates for catalytic oxidation processes. The catalytic properties of Mn are attributed to its ability to form oxides with different oxidation states, Mn^{2+}/Mn^{3+} or Mn^{3+}/Mn^{4+}, and its oxygen storage capacity in the crystalline lattice [17].

Considering the latest aspects, in this work, mixed oxides of Mn and Cu were synthesized by the autocombustion method for the catalytic wet peroxide oxidation of violet crystal, with a more extended range of the pH than is normally employed in Fenton-type reactions. In addition, the evaluation of a possible cooperative effect within these transition metals was studied.

2. Results

2.1. Mixed Oxide Catalyst

In order to evaluate the degradation of CV, five solids were synthesized, defined by the following nomenclature: Mn, Cu, Mn–Cu (1:1), Mn–Cu (1:2), and Mn–Cu (2:1), with the stoichiometric ratio (Table 1).

Table 1. Stoichiometric ratio and nomenclature of oxide-mixed catalyst.

Catalyst	Mn/Cu	Al	Mg	Glycine
Mn–Cu (1:1)	1	20	30	100
Mn–Cu (1:2)	0.5	20	30	100
Mn–Cu (2:1)	2	20	30	100
Mn	0	20	30	100
Cu	5	20	30	100

2.2. Characterization of Catalyst

Table 2 summarizes the results of the chemical and structural properties of mixed manganese–copper oxides. The X-ray diffraction (XRD) profiles (Figure 1) provided evidence for the presence of the husmanite phase (Mn_3O_4) and copper oxide (CuO) for the oxides that present as active phase Mn and Cu, respectively, and periclase (MgO) in all the oxides. The size of particle reduction of solid Mn after the incorporation of Cu (MnCu) revealed a possible effect of Cu distribution on manganese oxides, being more important the change in the size of particle in the MnCu (1:2) oxide.

Table 2. Particle size of the mixed solids, according to the crystalline phase determined from the XRD signals. Ratios determined by XRF.

Catalyst	Crystaline Phase	FWHM		2θ		Crystal Size (nm) *		Ratios		
		Mn	Cu	Mn	Cu			Cu/Al	Mn/Al	Mg/Al
Mn	Mn_3O_4	0.2525	-	44.4	-	6.59	-	-	0.4	1.8
Cu	Cu_2O	-	0.8762	-	42.5	-	1.89	0.4	-	1.8
Mn–Cu (1:1)	Mn_3O_4; Cu_2O	0.2800	0.3244	44.4	42.5	5.51	5.09	0.2	0.2	1.7
Mn–Cu (2:1)	Mn_3O_4; Cu_2O	0.9153	0.8840	44.4	42.5	1.82	1.87	0.1	0.3	1.8
Mn–Cu (1:2)	Mn_3O_4; Cu_2O	1.2066	1.2212	44.4	42.5	1.38	1.35	0.3	0.1	1.8

* Crystal size was determined with the angles 2θ of 44.35 and 42.5 for the manganese and copper metals, respectively. FWHM: full width at half maximum.

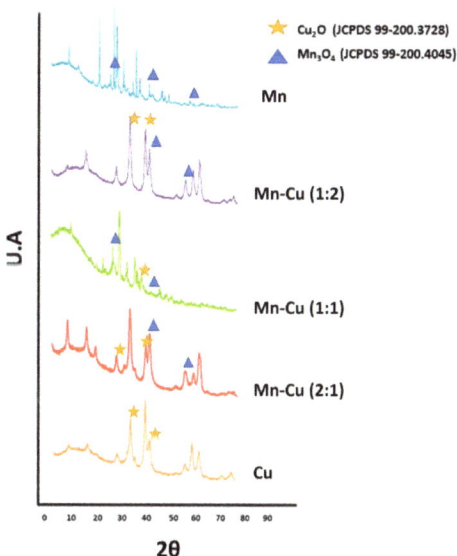

Figure 1. XRD patterns of catalyst.

Regarding the chemical composition by X-ray fluorescence XRF (Table 2) indicated through molar relationships between the elements, the catalysts revealed a similar and consistent chemical composition with the nominal ratio used in the synthesis, highlighting that the autocombustion method allows solids to be obtained without loss of the nominal value.

Concerning the surface analysis, the solids had isotherms of type IV (IUPAC classification) with a hysteresis H3 type (Figure 2), which corresponds to the solids with rough surfaces that are in aggregates of particles in the form of plates and that give rise to pores in the form of slits [18,19].

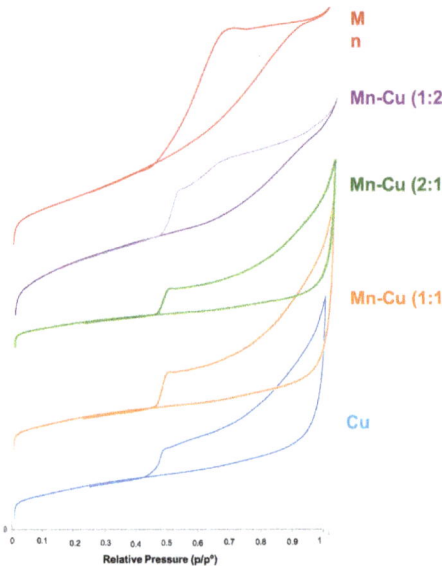

Figure 2. Nitrogen adsorption/desorption isotherms for catalyst.

BET areas are in Table 3. Table 3 also records the maximum temperature values of reduction in the analysis by TPR-H_2. Comparing the reduction temperatures of the solid Mn with the solids that have copper, the displacement at lower temperatures for mixed oxides MnCu indicated that reduction of manganese oxides is easier when Cu is added, with a further shift observed in the MnCu (1:2) oxide (Figure 3). This result can be related to the increased mobility of oxygen by the presence of a second metal of transition (Cu) and the particle size reduction that can facilitate the redox behavior in solids [15].

Table 3. Superficial area BET and maximum temperature values of reduction by TPR-H_2.

Catalyst	BET Area [m^2/g]	T Max (°C) TPR-H_2
Mn	71	505
Cu	112	211
Mn–Cu (1:2)	92	367
Mn–Cu (1:1)	55	194
Mn–Cu (2:1)	62	412

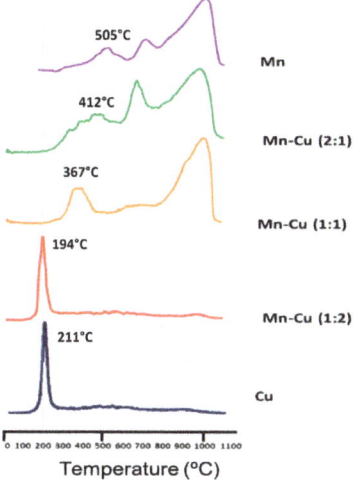

Figure 3. TPR-H$_2$ profiles for catalysts.

2.3. Catalytic Activity

Catalytic activity in CV oxidation is shown in Figure 4, where it can be appreciated that the activity was significantly influenced by the mixed oxide synthesis method, and a cooperative effect was observed between the two metals; the solid Cu–Mn (1:2) was the most active. These results are associated with the good redox properties of Mn and Cu, which are enhanced when they are together. Likewise, Figure 4 shows that the activity was directly related to the presence of a catalyst and not to the medium of reaction (H$_2$O$_2$ and air).

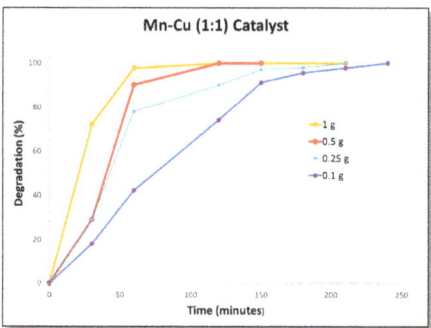

Figure 4. Effect of the catalyst loading in the catalytic wet peroxide oxidation (CWPO) of crystal violet (CV).

Additionally, with solid Mn–Cu (1:1), the effect of catalyst loading was evaluated (1, 0.5 and 0.25 g), finding as a result that only 0.25 g of catalyst yielded a degradation of 100% CV in 150 min (Figure 5), a selectivity in TOC of 70%, and an elimination in COD of 59% (Table 4). The results clearly indicated the great potential of these solids in respect to other catalysts reported in literature [20,21], under environmental conditions and low charges of active metal.

In order to clarify changes in the molecular and structural characteristics through CV oxidation as a function of time (Figure 5), GC-MS was used to further identify the intermediate products formed during the reaction (Figure 6). Equipment AGILENT 5975B VL MSD was used. At the beginning, the results showed evidence of the color degradation of the aromatic fragment in the molecule and the appearance of different intermediaries due to breaks by •OH radicals. Finally, the gradual breakdown of aromatic intermediate compounds led to the formation of carboxylic acids, before the conversion to carbon dioxide.

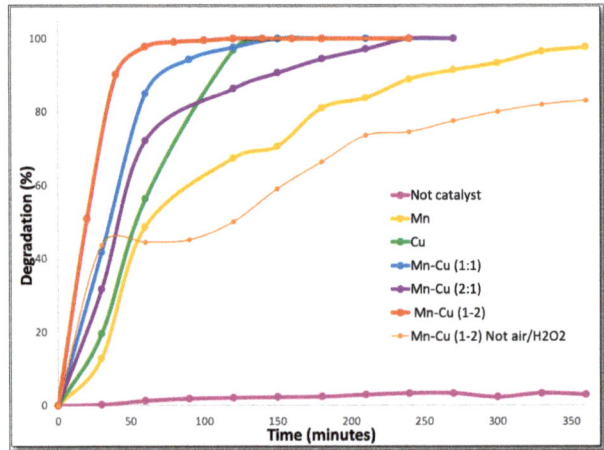

Figure 5. Catalytic activity for CWPO of CV by Mn–Cu catalyst.

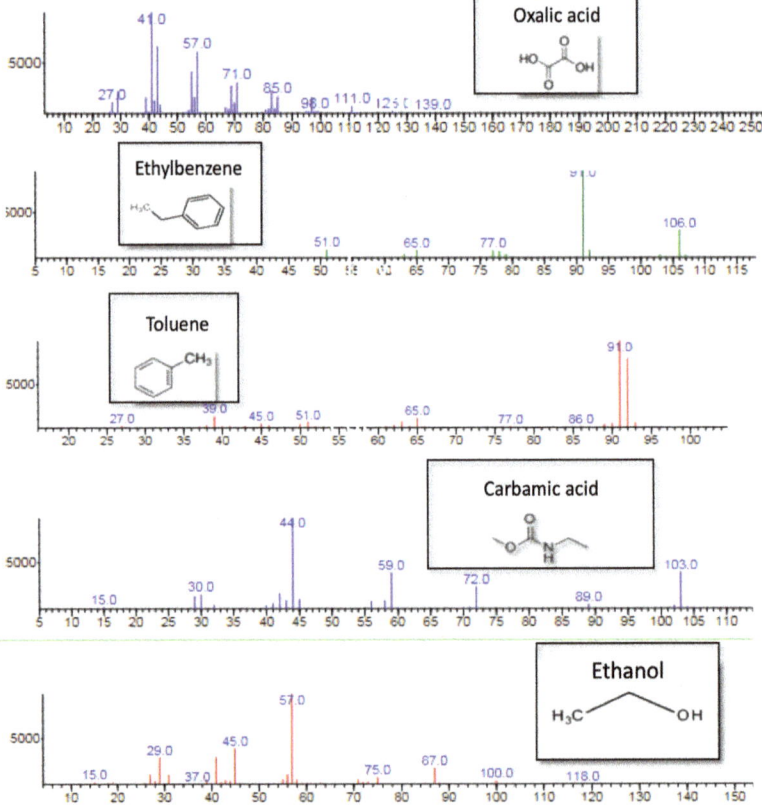

Figure 6. GC-MS chromatogram for CWPO in CV after 80 min of reaction.

Table 4. Catalytic activity in CV degradation of Mn–Cu catalyst.

Catalyst	Mn	Cu	Mn–Cu		
			1:1	1:2	2:1
Removal time 100% (min)	360	150	150	60	240
%TOC [1] (±2)	49	68	70	74	52
%COD [2] (±1)	52	60	59	71	50

[1] % TOC: total organic carbon; [2] % COD: chemical oxygen demand.

Concerning lixiviation during the reaction process, at the end of the catalytic test, the stability of the active phases was evaluated by taking the solutions after the final reaction time, filtering them to separate the solid from the aqueous phase, and assessing the possible leaching of the catalyst metals during the reaction. The analysis was done by atomic absorption spectrophotometry technique using equipment Agilent 280.

The nonexistence of leaching of metals (Mn and Cu), measured by atomic adsorption, indicated that the active phase was stable under the reaction conditions and avoided the generation of additional pollutant [22].

3. Discussion

Regarding the XRD patterns (Figure 1), the non-formation of Al_2O_3 suggested that Al was part of the crystal lattice, and allowed better Cu dispersion as the active phase. Also, peaks observed in 2θ values of 44.6 and 66.3 corresponded to the spinel phase (Al_2MgO_4) and regarded the mixed oxides that were generated.

One of the principal signals for copper is in 2θ = 37.5 as the copper oxide (CuO) crystalline phase, followed by two minor peaks in 2θ = 36 and 42.5, which are attributed to the Cu_2O cuprite (JCPDS 99-200-3728). These phases are essential in the Fenton oxidation process, since the Cu (II)/Cu (I) redox couples activate H_2O_2 molecules for the production of hydroxyl radical [23,24].

In contrast to Cu, the manganese pattern Mn (Figure 1) exposed different peaks associated with its multiple oxidation states and recorded signals corresponding to the periclase (MgO) phase. In addition, signals were related with spinel-type phases, which were preferential due to the high temperatures reached during the synthesis and subsequent calcination. This result showed the effectiveness of autocombustion as a novel method when compared with similar phases with traditional methods, such as co-precipitation [25].

Signals of MnO_2 at 2θ = 21, 28, and 36 are shown (Figure 1). MnO at 2θ = 26.8 and Mn_3O_4 at 2θ = 0.5 and 44.35 were related with these oxidative states as the most active phases. During the oxidation process, manganese oxide (III) and the husmanite phase Mn_3O_4 (JCPDS 99-200-4045) is highlighted as the most important phase in the decomposition of hydrogen peroxide during degradation processes [26–28].

The diffraction signals of the mixed catalyst of Mn–Cu (x:y) displayed the periclase phase and the formation of spinels, as occurred with the oxides separately. The crystalline phases that were related to the solids and standard signals evaluated previously, presented a favorable relationship with oxidizing agents such as oxygen and hydrogen peroxide for the production of radicals in the degradation of pollutants, evidenced by the most intense peaks belonging to Cu_2O, with values 2θ = 42.5 and 2θ = 44.35, respectively.

The increase of FWHM of XRD in mixed solids Mn–Cu (1:2) and Mn–Cu (2:1), with respect to Cu, Mn, and Mn–Cu (1: 1) solids, indicated that after the incorporation of Cu in the structure, a decrease in the crystallinity occurred. The latest result suggesting that the oxides are amorphous and highly dispersed structures. The decrease of the particle size of the solid of Mn, after the incorporation of the metal Cu (Mn–Cu), revealed a possible effect of distribution of cooper in the manganese oxides, being more significant than the change of the particle size in the oxide Mn–Cu (1:2).

The addition of Mn to the solids with copper led to lower surface areas being obtained, possibly due to processes of agglomeration of this metal on the surface. For the solid of Mn the particle size was greater in relation to the catalysts that present Cu, concluding that the adding of Cu as an active phase produced a decrease in particle size, which could result in a better distribution in the catalyst. The crystal size turned out to be better for the solid Mn–Cu (1:2), indicating high distribution on the surface of the active phase [29,30].

The average particle size (Table 2) was calculated from the Scherrer equation. For Mn solid, the particle size was greater in relation to the catalysts with Cu, so when adding Cu as an active phase there was a decrease in particle size, which could result in a better distribution in the catalyst. The smaller particle size of solid Mn–Cu (1:2) and Mn–Cu (2:1) indicated a good distribution of the active phase on the surface and therefore a possible better catalytic activity, which is evident in the reaction [27].

Regarding the results from the surface area BET, type IV isotherms characteristic of mesoporous solids were observed (Figure 2). In addition, according to the (IUPAC), the hysteresis type (H3) representative of a structure containing pores with different sizes and non-uniform shapes, as well as a large internal surface area, indicated the possible efficiency in the activity of the reaction [18,19]. The BET surface areas of the solids were summarized in Table 3. The Mn–Cu (1:2) sample had a surface area of 92 m^2/g, being smaller than the Cu solid (112 m^2/g), probably due to the partial obstruction of the pores and the addition of the active phase, where the introduction of Mn into the structure decreased the surface area and the pore volume of the material.

The resulting area was related to the evolution of gases such as CO_2 and H_2O, which were generated during the combustion process. Therefore, the solids with the greatest surface area were those with copper as the major phase. Thus, the mixed solid Mn–Cu (1:2) was the solid with the largest area compared to the other double-metal solids in the active phase.

According to the TPR-H_2 analysis (Figure 3), Cu catalyst showed a reduction peak between 200 °C and 300 °C, belonging to the reduction of cations from Cu^{+2} to Cu^0, and another reduction peak above 450 °C. According to the literature, the peak of low temperature can be attributed to the reduction of cations from Cu^{+2} to Cu^0, while the broad, but slightly pronounced peak is due to the hydrogenation of residual carbonates [31].

The Mn showed multiple peaks due to different oxidation states (Table 5) and finally, in binary samples Mn–Cu (x:y), comparing the reduction temperatures of the solid with only Mn or Cu, there was evidence of a shift to lower temperatures for the mixed oxides Mn–Cu, with a more pronounced shift in the Mn–Cu oxide (1:2) being observed, which indicated that the reduction of manganese oxides is easier when Cu is added (Table 3). This result can be attributed to the oxygen mobility increasing due to the presence of a second transition metal (Cu) and the decrease in particle sizes that can facilitate the redox behavior of the materials [32]. When Cu is added within the mixed oxides, lower reducibility temperatures are evident, which would favor the reaction to moderate conditions.

Table 5. Temperature range for reduction peaks in Mn.

Solid	Temperature °C	Reduction Signal
	<300	Mn_5O_8 to Mn_2O_3 (Mn^{+4} -> Mn^{+3})
Mn	350–450	Mn_2O_3 to Mn_3O_4 (Mn^{+3} -> Mn^{+2})
	500–600	Mn_3O_4 a MnO
	>600	Mn spinel phase

Catalytic Activity

In order to establish the ideal catalyst ratio for the CWOP of CV, the effect of the catalyst load was assessed with solid Mn–Cu (1:1). This solid was used to determine a cooperative effect between the metals Mn and Cu for the oxidation of the contaminant. At the same time, it established a possible reduction for raw material, increased the available amount of solid for subsequent tests, and ensured a good performance in the reaction times. According to the literature, the quantities chosen were 1 g, 0.5 g, 0.25 g, and 0.1 g

As a result, only 0.25 g of catalyst (Figure 4) achieved a degradation of 100% CV in 1 h, selectivity in COT of 74%, and elimination in the COD of 70% (Table 4). The results highlight the great potential of these solids compared with catalysts reported previously in the literature [1,2]. Although with any selected amount of catalyst 100% degradation is obtained, increasing the catalytic dose leads to considerable diminution in degradation times, thanks to the increase of active sites of the solid. Then, the highest efficiency was achieved with 1 g of catalyst in a time of 60 min. However, in consideration of both efficiency and costs, a catalyst amount of 0.25 g was selected, with short reaction times of approximately 110 min, with Mn–Cu catalyst (1:1), achieving lower raw material expense, compared with that used for each reaction. Taking into account the ratio of degradation time and the amount of catalyst, with 75% less catalyst compared to 1 g, the increase in the removal time was 45%. Due to these results, a value of 0.25 g was established as the amount of catalyst in the subsequent reactions.

The catalytic wet peroxide oxidation of CV using 0.25 g of catalyst is shown in Figure 5, where the activity was significantly influenced by the mixed oxide synthesis method, and a cooperative effect between the two metals was observed. The results showed that the catalytic activities of the sample Mn–Cu (1:2) were higher than those of the samples of Mn–Cu (1:1) > Mn–Cu (2:1) > Cu > Mn, with degradation times of only 80 min, presenting a clear influence of the molar ratio on the catalytic activity. In this way, a cooperative effect was evidenced, since there was a synergic phenomenon with the molar ratio, and a correlation in which the use of mixed active phases also optimized the physicochemical characteristics and the catalytic properties, in comparison with the system exclusively of Mn or Cu. Additionally, a clear influence was evidenced in shorter times of degradation of the CV in those solids whose Cu ratio was higher, compared to those that had Mn, for which Cu plays an important role in radicals production.

This behavior is related to the TPR-H_2 profiles (Figure 3), where the temperatures with minor reduction, corresponding to the solid of the best catalytic activity, validated the efficiency under moderate conditions proposed for the degradation of the CV dye.

Additionally a catalytic test was done without catalyst in order to determine if only H_2O_2/air is enough for the oxidation of CV. The percentage of degradation did not exceed 3%, so the presence of catalyst is necessary, confirming the catalytic character of the process and not the reaction medium. Likewise, to see the feasibility of the technique, a test was generated without the addition of H_2O_2 or air, and presented a percentage of degradation greater than 50% in 350 min, three times longer than the result with the oxidizing agent. In consequence, the importance of an oxidizing agent such as air and H_2O_2 is evident.

A supplementary analysis was done by UV-Vis spectrophotometry. The changes of the UV-visible spectra are shown in Figure 7, where the behavior of the Mn–Cu catalyst (1:2) is highlighted. The signal decrease at 600 nm showed the rupture of the aromatic rings and the formation of aliphatic and easily biodegradable molecules, as evidenced by the appearance of the peaks at 380 and 220 nm, related with the appearance of intermediate compounds during the reaction.

These results confirmed the decomposition of the molecule and the development of by-products that were analyzed by GC-MS (Figure 6). The results in GC-MS for the different characteristic peaks in the spectra allowed the substances to be classified as oxalic acid, ethylbenzene, toluene, carbamic acid, and ethanol, which are probably due to the triphenylmethane breakdown. All these breaks prove that the CV can be degraded into non-toxic small molecules and ensures the elimination of the contaminant in effluents from the proposed method [33].

Based on previous studies about the degradation of CV, it occurs through N-demethylation, chromophore excision and rupture of the trajectories of the ring structure, until the final formation of the desired products. According to the results, it can be stated that the oxidation route was N-demethylation, as a consequence of the reaction caused by the addition of hydroxyl, decomposition of the conjugated structures, elimination of the benzene ring, and the ring opening in smaller molecules [34].

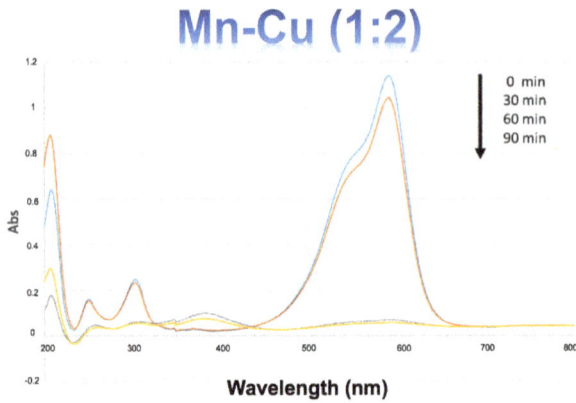

Figure 7. UV-Vis spectra for CWPO of CV with Mn–Cu (1:2) catalyst.

4. Materials and Methods

4.1. Catalyst Synthesis

The mixed oxides were synthesized from hydrated nitrate solutions by the self-combustion method. For the synthesis of Mn–Cu catalysts, $Mn(NO_3)_2 \cdot 6H_2O$, $Mg(NO_3)_2 \cdot 6H_2O$, $Al(NO_3)_3 \cdot 9H_2O$, and $Cu(NO_3)_2 \cdot 3H_2O$ (Merck (Darmstadt, Germany), 95.0% purity), reagents which act as oxidants were used, and glycine [CH_2NH_2COOH] (Merck (Darmstadt, Germany), reagent analytical degree) was used as combustible. Molar ratios for the preparation were Mn = 5, Mn/Cu 1:1 = 1, Mn/Cu 2:1 = 2, Mn/Cu 1:2 = 0.5, and Cu = 5, and a ratio (nitrates/glycine) = 0.8 [35,36]. The aqueous solution obtained was subjected to a slow evaporation process at a temperature of 110 °C, during 120–140 min, and with constant agitation between 200 and 300 rpm, until gel formation. The gel was heated to a temperature of 500 °C to provide a thermal shock and carry out the ignition process, which happened in a matter of seconds. Once the solid was obtained, calcination was carried out at a temperature of 700 °C for 14 h, where the elimination of glycine residues, conformation of the grains, and formation of a well-consolidated crystalline structure were ensured (Figure 8).

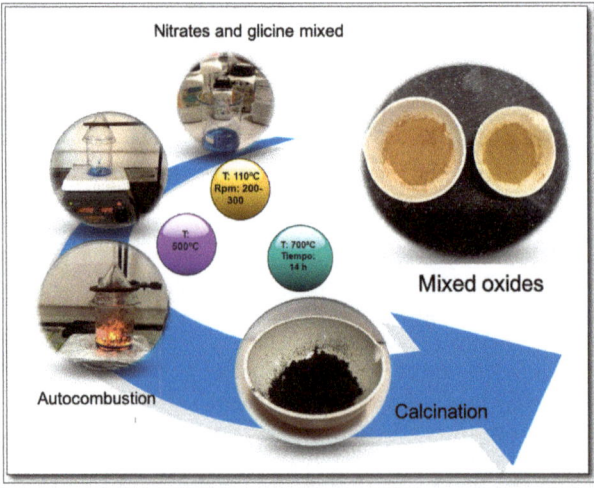

Figure 8. Procedure for catalysts synthesis using the autocombustión technique.

4.2. Characterization

The catalysts were characterized by X-ray Diffraction XRD (powder sample), using a Panalytical X'Pert PRO MPD (Almelo, The Netherlands) diffractometer equipped with a copper anode ($\lambda = 1.5406$ Å), using an angular velocity of 1°/min and a step size of 0.02° θ to evaluate the crystalline phases present. The chemical analysis of catalysts was carried out using the X-ray fluorescence XRF technique on a Philips MagiX Pro PW2440 device (PHILIPS / PANALYTICAL, AUSTIN, TX, USA). The oxidative–reductive properties of the materials were measured by means of TPR-H_2 in a CHEMBET 3000 QUANTACHROME (Anton-Paar, Boynton Beach, FL, USA) equipment, where a reduction was carried out with a gaseous mixture of H_2/Ar, analyzing the processes of reducibility as a function of the temperature. The textural characteristics were determined through N_2 adsorption–desorption isotherms using ASAP 2020 Micromeritics equipment (Micromeritics Instrument Corp. Norcross, GA, USA).

4.3. Catalytic Evaluation

In order to evaluate the efficiency of the catalyst in the elimination of CV, a semi-batch reactor was used, adding 200 mL of water at a concentration of CV of 50 ppm. It was provided with continuous agitation of 500 rpm, a temperature of 25 °C, constant flow of peroxide of hydrogen 0.1 M 2 mL/h, and air flow of 2 mL/min (Figure 9).

Figure 9. Reactor used for the catalytic degradation of CV.

To monitor the elimination of the CV, a Thermo Scientific GENESYS 20 spectrophotometer (Thermo Fisher Scientific Inc, Waltham, MA, USA) was used. The degradation process was measured at a wavelength of 590 nm. In order to determine the selectivity, the amount of total organic carbon (TOC) still present in each reaction time was measured using a Shimadzu Model TOC-L CPH analyzer (Shimadzu, Columbia, MD, USA. The COD was determined through an adaptation of the method approved by EPA 410.4; in a HANNA HI839800 reactor (Hanna Instruments SAS, Madrid, Spain) which determined the amount of oxygen that was required to oxidize the organic matter still present in the samples during the degradation treatment. The reaction was also followed by the GC-MS, which identified the formation of several by-products by the oxidation of the CV during the Fenton-type reaction. Intermediates were identified using a program from the NIST14.L library, with settings around 80%. The reaction mechanisms proposed in this research should be useful for the future application of dye treatment technologies.

5. Conclusions

The catalytic activity of the Mn–Cu solids was highly influenced by the association of copper and manganese, exposing a short time of removal of the contaminant (CV) at environmental conditions, where the catalyst Mn–Cu (1:2), with an elimination time of 80 min, was the catalyst with greater efficiency when compared with the other synthesized solids.

The implementation of the method with hydrogen peroxide in heterogeneous phase was efficient in the oxidation of the pollutant, where its catalytic degradation was observed by obtaining by-products (carboxylic acids), until mineralization was achieved.

Additionally, the method showed that no new pollutants were generated. In addition, the mild conditions of temperature and pressure, low airflow and hydrogen peroxide flow are promising for large-scale pilot and industrial implementation, where effluents are handled with high concentrations, being preferred to the techniques exposed in the reference literature, with the aim of reducing pollution and potential toxicity.

The stability of Mn–Cu mixed oxide catalysts during the Fenton reaction process indicated that there was no dissolved metal content in the final water samples, so no additional contamination would be present, excluding some indication of homogeneous catalysis phenomena and indicating, in turn, that the specified operating conditions were suitable within the catalytic system.

Author Contributions: Formal analysis, A.P.; Investigation, S.M.; Methodology, D.L.L. and J.A.B.; Project administration, C.A.C.; Resources, S.M.; Supervision, A.P.; Writing—original draft, P.F.R.; Writing—review & editing, A.M.C.

Funding: This research was funded by Pontificia Universidad Javeriana. Bogotá. Colombia. Proyecto ID 00007181.

Conflicts of Interest: The authors declare no conflict of interest.

References

1. Yin, J.; Cai, J.; Yin, C.; Gao, L.; Zhou, J. Degradation performance of crystal violet over CuO/AC and CeO_2-CuO/AC catalysts using microwave catalytic oxidation degradation method. *J. Environ. Chem. Eng.* **2016**, *4*, 958–964. [CrossRef]
2. Singh, K.P.; Gupta, S.; Singh, A.K.; Sinha, S. Optimizing adsorption of crystal violet dye from water by magnetic nanocomposite using response surface modeling approach. *J. Hazard. Mater.* **2011**, *186*, 1462–1473. [CrossRef]
3. Rezaei, S.; Tahmasbi, H.; Mogharabi, M.; Firuzyar, S.; Ameri, A.; Khoshayand, M.R.; Faramarzi, M.A. Efficient decolorization and detoxification of reactive orange 7 using laccase isolated from Paraconiothyrium variabile, kinetics and energetics. *J. Taiwan Inst. Chem. Eng.* **2015**, *56*, 113–121. [CrossRef]
4. Santos-Juanes, L.; Sánchez, J.G.; López, J.C.; Oller, I.; Malato, S.; Pérez, J.S. Dissolved oxygen concentration: A key parameter in monitoring the photo-Fenton process. *Appl. Catal. B Environ.* **2011**, *104*, 316–323.
5. Inchaurrondo, N.S.; Massa, P.; Fenoglio, R.; Font, J.; Haure, P. Efficient catalytic wet peroxide oxidation of phenol at moderate temperature using a high-load supported copper catalyst. *Chem. Eng. J.* **2012**, *198–199*, 426–434. [CrossRef]
6. Carrillo, A.M.; Carriazo, J.G. Cu and Co oxides supported on halloysite for the total oxidation of toluene. *Appl. Catal. B Environ.* **2015**, *164*, 443–452. [CrossRef]
7. Deganello, F.; Marcì, G.; Deganello, G. Citrate–nitrate auto-combustion synthesis of perovskite-type nanopowders: A systematic approach. *J. Eur. Ceram. Soc.* **2009**, *29*, 439–450. [CrossRef]
8. Castaño, M.H.; Molina, R.; Moreno, S. Mn–Co–Al–Mg mixed oxides by auto-combustion method and their use as catalysts in the total oxidation of toluene. *J. Mol. Catal. A Chem.* **2013**, *370*, 167–174. [CrossRef]
9. Ovejero, G.; Rodríguez, A.; Vallet, A.; García, J. Ni/Fe-supported over hydrotalcites precursors as catalysts for clean and selective oxidation of Basic Yellow 11: Reaction intermediates determination. *Chemosphere* **2013**, *90*, 1379–1386. [CrossRef] [PubMed]
10. Chen, C.-C.; Chen, W.C.; Chiou, M.R.; Chen, S.W.; Chen, Y.Y.; Fan, H.J. Degradation of crystal violet by an FeGAC/H_2O_2 process. *J. Hazard. Mater.* **2011**, *196*, 420–425. [CrossRef]
11. Lousada, C.M.; Yang, M.; Nilsson, K.; Jonsson, M. Catalytic decomposition of hydrogen peroxide on transition metal and lanthanide oxides. *J. Mol. Catal. A Chem.* **2013**, *379*, 178–184. [CrossRef]
12. Ribeiro, R.S.; Silva, A.M.; Figueiredo, J.L.; Faria, J.L.; Gomes, H.T. Catalytic wet peroxide oxidation: A route towards the application of hybrid magnetic carbon nanocomposites for the degradation of organic pollutants. A review. *Appl. Catal. B Environ.* **2016**, *187*, 428–460. [CrossRef]
13. Matilainen, A.; Sillanpää, M. Removal of natural organic matter from drinking water by advanced oxidation processes. *Chemosphere* **2010**, *80*, 351–365. [CrossRef] [PubMed]

14. Barrault, J.; Bouchoule, C.; Tatibouët, J.M.; Abdellaoui, M.; Majesté, A.; Louloudi, I.; Papayannakos, N.; Gangas, N.H. Catalytic wet peroxide oxidation over mixed (Al-Fe) pillared clays. In *Studies in Surface Science and Catalysis*; Corma, A., Melo, F.V., Mendioroz, S., Fierro, J.L.G., Eds.; Elsevier: Amsterdam, The Netherlands, 2000; pp. 749–754.
15. Huang, L.; Zhang, F.; Wang, N.; Chen, R.; Hsu, A.T. Nickel-based perovskite catalysts with iron-doping via self-combustion for hydrogen production in auto-thermal reforming of Ethanol. *Int. J. Hydrogen Energy* **2012**, *37*, 1272–1279. [CrossRef]
16. Cesar, M.A. *Quimisorción*; Universidad de Juarez: Ciudad Juárez, Mexico, 2011.
17. Mossino, P. Some aspects in self-propagating high-temperature synthesis. *Ceram. Int.* **2004**, *30*, 311–332. [CrossRef]
18. Liu, L.; Song, Y.; Fu, Z.; Ye, Q.; Cheng, S.; Kang, T.; Dai, H. Enhanced catalytic performance of Cu- and/or Mn-loaded Fe-Sep catalysts for the oxidation of CO and ethyl acetate. *Chin. J. Chem. Eng.* **2017**, *25*, 1427–1434. [CrossRef]
19. Rouquerol, F.; Rouquerol, J.; Sing, K. Adsorption by Powders and Porous Solids. 2012. Available online: http://linux0.unsl.edu.ar/~{}rlopez/cap3new.pdf (accessed on 27 May 2019).
20. Wu, J.; Gao, H.; Yao, S.; Chen, L.; Gao, Y.; Zhang, H. Degradation of Crystal Violet by catalytic ozonation using Fe/activated carbon catalyst. *Sep. Purif. Technol.* **2015**, *147*, 179–185. [CrossRef]
21. Han, J.; Zeng, H.; Xu, S.; Chen, C.; Liu, X. Catalytic properties of CuMgAlO catalyst and degradation mechanism in CWPO of methyl orange. *Appl. Catal. A Gen.* **2016**, *527*, 72–80. [CrossRef]
22. Ahmad, M.; Chen, S.; Ye, F.; Quan, X.; Afzal, S.; Yu, H.; Zhao, X. Efficient photo-Fenton activity in mesoporous MIL-100(Fe) decorated with T ZnO nanosphere for pollutants degradation. *Appl. Catal. B Env.* **2019**, *245*, 428–438. [CrossRef]
23. Lu, H.; Kong, X.; Huang, H.; Zhou, Y.; Chen, Y. Cu–Mn–Ce ternary mixed-oxide catalysts for catalytic combustion of toluene. *J. Environ. Sci.* **2015**, *32*, 102–107. [CrossRef]
24. Chi, H.; Andolina, C.M.; Li, J.; Curnan, M.T.; Saidi, W.A.; Zhou, G.; Yang, J.C.; Veser, G. Dependence of H_2 and CO_2 selectivity on Cu oxidation state during partial oxidation of methanol on Cu/ZnO. *Appl. Catal. A Gen.* **2018**, *556*, 64–72. [CrossRef]
25. Molina, R.; Castaño, M.H.; Moreno, S. Catalizadores de manganeso sintetizados por autocombustión y coprecipitación y su empleo en la oxidación del 2-propanol. *Cienc. Químicas* **2015**, *39*, 26–35. [CrossRef]
26. Zhao, H.; Fang, K.; Zhou, J.; Lin, M.; Sun, Y. Direct synthesis of methyl formate from syngas on CueMn mixed oxide catalyst. *Int. J. Hydrogen Eng.* **2016**, *41*, 8819–8828. [CrossRef]
27. Răciulete, M.; Layrac, G.; Papa, F.; Negrilă, C.; Tichit, D.; Marcu, I.C. Influence of Mn content on the catalytic properties of Cu-(Mn)-Zn-Mg-Al mixed oxides derived from LDH precursors in the total oxidation of methane. *Catal. Today* **2018**, *306*, 276–286. [CrossRef]
28. Aguilera, D.A.; Perez, A.; Molina, R.; Moreno, S. Cu–Mn and Co–Mn catalysts synthesized from hydrotalcites and their use in the oxidation of VOCs. *Appl. Catal. B Environ.* **2011**, *104*, 144–150. [CrossRef]
29. Li, Z.; Wang, H.; Wu, X.; Ye, Q.; Xu, X.; Li, B.; Wang, F. Novel synthesis and shape-dependent catalytic performance of Cu–Mn oxides for CO oxidation. *Appl. Surf. Sci.* **2017**, *403*, 335–341. [CrossRef]
30. Zhao, H.; Fang, K.; Dong, F.; Lin, M.; Sun, Y.; Tang, Z. Textual properties of Cu–Mn mixed oxides and application for methyl formate synthesis from syngas. *J. Ind. Eng. Chem.* **2017**, *54*, 117–125. [CrossRef]
31. Jabłońska, M.; Chmielarz, L.; Węgrzyn, A.; Góra-Marek, K.; Piwowarska, Z.; Witkowski, S.; Bidzińska, E.; Kuśtrowski, P.; Wach, A.; Majda, D. Hydrotalcite derived (Cu, Mn)–Mg–Al metal oxide systems doped with palladium as catalysts for low-temperature methanol incineration. *Appl. Clay Sci.* **2015**, *114*, 273–282. [CrossRef]
32. Perdomo, C.; Perez, A.; Molina, R.; Moreno, S. Storage capacity and oxygen mobility in mixed oxides from transition metals promoted by cerium. *Appl. Surf. Sci.* **2016**, *383*, 42–48. [CrossRef]
33. Fan, H.J.; Huang, S.T.; Chung, W.H.; Jan, J.L.; Lin, W.Y.; Chen, C.C. Degradation pathways of crystal violet by Fenton and Fenton-like systems: Condition optimization and intermediate separation and identification. *J. Hazard. Mater.* **2009**, *171*, 1032–1044. [CrossRef]
34. Fida, H.; Zhang, G.; Guo, S.; Naeem, A. Heterogeneous Fenton degradation of organic dyes in batch and fixed bed using La-Fe montmorillonite as catalyst. *J. Colloid Interface Sci.* **2017**, *490*, 859–868. [CrossRef] [PubMed]

35. Hwang, C.-C.; Tsai, J.S.; Huang, T.H.; Peng, C.H.; Chen, S.Y. Combustion synthesis of Ni–Zn ferrite powder—influence of oxygen balance value. *J. Solid State Chem.* **2005**, *178*, 382–389. [CrossRef]
36. Hua, Z.; Cao, Z.; Deng, Y.; Jiang, Y.; Yang, S. Sol–gel autocombustion synthesis of Co–Ni alloy powder. *Mater. Chem. Phys.* **2011**, *126*, 542–545. [CrossRef]

© 2019 by the authors. Licensee MDPI, Basel, Switzerland. This article is an open access article distributed under the terms and conditions of the Creative Commons Attribution (CC BY) license (http://creativecommons.org/licenses/by/4.0/).

Article

Wet Peroxide Oxidation of Chlorobenzenes Catalyzed by Goethite and Promoted by Hydroxylamine

David Lorenzo, Carmen M. Dominguez *, Arturo Romero and Aurora Santos

Chemical Engineering and Materials Department, Facultad de Ciencias Químicas, Universidad Complutense de Madrid, Ciudad Universitaria S/N, 28040 Madrid, Spain; dlorenzo@ucm.es (D.L.); aromeros@ucm.es (A.R.); aursan@ucm.es (A.S.)
* Correspondence: carmdomi@ucm.es; Tel.: +34-913944170

Received: 31 May 2019; Accepted: 18 June 2019; Published: 20 June 2019

Abstract: In this work, the abatement of several chlorobenzenes commonly found as pollutants in the aqueous phase has been carried out by catalytic wet peroxide oxidation using goethite as the catalyst and hydroxylamine as the promotor. Spiked water with monochlorobenzene and different positional isomers of dichlorobenzene, trichlorobenzene, and tetrachlorobenzene, at concentrations ranging from 0.4 to 16.9 mg L^{-1} was treated. Runs were carried out batch-way, at room conditions, without headspace. The heterogeneous catalyst was commercial goethite, with a specific surface area (S$_{BET}$) of 10.24 m^2 g^{-1} and a total iron content of 57.3 wt%. Iron acts as a catalyst of hydrogen peroxide decomposition to hydroxyl radicals. Hydroxylamine (in a range from 0 to 4.9 mM) was added to enhance the iron redox cycle from Fe (III) to Fe (II), remarkably increasing the radical production rate and therefore, the conversion of chlorobenzenes. Iron was stable (not leached to the aqueous phase) even at the lowest pH tested (pH = 1). The effect of pH (from 2 to 7), hydrogen peroxide (from 1 to 10 times the stoichiometric dosage), hydroxylamine, and catalyst concentration (from 0.25 to 1 g/L) was studied. Pollutant removal increased with hydroxylamine and hydrogen peroxide concentration. An operating conditions study demonstrated that the higher the hydroxylamine and hydrogen peroxide concentrations, the higher the removal of pollutants. The optimal pH value and catalyst concentration was 3 and 0.5 g L^{-1}, respectively. Operating with 2.4 mM of hydroxylamine and 10 times the stoichiometric H$_2$O$_2$ amount, a chlorobenzenes conversion of 90% was achieved in 2.5 h. Additionally, no toxic byproducts were obtained.

Keywords: chlorobenzenes; goethite; catalytic wet peroxide oxidation; hydroxylamine; iron redox cycle

1. Introduction

A drawback of industrial activities is the generation of toxic wastes. In the last decades, these wastes were often dumped without environmental concern near the production sites, resulting in an important soil and groundwater contamination in the nearby area [1,2]. Among these toxic wastes, the most concerning are those known as persistent organic pollutants (POPs) since they are poorly biodegradable and frequently pose a notable risk for health and the ecosystems [3–5]. Therefore, the development of effective techniques for the removal of these organic compounds is mandatory.

Compounds with chlorine atoms are among the more toxic organic compounds frequently appearing in soil and groundwater [6,7]. Between these chlorinated organic compounds (COCs), chlorobenzenes (nCBs, (n = mono, di, tri, tetra)) are commonly chlorinated pollutants in soil and groundwater [8]. Their occurrence in the environment is because they are often used in the manufacture of pesticides, dyes, and other widely used chemicals [9–13].

Monochlorobenzene (CB) is mainly used as a raw material in the production of nitrochlorobenzenes, diphenyl oxide, and diphenyldichlorosilane, and as a solvent in the production of isocyanates and in the dyes industry [14]. On the other hand, 1,4 dichlorobenzene (DCB), and 1,2,3 and 1,2,4 trichlorobenzene (TCBs) are raw materials in the synthesis of pesticides, resins, and in the production of several fine chemicals, particularly herbicides, pigments, and dyes [9,12,13,15–17]. Some pesticides as hexachlorocyclohexanes and heptachlorocyclohexanes are transformed in trichlorobenzenes and tetrachlorobenzenes, respectively, at alkaline pH [18].

The technologies developed for the treatment of chlorobenzenes in aqueous phase include reductive and oxidative treatments. Reductive treatments, mainly using zerovalent iron, are effective but the times required to achieve high pollutant conversion are often too long. The use of iron nanoparticles reduces this time [19–21] from iron microparticles [22], although they are more expensive, and their stability needs to be improved. Dominguez et al. found that the dechlorination rate increased with the chlorine content of the organic molecule and that the non-aromatic chlorinated organic compounds, as hexachlorocyclohexanes (HCHs), were more rapidly eliminated in the presence of zero valent iron microparticles via dechloroelimination, than chlorobenzenes, while the last ones were dechlorinated via hydrogenolysis [22].

Among the oxidative technologies, advanced oxidation processes (AOPs) stand out [23]. These methods are suitable for the treatment of water-soluble contaminants and their applicability has already been documented for water, groundwater, and wastewater decontamination [24]. Among the AOPs treatments proposed for nCBs abatement are the photo-oxidation process, coupling UV, and H_2O_2 [23,25]. These treatments have achieved conversions up to 90% in several hours but imply high operation costs [24]. Santos et al. used persulfate activated by alkali for the removal of nCBS in groundwater polluted with lindane wastes [26]. The complete conversion of nCBS was obtained in 15 days. This technology could be adapted for an in situ remediation process due to the high stability of persulfate in the subsurface.

Fenton's reagent (iron + H_2O_2) has been also tested in nCBs oxidation in aqueous phase, using several sources of iron. Sedlak and Andren reported the effectivity of adding a Fe (II) salt to hydrogen peroxide in the removal of nCBS from wastewaters [27]. Kuang et al. used iron nanoparticles and Pagano et al. used iron powder as a source of Fe^{2+} ions [28,29]. In all these works, soluble iron cation catalyzed the decomposition of hydrogen peroxide in homogenous phase.

Despite the classical Fenton process, using iron in solution at low pH, is an efficient process [30], but has a major drawback: the catalyst is lost after each reaction cycle. Additionally, additional treatments are usually required (i.e., neutralization, separation, and management of the iron hydroxide sludge generated) [31–33].

To overcome these drawbacks, several heterogeneous catalysts, that can be easily recovered at the end of the process, have been tested [31], giving rise to the process known as Fenton-like or catalytic wet peroxide oxidation (CWPO). Many researchers studied this process using mainly iron minerals [31,34–37]. Those that attract the most attention, due to their availability and low cost, are the naturally occurring iron materials, such as magnetite, hematite, ilmenite, goethite, etc. [34]. The only catalytic species considered in literature when these minerals are employed is iron [31,32,34–37], responsible for the production of hydroxyl radicals by reaction with the oxidant (hydrogen peroxide). These kinds of heterogeneous catalysts, specifically hematite, were successfully applied in the oxidation of nCBS (1,3,5 TCB and 1,2,3,4 TetraCB) with hydrogen peroxide [38]. The main drawback of these heterogeneous catalysts is the lower rate of H_2O_2 decomposition and therefore, hydroxyl radical (·OH) generation vs. the classical Fenton process which leads to longer reaction times and restricts their full-scale applications [32,39]. However, heterogeneous catalysts based on iron oxides prepared in the laboratory often promote the iron leaching at acidic pH and the process turns from heterogeneous to homogeneous (Fenton's reagent) [34].

The use of reducing agent compounds, such as hydroxylamine, to enhance the redox iron cycle in the Fenton and Fenton-like reactions, and therefore, the pollutant oxidation performance, has been recently explored [40–42]. Some authors reported the use of hydroxylamine to accelerate the Fenton reaction applied to the abatement of different organics such as benzyl alcohol [42], methylene blue [43], and carbamazepine [44].

A very recent work reported the combined use of goethite (a naturally occurring iron material) and hydroxylamine to enhance the oxidation of alachlor with hydrogen peroxide, finding very interesting results [32]. However, to the best of our knowledge, neither goethite nor hydroxylamine have been applied for the degradation of pollutants as nCBs, which is the goal of the present work. Moreover, more information is required on the role of hydroxylamine, as well as its oxidation byproducts.

Therefore, in the current work, the use of goethite, as an inexpensive heterogeneous source of supported iron (Fe III), coupled with hydroxylamine, has been explored for the first time to promote the oxidation of nCBS with hydrogen peroxide. Moreover, the effect of the main operating variables influencing the performance of the process was evaluated.

2. Results

2.1. Blank Experiments

Seven experiments, gathered as B1a–g, in Table 1, were carried out. Each one was carried out at a different initial pH value within the range of 1 to 7, using the same goethite concentration (0.5 g L^{-1}). After 48 h at room temperature, the concentration of the iron leached was measured in the filtered aqueous phase. At pH = 1, a concentration of 0.52 mg L^{-1} of total iron in the aqueous phase was found, whereas concentrations below the detection limit were measured at higher pHs. Therefore, it is assumed that the iron leached from the heterogeneous catalyst is negligible even at the lowest pH used.

Table 1. Experimental conditions of the runs carried out. Variables: initial concentration of hydroxylamine (C_{HA}), theoretical stoichiometric amount of hydrogen peroxide for the complete mineralization of the chlorobenzenes (nCBs; n = mono, di, tri, tetra). ($C_{H_2O_2,0}/C^{st}_{H_2O_2,0}$), goethite concentration (C_{GOE}), pH using either milliQ or spiked water with chlorinated organic compounds (COCs).

	C_{HA} mM	$C_{H_2O_2,0}/C^{st}_{H_2O_2,0}$	C_{GOE} g L^{-1}	pH_0	$\Sigma COCs$ mg L^{-1}
B1 a–g*	0	0	0.5	1–7	
B2	2.4	0	0.5	3	0
B3	2.4	5	0	3	
B4	2.4	5	0.5	3	
S1 a, b, c	2.4	5	0.5	7, 5, 2	31.73
S2	0				
S3	0.6				
S4	1.2	5	0.5	3	31.73
S5	2.4				
S6	4.9				
S7	2.4	1	0.5	3	31.73
S8	2.4	10	0.5	3	31.73
S9	2.4	5	0.25	3	31.73
S10	2.4	5	1	3	31.73

* A total of seven experiments were carried out in the run gathered as B1.

To investigate the reaction between hydroxylamine (HA) and hydrogen peroxide in the presence or absence of goethite, the experiments B2, B3, and B4 were carried out at pH 3. HA conversions at 300 min of reaction time, calculated from Equation (1), are plotted in Figure 1. As can be seen, the conversion of HA was almost negligible in the presence of the catalyst without H_2O_2 (B2), whereas around 0.25 of HA conversion was obtained when hydrogen peroxide (without catalyst addition) was used (B3). On the contrary, the consumption of HA was almost total when goethite and hydrogen peroxide were added simultaneously (B4). The evolution of HA and hydrogen peroxide with reaction time in this run (B4) is plotted in Figure 2. HA was rapidly consumed in the first stage, achieving more than 90% of the conversion at 300 min. After 24 h of reaction time the amount of HA detected in the media was negligible. Therefore, this compound does not remain in the media for longer times. However, the hydrogen peroxide conversion at the end of the experiment (300 min) was only around 0.2, which suggests that HA is the limiting reagent.

$$X_{HA} = 1 - \frac{C_{HA}}{C_{HA,0}} \quad (1)$$

where X_{HA} is the conversion of HA, C_{HA} is the remaining concentration of HA at a given reaction time, and $C_{HA,0}$ is the initial concentration of HA.

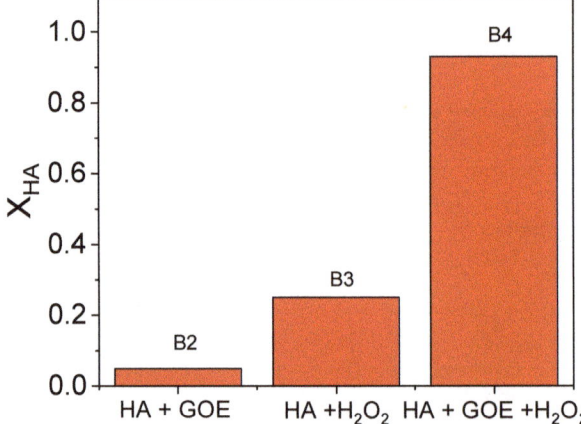

Figure 1. Hydroxylamine conversion (X_{HA}) in runs B2, B3, and B4 after 300 min of reaction time ($C_{GOE} = 0.5$ g L^{-1}, $C_{HA,0} = 2.4$ mM, and $C_{H_2O_2,0} = 12.35$ mM, pH = 3).

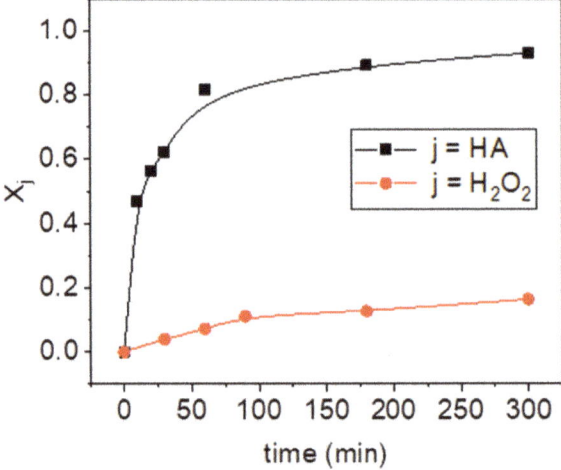

Figure 2. Hydroxylamine (HA) and hydrogen peroxide conversion (X_j) vs. reaction time (run B4) ($C_{GOE} = 0.5$ g L^{-1}, $C_{HA} = 2.44$ mM, and $C_{H_2O_2} = 12.35$ mM, pH = 3).

It has been described that the oxidation of HA produced inorganic anions (i.e., nitrates and nitrites) [42,45]. In the present work, only nitrates were identified by ion chromatography, which agrees with the results obtained by Chen et al. [42]. It should be noted that this result was expeceted since hydrogen peroxide was used in high stoichiometric excess.

The profile of the molar yield of HA to nitrates with reaction time, plotted in Figure 3, has been calculated using Equation (2). As can be seen, an asymptotic value on nitrate concentration was achieved after 30 min of reaction time. However, the hydroxylamine conversion at that time was only around 0.5. The consumption of this compound continued increasing until it reached almost the complete conversion. This fact could be explained if nitrates are intermediate byproducts of HA oxidation. The products proposed in the literature for the oxidation of hydroxylamine are, besides nitrates, N_2 and N_2O [42]. Moreover, it was found that the formation of N_2O was preferred at pH = 3 (the pH used in run B4), which would explain the nitrogen mismatch.

$$Y_{NO_3^-} = \frac{C_{NO_3^-}}{C_{HA,0} \cdot X_{HA}} \tag{2}$$

where $Y_{NO_3^-}$ is the yield of the oxidation of HA to nitrate. It is calculated as the ratio between the nitrate concentration ($C_{NO_3^-}$) measured at a given reaction time (quantified by ion chromatography (IC)), and the concentration of HA reacted.

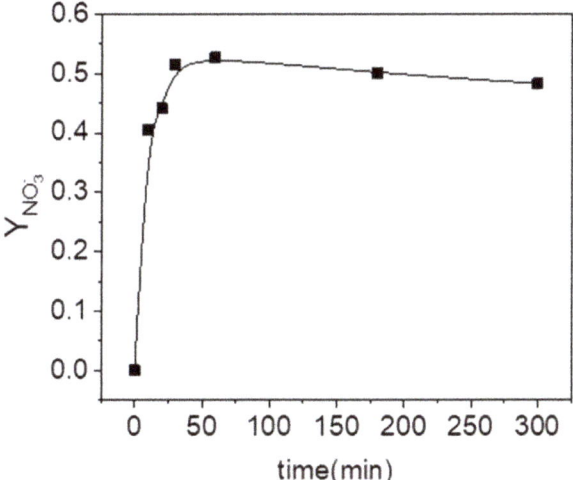

Figure 3. Nitrogen yield to nitrates ($Y_{NO_3^-}$) in the oxidation of hydroxylamine ($C_{GOE} = 0.5$ g L^{-1}, $C_{HA} = 2.44$ mM, and $C_{H_2O_2} = 12.35$ mM, pH = 3).

2.2. CWPO Experiments

A systematic study of the pH, HA, H_2O_2, and catalyst concentrations on nCBS (n = mono, di, three, and tetra) removal in the aqueous phase has been carried out. It should be pointed out that there was neither evaporation nor reaction of nCBs during the 24 h (control experiments).

2.2.1. Study of pH Effect

The effect of the initial pH was evaluated within the range 7–2, by comparison of nCBs conversion obtained at 300 min in runs S1 a, b, c, carried out at pH = 7, 5, and 2, respectively, and S5, carried out at pH = 3. The operation conditions for these runs are summarized in Table 1. Taking the CB conversion vs. pH as representative of the profile for the other nCBS, it was noticed that its conversion is greatly increased by acidifying the pH from 7 ($X_{CB} = 0.05$) to 3 ($X_{CB} = 0.76$), as can be seen in Figure 4. At pH 5, the conversion of CB obtained at 300 min was about 0.2. The same trend was noticed for the other pollutants studied: the lower the initial pH used, the higher the conversion of chlorobenzenes was found. However, when using lower pHs (pH = 2), only slight differences were found in nCBs conversion over time.

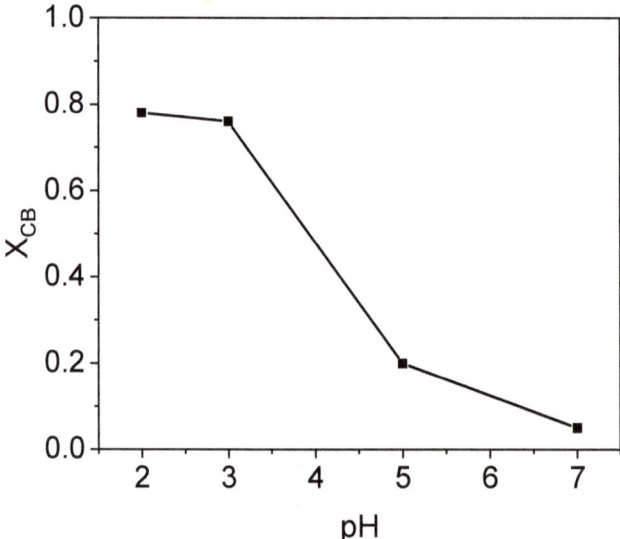

Figure 4. Conversion of chlorobenzene X_{CB} vs. initial pH at 300 min using $C_{GOE} = 0.5$ g L^{-1} and $(C_{H_2O_{2,0}})/(C^{st}_{H_2O_2}) = 5$ (12.43 mM), and $C_{GOE} = 2.4$ mM.

The influence of pH on nCBs abatement found here can be related to the oxidation mechanism of hydroxylamine, which is affected by the pH of the solution [42]. The pH value selected in this work, using the system goethite-H_2O_2 and hydroxylamine, agrees with the optimal pH value reported in the literature for the homogeneous Fenton reaction [24,30]. In fact, it is widely documented that the highest rate for the hydroxyl radical generation takes place at this pH range using Fenton's reagent. Chen et al. [42] studied the abatement of benzyl alcohol by homogeneous Fenton reagent (H_2O_2) enhanced with HA and found that the optimal pH range was also in the acidic range (3–4).

It has been described that at pH values above 6, hydroxylamine can act as a strong termination agent [42]. Consequently, the initial pH selected for further experiments was 3.

The nCBs conversion was checked when 1 mg L^{-1} of homogeneous iron, Fe (II) (the maximum iron leached value found in blank runs, B1a, pH = 1, t = 48 h was 0.52 mg L^{-1}), was added to the aqueous phase as a catalyst instead of goethite, using the same H_2O_2 and HA concentrations, as well as the pH used in run S5. At these operating conditions, nCBs conversion (<0.05) at 300 min was found negligible. Therefore, it can be concluded that the Fenton reaction studied here was heterogeneously promoted by the iron in the surface of the solid catalyst, in agreement with that reported in the literature [32].

When using an initial pH of about 3, a decrease in the pH value was noticed during the runs, because of the generation of short chain organic acids [46]. A minimum value pH of about 2.3 was achieved. However, in the pH range 2–3, a significant effect of the pH on the nCBS conversion was not noticed, as was shown in Figure 4.

2.2.2. HA Concentration Effect

To study the influence of the HA addition on the pollutant's abatement, several HA concentrations (from 0 to 4.9 mM, runs S2 to S6) have been used to treat the contaminated water, keeping a constant concentration of goethite (0.5 g L^{-1}) and hydrogen peroxide (12.34 mmol L^{-1}). As can be deduced from Table 2, where the initial concentration of each chlorinated compound is included, the selected concentration of the oxidant is 5 times the theoretical stoichiometric amount considering the complete mineralization of the nCBs in the polluted water.

Table 2. Initial concentration of chlorobenzene compounds (nCBs) in the spiked water.

COCs	Chemical Formula	Acronym	Initial Concentration (mg L^{-1})	Initial Concentration (mM)	$C^{st}_{H_2O_2}$ (mM) [a]
chlorobenzene	C_6H_5Cl	CB	16.29	0.145	2.030
1,3-dichlorobenzene	$C_6H_4Cl_2$	1,3 DCB	0.33	0.002	0.026
1,4-dichlorobenzene	$C_6H_4Cl_2$	1,4 DCB	4.77	0.032	0.416
1,2-dichlorobenzene	$C_6H_4Cl_2$	1,2 DCB	4.64	0.032	0.416
1,2,4-trichlorobenzene	$C_6H_3Cl_3$	1,2,4 TCB	3.86	0.021	0.252
1,2,3-trichlorobenzene	$C_6H_3Cl_3$	1,2,3 TCB	0.75	0.004	0.048
1,2,4,5 tetrachlorobenzene	$C_6H_2Cl_4$	TetraCB-a	0.41	0.002	0.020
1,2,3,4 tetrachlorobenzene	$C_6H_2Cl_4$	TetraCB-b	0.68	0.003	0.030

[a] Theoretical stoichiometric dosage of H_2O_2 required for the mineralization of each compound.

The normalized remaining concentration of mono CB with reaction time in runs S2 to S6 is plotted in Figure 5. The normalized profiles of the positional isomers of DCB (1,2-DCB, 1,3-DCB, and 1,4-DCB), TCBs (1,2,3-TCB and 1,2,4-TCB), and TetraCBs (TetraCB-a and TetraCB-b) concentrations are represented in Figures 6 and 7a–d. The normalized concentration of each chlorinated compound has been calculated using Equation (3):

$$1 - X_j = C_j / C_{j,0} \tag{3}$$

where X_j is the conversion of the pollutant j, C_j and $C_{j,0}$ are the concentration of pollutant j at a given reaction time and at zero time, respectively.

As can be seen in these figures, in the absence of hydroxylamine (run S2), the reaction rate was very slow, and the concentration of the pollutants hardly decreased with reaction time. On the contrary, when hydroxylamine was added to the reaction medium, the abatement of the pollutants was greatly increased. This could be explained considering that the presence of hydroxylamine facilitates the reduction of Fe (III) to Fe (II) in the catalyst surface, accelerating the decomposition of H_2O_2 to generate hydroxyl radicals, which are responsible for the oxidation of organic matter in the solution, in agreement with that previously reported in the literature [32]. As the iron redox cycle is enhanced by the presence of hydroxylamine and nitrate being the only inorganic product detected in the reaction media, the mechanism in Equations (4–6) can be proposed to explain the iron redox cycle in the presence of hydroxylamine, in accordance with the mechanism reported in literature [47].

$$Fe^{3+} + NH_2OH \rightarrow NH_2O \cdot + Fe^{2+} + H^+ \tag{4}$$

$$Fe^{3+} + NH_2O \cdot \rightarrow NHO \cdot + Fe^{2+} + H^+ \tag{5}$$

$$5Fe^{3+} + 2H_2O + NH_2O \cdot \rightarrow NO_3^- + 5Fe^{2+} + 6H^+ \tag{6}$$

In Figure 4, Figure 5, and Figure 6 it can be noted that a fast depletion of the contaminants took place during the first 5 h of the reaction. Following, the pollutant oxidation rate decreased, which can be related to the consumption of HA in the fast stage. In fact, as noted in Figure 2, the almost complete conversion of HA was obtained at 5 h when using the same oxidant and catalyst concentration employed in run S5 (run B4). Therefore, the slow stage in the depletion of COCs observed in Figure 5, Figure 6, and Figure 7 is a consequence of the absence of HA in this period. Concluding, the presence of HA improves the reduction of Fe (III) to Fe (II) over the goethite surface, enhancing the hydroxyl radical formation and the pollutant abatement.

Attending to the results shown in Figures 5–7, it can be noticed that the higher the concentration of hydroxylamine used, the higher the conversions of nCBs. However, the differences seem to be narrower when HA concentration raised from 2.4 to 4.9 mM. This fact can be explained considering that HA snot only participate in the iron redox cycle but also competes with the pollutant for the oxidant.

Comparing the profiles obtained for the different COCs (Figures 5–7), it can be concluded that the oxidation rates followed the subsequent order: CBs > DCBs >> TCBS = TetraCBs. As can be seen in Figures 6 and 7, the positional stereoisomers of DCBs, TCBs, and TetraCBs showed similar normalized concentration profiles with reaction time, indicating that the catalyst is not regioselective.

On the other hand, the maximum conversion achieved of H_2O_2 was about 10%, obtained in S6, when a concentration of 4.9 mM of HA was used (data not shown). Therefore, it can be assumed that, at the operating conditions, the limiting reagent was HA, as previously proposed, and in this case, the concentration of hydrogen peroxide can be considered almost constant with reaction time. To confirm that at the operating conditions HA was the limiting reagent, two new reaction vials were prepared at the conditions of run S5 in Table 1. At a reaction time of 2.5 h, a volume of 0.2 mL of the aqueous reaction media was replaced by an aliquot of 0.2 mL of a concentrated HA solution. The amount of HA added in the aliquot corresponds to an amount of 2.4 mM in the final aqueous volume in the vial. At 2.5 and 5 h after the HA aliquots were added, the vials were sacrificed and the pollutants in the aqueous phase were analyzed. Results obtained are shown as open symbols in Figures 5–7. As can be seen, the further addition of HA after 2.5 h produced a significant rise in the pollutant conversions, confirming that HA is the limiting reagent. Further, the concentration of hydrogen peroxide was found to be almost constant with the reaction time.

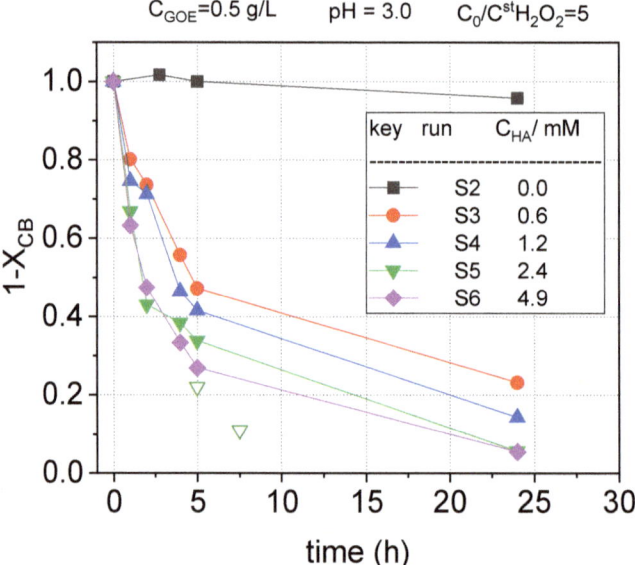

Figure 5. Remaining fractional chlorobenzene concentration $(1 - X_{CB})$ vs. reaction time at $pH_0 = 3$, $C_{GOE} = 0.5$ g L^{-1}, and $(C_{H_2O_{2,0}})/(C^{st}_{H_2O_2}) = 5$ (12.43 mM), at room temperature (controlled at 22 °C). Conversions obtained after addition of further aliquots of HA (+2.4 mM) in run S5 at 2.5 h are shown with open symbols.

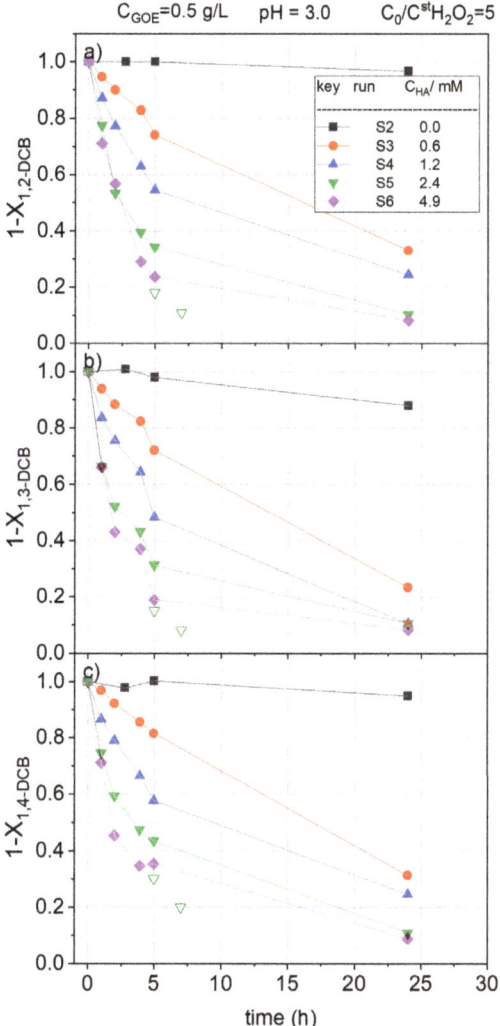

Figure 6. Remaining fractional concentration of (**a**) 1,2-dichlorobenzene $(1 - X_{1,2-DCB})$; (**b**) 1,3-dichlorobenzene $(1 - X_{1,3-DCB})$; and (**c**) 1,4-dichlorobenzene $(1 - X_{1,4-DCB})$ vs. reaction time at $pH_0 = 3$, $C_{GOE} = 0.5$ g L^{-1}, and $(C_{H_2O_2,0})/(C^{st}_{H_2O_2}) = 5$ (12.43 mM), at room temperature (controlled at 22 °C). Conversions obtained after addition of further aliquots of HA (+2.4 mM) in run S5 at 2.5 h are shown with open symbols.

To evaluate if the removal of nCBs was linked to the loss of the chlorine atoms from the pollutant molecule, the release of chloride to the reaction media was analyzed. The profile of chloride concentration in the aqueous phase in run S5 was measured and the results are shown in Figure 8. The predicted chloride concentration was calculated assuming that the conversion of the chlorinated pollutant corresponds to the loss of all the chlorine atoms of the molecule, as indicated in Equation (7).

$$C^{stc}_{Cl^-} = \sum C_{j,0} \cdot X_j \cdot n \cdot 35.5 \qquad (7)$$

145

where $C_{Cl^-}^{stc}$ is the stoichiometric amount of chloride released assuming the complete dechlorination of each nCBs (n = 1, 2, 3, 4) in mg L^{-1}; $C_{j,0}$ is the initial concentration of the pollutant j (in mM, summarized in Table 2); X_j is the experimental conversion of each nCBs; and n is the number of chlorine atoms in the starting pollutant (1 for CB, 2 for DCBs, 3 for TCBs and 4 for TetraCB).

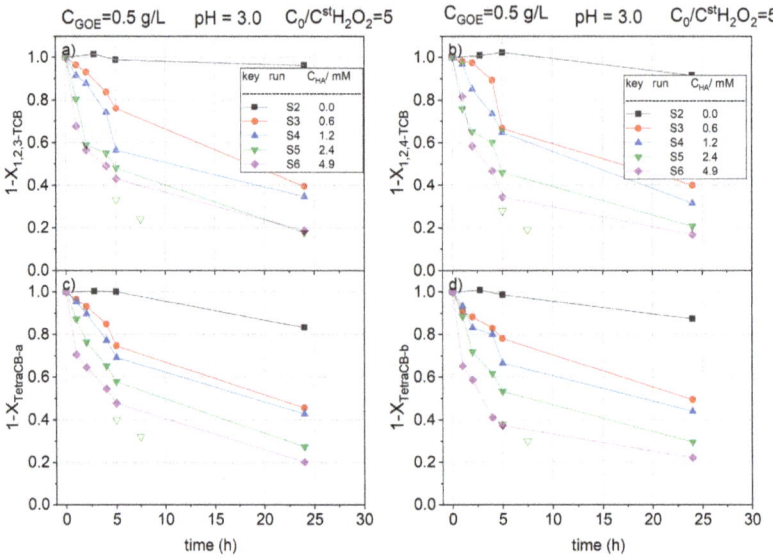

Figure 7. Remaining fractional concentration of (**a**) 1,2,3-trichlorobenzene $(1 - X_{1,2,3-TCB})$; (**b**) 1,2,4-trichlorobenzene $(1 - X_{1,2,4-TCB})$; (**c**) tetrachlorobenzene-a $(1 - X_{TetraCB-a})$; (**d**) tetrachlorobenzene-b $(1 - X_{TetraCB-b})$ vs. reaction time at pH$_0$ = 3, $C_{GOE} = 0.5$ g L^{-1}, and $(C_{H_2O_{2,0}})/(C_{H_2O_2}^{st}) = 5$ (12.43 mM), at room temperature (controlled at 22 °C). Conversions obtained after addition of further aliquots of HA (+2.4 mM) in run S5 at 2.5 h are shown with open symbols.

The profile of the chloride concentration predicted by Equation (7) with the data obtained is compared with that experimentally measured (Figure 8). As can be seen, both experimental and predicted values are similar, confirming that the dehalogenation of the organic molecules was achieved. As the toxicity of the organic molecules is related to its chlorine content, this a desirable result if the detoxification of the contaminated water is required. No aromatic compounds different than those present in the initial polluted water were detected by gas chromatography/mass spectrometry (GC/MS) analysis of the reaction samples. The only byproduct detected in the oxidation of nCBs with goethite was acetic acid, identified by ion chromatography.

Moreover, the iron leached was measured with the reaction time in all the runs in Table 1. In all the runs the maximum iron concentration in the aqueous phase was less than 0.6 mg/L. The presence of chlorides at the end of the reaction did not increase the amount of iron leached. To check for further reaction in the absence of the solid, a hot filtration test was carried out at the conditions of run S5 in Table 1. To do this, two GC vials of 20 mL were prepared with the corresponding goethite, oxidant and HA concentration employed in run S5. After 2.5 h of reaction time, the stirring was stopped, the liquid of each vial was filtered, an aliquot of the liquid was analyzed by GC/MS and two aliquots of 10 mL of the remaining aqueous phase were placed in two 10 mL GC vials to check the reaction progress in absence of the solid. Vials containing the filtered aqueous phase were sacrificed at 2.5 and 5 h and the pollutant concentration in the aqueous phase was measured. A negligible increase in pollutant conversion in the filtered media was noticed, confirming that the reaction took place in a heterogeneous way.

Considering the nCBs profiles shown in Figures 5–7, it can be deduced that serial reactions as tetrachlorobenzene to trichlorobenzene, trichlorobenzene to dichlorobenzene, and dichlorobenzene to monochlorobenzene did not take place.

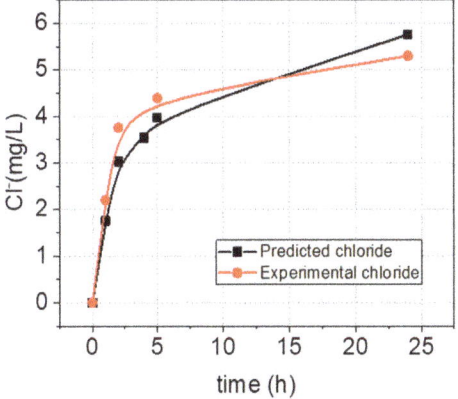

Figure 8. Predicted and experimental chloride concentration vs. reaction time (run S5), at $pH_0 = 3$ $C_{GOE} = 0.5$ g L^{-1}, and $(C_{H_2O_{2,0}})/(C^{st}_{H_2O_2}) = 5$ (12.43 mM), at room temperature (controlled at 22 °C).

2.2.3. Hydrogen Peroxide Concentration Effect

To study the effect of the oxidant on the COCs abatement, runs S5, S7, and S8 (Table 1) were carried out. The dose of hydrogen peroxide varied from 1 to 10 times the stoichiometric amount calculated for the complete mineralization of the starting nCBs (listed in Table 2). The normalized concentration profiles of remaining CB, DCBs, TCBs, and TetraCBs concentration vs. reaction time, are plotted in Figures 9–11, respectively.

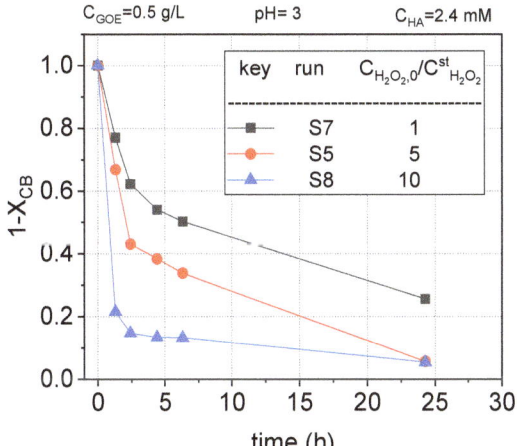

Figure 9. Remaining fractional chlorobenzene concentration $(1 - X_{CB})$ vs. reaction time using different doses of hydrogen peroxide in a range of 1–10 times the theoretical stoichiometric amount for the complete mineralization of the nCBs at $pH_0 = 3$, $C_{GOE} = 0.5$ g L^{-1}, and $C_{HA} = 2.4$ mM, at room temperature (controlled at 22 °C).

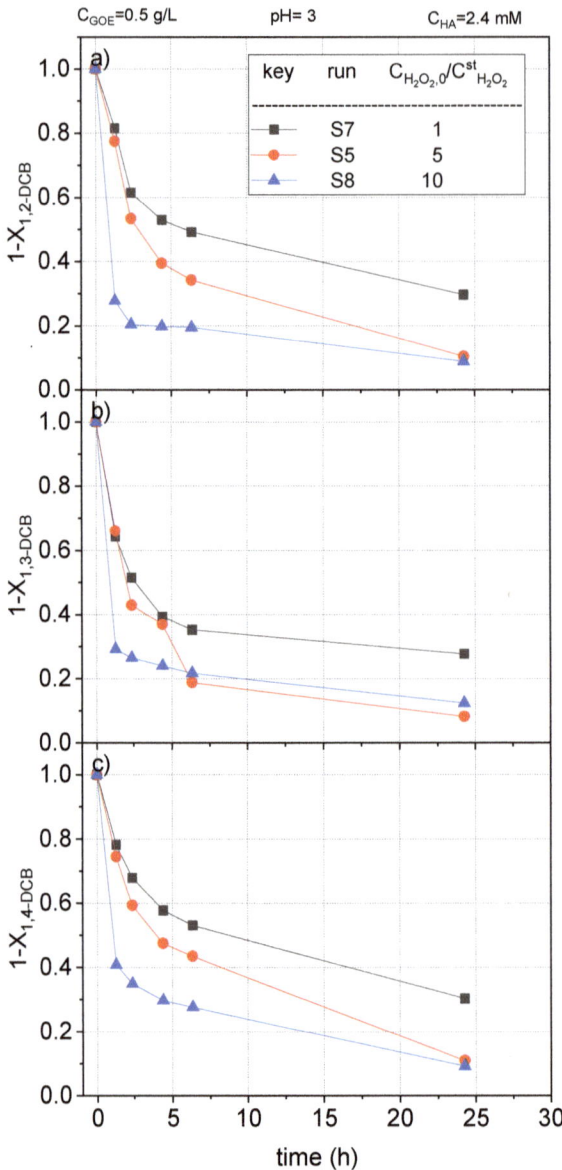

Figure 10. Remaining fractional concentration of (**a**) 1,2-dichlorobenzene $(1 - X_{1,2-DCB})$; (**b**) 1,3-dichlorobenzene $(1 - X_{1,3-DCB})$; and (**c**) 1.4-dichlorobenzene $(1 - X_{1,4-DCB})$ vs. reaction time using different doses of hydrogen peroxide in a range of 1–10 times the theoretical stoichiometric amount for the complete mineralization of the CBs at $pH_0 = 3$, $C_{GOE} = 0.5$ g L^{-1}, and $C_{HA} = 2.4$ mM, at room temperature (controlled at 22 °C).

As can be observed, the higher concentration of hydrogen peroxide, the higher conversion of nCBs, due to a higher production of hydroxyl radicals. This trend was maintained regardless of the type of chlorobenzene. Thus, when a concentration of hydrogen peroxide 10 times the stoichiometric one was used, the reaction time was greatly decreased. For instance, in the case of CB (Figure 9), a conversion of the pollutant above 0.9 was reached in 2 h. On the other hand, more than 24 h of reaction time was needed to reach this conversion value when the stoichiometric dose of H_2O_2 was used. The samples obtained at 24 h in S7 were analyzed to quantify the remaining hydrogen peroxide and the leaches of iron. For a stoichiometric ratio for H_2O_2 of 1, the conversion factor was roughly 85%.

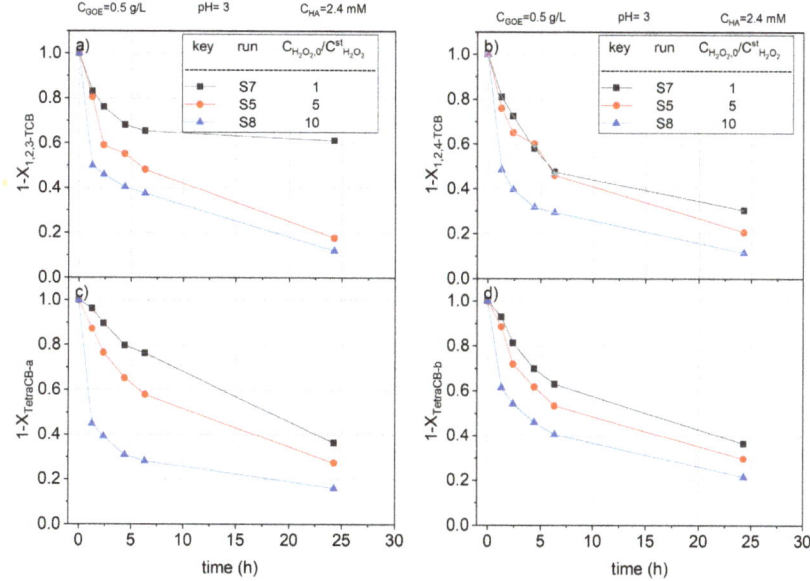

Figure 11. Remaining fractional concentration of (**a**) 1,2,3-trichlorobenzene $(1 - X_{1,2,3-TCB})$; (**b**) 1,2,4-trichlorobenzene $(1 - X_{1,2,4-TCB})$; (**c**) tetrachlorobenzene-a $(1 - X_{TetraCB-a})$; and (**d**) tetrachlorobenzene-b $(1 - X_{TetraCB-b})$ vs. reaction time using different doses of hydrogen peroxide in a range of 1–10 times the theoretical stoichiometric amount for the complete mineralization of the CHBs at $pH_0 = 3$, $C_{GOE} = 0.5$ g L^{-1}, and $C_{HA} = 2.4$ mM, at room temperature (controlled at 22 °C).

2.2.4. Effect of Goethite Concentration

Finally, the influence of the goethite concentration on nCBs removal was also evaluated (runs S5, S9 and S10 of Table 1). The concentration of HA was kept as 2.4 mM, the initial pH was set to 3, and the dose of hydrogen peroxide was fixed at 5 times the stoichiometric amount.

It was noticed that the addition of the heterogeneous catalyst produced a sludge with an orange color (the higher the concentration of goethite, the higher the color intensity). The reaction samples were filtered using nylon filters to remove the heterogenous catalyst, obtaining a colorless aqueous solution. As commented before, the iron leaching at these conditions was negligible.

The fractional remaining concentration of nCBs measured in runs S5, S9, and S10 are plotted in Figures 12–14. As expected, when 0.25 g L^{-1} of goethite was used, the conversion of CB, DCB, TCBS, and TetraCB isomers was lower than those obtained with a catalyst concentration of 0.5 g L^{-1}. Surprisingly, an increase in the concentration of goethite from 0.5 to 1 g·L^{-1} did not produce an improvement in the removal of COCs. When the catalyst concentration is higher than a critical value in slurry reactors it is common that the slurry formed is not well mixed and the catalyst agglomerates at the bottom of the reactor, resulting in a defective contact between the solid and the aqueous phase. In our work, this effect was noticed at values of 1 g/L or higher of the iron mineral. This fact has also been reported in literature, where an optimal catalyst concentration in slurry systems was found [48]. However, it should be considered that the optimal value for catalyst loading to avoid agglomeration and defective liquid–solid contact will depend on the reactor construction and the agitation system used.

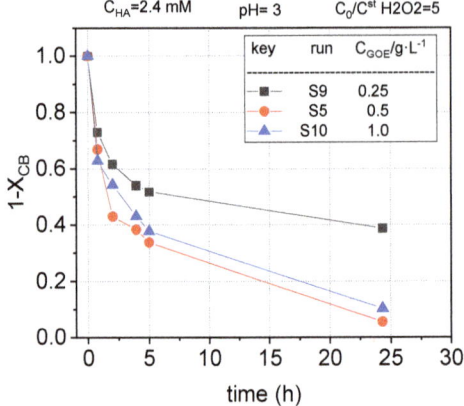

Figure 12. Remaining fractional chlorobenzene concentration $(1 - X_{CB})$ at different goethite concentrations vs. reaction time at $pH_0 = 3$, $C_{HA} = 2.4$ mM and $(C_{H_2O_2,0})/(C^{st}_{H_2\,O_2}) = 5$ (12.43 mM), at room temperature (controlled at 22 °C).

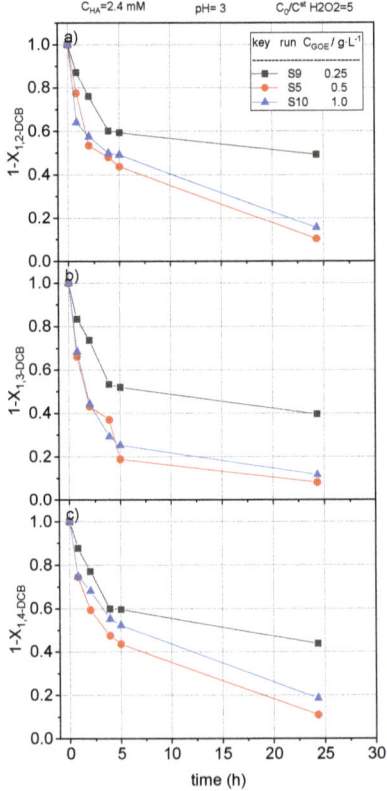

Figure 13. Remaining fractional concentration of (**a**) 1,2-dichlorobenzene $(1 - X_{1,2-DCB})$; (**b**) 1,3-dichlorobenzene $(1 - X_{1,3-DCB})$; and (**c**) 1,4-dichlorobenzene $(1 - X_{1,4-DCB})$ vs. reaction time at $pH_0 = 3$, $C_{HA} = 2.4$ mM and $(C_{H_2O_2,0})/(C^{st}_{H_2\,O_2}) = 5$ (12.43 mM), at room temperature (controlled at 22 °C).

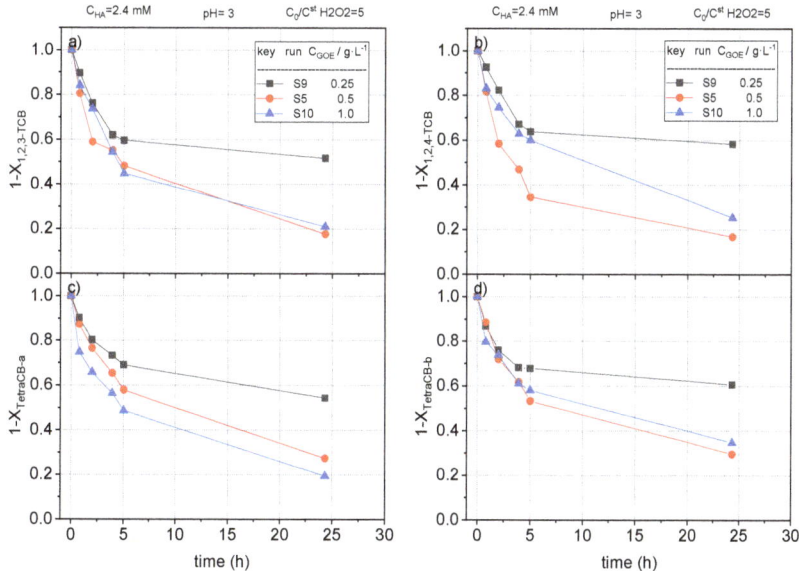

Figure 14. Remaining fractional concentration of (**a**) 1,2,3-trichlorobenzene $(1 - X_{1,2,3-TCB})$; (**b**) 1,2,4-trichlorobenzene $(1 - X_{1,2,4-TCB})$; (**c**) tetrachlorobenzene-a $(1 - X_{TetraCB-a})$; and (**d**) tetrachlorobenzene-b $(1 - X_{TetraCB-b})$ vs. reaction time at $pH_0 = 3$, $C_{HA} = 2.4$ mM and $(C_{H_2O_{2,0}})/(C^{st}_{H_2O_2}) = 5$ (12.43 mM), at room temperature (controlled at 22 °C).

3. Methods

3.1. Reagents and Catalyst

Chlorobenzene (CB), 1,2-dicholorobenzene (1,2-DCB), 1,3-dichlorobenzene (1,3-DCB), 1,4-dicholorobenzene (1,3-DCB), 1,2,3-trichlorobenzene (1,2,3-TCB), 1,2,3,4-tetrachlorobenzene (1,2,3,4-TetraCB), 1,2,3,5-tetrachlorobenzene (1,2,3,5-TetraCB), and 1,2,3,4-tetrachlorobenzene (1,2,3,4-TetraCB), all of analytical quality, were purchased from Sigma-Aldrich (Darmstadt, Germany) and dissolved in n-hexane to prepare the standards used in the calibration curves of these compounds. Bicyclohexyl, also purchased from Sigma-Aldrich (Darmstadt, Germany), was selected as the internal standard (ISTD). The polluted water used in this work was prepared by spiking milli-Q water with chlorobenzenes (nCBs, being n = mono, di, tri, and tetra). The concentration of each compound, summarized in Table 2, was selected to simulate the concentration range found in the groundwater of the hot spots where these compounds were usually dumped [49]. As can be seen, 1,3 DCB has a lower concentration than 1,2 and 1,4 DCB in accordance with the occurrence of these compounds noticed in groundwaters. The concentration of COCs used in the spiked water also considered that the solubility of each nCB in aqueous phase decreases with the number of chlorine atoms in the molecule. The theoretical stoichiometric dosage of H_2O_2 required for the mineralization of each compound is also given in this table. As can be seen, the total amount of H_2O_2 required for the complete oxidation of COCs was 3.24 mmol L^{-1} (0.11 g L^{-1}).

Hydrogen peroxide (33 wt%), employed as an oxidant, was supplied by Sigma-Aldrich; hydroxylamine, used to enhance the reduction of Fe (III) to Fe (II) in the Fenton redox cycle, was added as hydroxylammonium sulfate (HAS), being supplied by Acros Organics (Geel, Belgium).

Goethite, used as a catalyst in CWPO reactions, was supplied by Sigma-Aldrich (Darmstadt, Germany). Goethite is a soil mineral compounded by Fe(III), being one of the most thermodynamically stable iron oxides. It presents an orthorhombic structure, where the iron atoms are in the center of an the octahedral formed by O^{2-} and OH^- anions, and the molecular formula is Fe(III)OOH [50–52].

A specific surface area (S_{BET}) of 10.24 m² g^{-1} was measured by the N_2 adsorption–desorption isotherms at 77 K (Figure 15). The total iron content of the catalyst was determined by spectroscopy of atomic emission at 259.94 nm. Previously, 0.5 g of the catalyst was dissolved in 15 mL of chlorhydric acid (Sigma-Aldrich, Darmstadt, Germany 35 wt%), finding that the iron mass percentage in the goethite was 57.3 wt%. Other metals were also measured by atomic emission spectrometry, but only a small amount of potassium (<0.01%) was found. Therefore, the Fe(III)OOH percentage in goethite can be calculated as 91.3 wt%. The corresponding mass percentages of oxygen and hydrogen happens to be 33.2 wt% and 1.03 wt%, respectively. About 8.6 wt% of the mineral mass could not be identified but it did not correspond with transition metals.

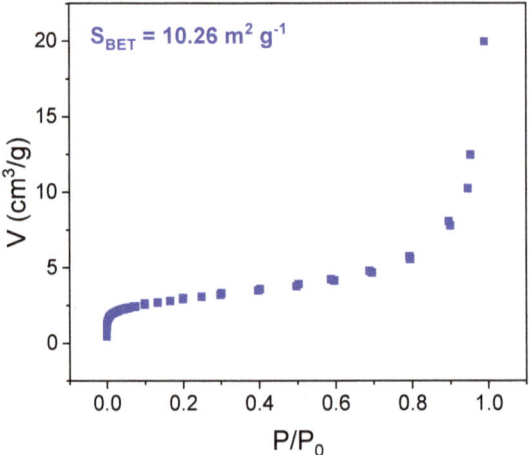

Figure 15. N_2 adsorption–desorption isotherm of goethite at 77 K, used as a solid catalyst.

Other reagents used in the present work were: titanium oxysulfate, ferric sulfate heptahydrate, n-hexane, and cyclohexanone (Sigma-Aldrich), carbonate, bicarbonate, H_2SO_4, acetone, nitrite, and nitrate (for Ion Chromatography analyses).

3.2. Blank Experiments

The leaching of iron from the heterogeneous catalyst (0.5 g L^{-1} of goethite) was evaluated in the seven runs gathered as B1a–g in Table 1. A concentrated solution of hydroxylamine sulfate (HAS) in water (400 g/L) was used to achieve the desired hydroxylammonium (HA, Sigma-Aldrich Darmstadt, Germany) concentration in the reaction media. It must be considered that each mol of HAS yields 2 moles of HA. A pH range from 1 to 7 was investigated and leached iron was measured in the batch reactor after 48 h at room temperature. Previously to the leached iron quantification, the goethite was separated from the aqueous phase using a nylon filter (0.22 μm).

In order to discriminate the interaction among the oxidant, the reductant, HA, and the catalyst in the absence of pollutants, a set of control experiments were carried out using milli-Q water without pollutants. The experimental conditions of these runs (B2, B3, and B4) are summarized in Table 1. These runs were carried out at room temperature (22 °C) in a well-mixed batch reactor of 250 mL using a magnetic plate, IKA C-MG HS 7 (Staufen, Germany). A stirring speed of 500 rpm was selected for the blank runs because almost the same HA conversions were obtained in run B4 at 400, 500, and 600 rpm.

Reaction samples were taken at different times to follow the evolution of the reagents in the aqueous phase.

3.3. CWPO Experiments

The effect of the initial pH value, as well as the oxidant (H_2O_2), reductant (HAS), and catalyst (goethite) concentration on the nCBs removal was investigated (runs S1 to S10, Table 1). Runs were carried out by duplicate with the standard deviation being lower than 5% in all cases.

A volume of 0.4 mL of an aqueous solution containing AS and H_2O_2 was added to 19.6 mL of the polluted water in a cap-sealed vial (20 mL) without headspace, in order to minimize the evaporation of CBs during the reaction time. The concentration of reagents (H_2O_2 and HAS) in the 0.4 mL volume was selected to achieve the corresponding concentration of each run (conditions collected in Table 1) in the final reaction volume (20 mL). The corresponding amount of catalyst was added at zero time. Several vials were prepared for each reaction and a vial was sacrificed at each reaction time. The vials were agitated using a rotatory agitator (Labolan, ref 51752, Navarra, Spain). The agitation speed was fixed to 90% of the maximum speed of the rotatory agitator employed. No differences were found in the pollutant conversion obtained in run S5 in the range of 80% to 100% of the maximum rotatory speed

As the addition of HAS acidified the pH of the spiked water, the pH of the polluted water and the HAS solution was adjusted before these solutions were mixed. To do this, H_2SO_4 and NaOH were used when necessary.

As can be seen in Table 1, the influence of pH was tested in runs S1 a, b, c and S5. Runs S2–S6 were carried out to study the influence of the hydroxylamine concentration (in the range of 0.6 to 4.9 mmol L^{-1}) on the nCBs abatement. Runs S5, S7, and S8 were carried out using different doses of hydrogen peroxide in a range of 1–10 times the theoretical stoichiometric amount, calculated for the complete mineralization to carbon dioxide, water and chloride of the nCBs (Table 2). Finally, the influence of the solid catalyst concentration on the removal of CBs was tested in runs S5, S9, and S10 within the range 1–0.25 g L^{-1}. Operation conditions of run S5 were selected as the central values for the study of other variables than H_2O_2.

A blank experiment with the polluted water in the absence of the reagents (H_2O_2 and HA) and the catalyst (goethite) was also carried out (by triplicate).

3.4. Analytical Methods

Firstly, COCs were identified by gas chromatography (Agilent 6890N, Santa Clara, CA, USA) coupled to a mass spectrometry detector (GC/MS). The concentration of nCBs in the aqueous phase during the experiments was determined by extracting 10 mL of the reaction mixture with 5 mL of n-hexane by the agitation during 2 min and after 10 min of settlement. It was experimentally confirmed that more than 95% of the COCs in the aqueous phase were extracted to the organic phase by this procedure. Both phases were separated by decantation and the concentration of nCBs in the organic phase was measured using a gas chromatograph coupled with a flame ionization detector (GC/FID) and an electron capture detector (ECD). On the other hand, the catalyst was separated from the aqueous phase using a nylon filter (0.22 µm). After that, the concentration of hydrogen peroxide, chloride, carboxylic acids, and iron was determined.

GC/MSD analysis: COCs were identified by gas chromatography (Agilent 6890N, Santa Clara, CA, USA) coupled to a mass spectrometry detector (Agilent MSD 5975B, Santa Clara, CA, USA), which operates under a vacuum. A column HP-5MS (30 m × 0.25 mm ID × 0.25 µm) was used as stationary phase. A flow rate of 1.7 mL min^{-1} of helium was used as the carrier gas and 1 µL of liquid samples was injected. The GC injection port temperature was set to 250 °C and a programed-temperature gradient was used for the GC oven, starting at 80 °C, increasing the temperature at a rate of 18 °C min^{-1} up to 180 °C, and then keeping it constant for 15 min.

GC/ECD-FID analysis: The quantification of nCBs was carried out using an Agilent 6890 gas chromatograph (Santa Clara, CA, USA) with FID and ECD detectors. An HP-5MS column (30 m × 0.25 mm ID × 0.25 µm) was also used. The same conditions as for the gas GC/MSD analysis (carrier gas, injector temperature, and oven program) were selected. The output flow of the capillary column was split (1:1), using FID and ECD detectors simultaneously. More details of the experimental procedure can be found elsewhere [18].

The concentration of hydrogen peroxide was determined by colorimetric titration with a BOECO S-20 UV–Vis (Hamburg, Germany) spectrophotometer at 410 nm (Eisenberg, 1943), while the pH evolution was measured using a Basic 20-CRISON pH (Barcelona, Spain) electrode. Ionic compounds such as carboxylic acids, nitrate, and chloride, coming from the oxidation of the starting compounds, were measured by ion chromatography (Metrohm 761 Compact IC, Gallen, Suiza) with anionic chemical suppression and a conductivity detector. A Metrosep (Gallen, Suiza) A SUPP5 5-250 column (25 cm length, 4 mm diameter) was used as stationary phase and an aqueous solution (0.7 mL min^{-1}) of Na_2CO_3 (3.2 mM) and $NaHCO_3$ (1 mM) as the mobile phase.

The total iron content in the aqueous solution was measured by spectroscopy of atomic emission (AES MP-4100 Agilent Technology, Santa Clara, CA, USA). Aqueous solutions with different concentration of ferric sulfate (Sigma-Aldrich Darmstadt, Germany) were measured at a weight length of 259.94 nm, a nebulizer pressure of 100 kPa, and a viewing position of 10 to calibrate.

The specific surface area (S_{BET}) value of the fresh catalyst was obtained from the N_2 adsorption/desorption isotherm at 77 K using a Micromeritics Tristar (Norcross, GA, USA) apparatus with the sample previously outgassed overnight at 150 °C to a reduced pressure $<10^{-3}$ torr in order to ensure a dry clean surface.

The HA concentration in the aqueous phase was determined by derivatization of this compound with cyclohexanone. Cyclohexanone reacts with HA in the aqueous phase to produce cyclohexanone oxime, which was quantified by GC/FID. The reaction was carried out adding 1.5 g of cyclohexanone to 3 mL of the aqueous phase. After 15 min of reaction time at 40 °C, the cyclohexanone oxime formed was quantified in the organic phase, composed of the non-reacted cyclohexanone and the cyclohexanone oxime, by GC/FID. To do this, a calibration curve using different concentrations of HA in water was previously accomplished.

4. Conclusions

It has been probed that goethite is an effective and stable catalyst in the CWPO of several COCs such as chlorobenzene, di, tri and tetrachlorobenzene isomers at room conditions. These chlorinated compounds are common pollutants found in aqueous phase because they are often used in the manufacture of a high number of chemicals.

It was confirmed that the addition of hydroxylamine (as hydroxylammonium sulfate) greatly enhances the chlorobenzenes conversion with time. This can be explained because hydroxylamine reduces Fe (III) in the goethite surface to Fe (II) and the reduced iron catalyzes faster (than Fe (III)) than the decomposition of hydrogen peroxide, yielding hydroxyl radicals. On the other hand, hydroxylamine competes with the pollutants for the oxidant and its consumption slows down the iron redox cycle and therefore, the pollutant removal rate. The only byproduct detected from hydroxylamine oxidation at the condition tested was nitrate ion.

The oxidation of COCs (around 90% at the selected conditions: 2.4 mM of HA, 10 times the stoichiometric H_2O_2 amount, 2.5 h) results in their dehalogenation (confirmed by the chlorine balance) and only short chain organic acids (mainly acetic acid) were detected as oxidation byproducts, meaning that this treatment leads to the detoxification of the polluted water. The oxidation rate follows the subsequent order: CBs > DCBs >> TCBS = TetraCBs, the catalyst not being regioselective. A positive effect of hydrogen peroxide and hydroxylamine concentrations on the reaction rate, and an optimal value of catalyst concentration (0.5 g L^{-1}) was found.

Author Contributions: A.S. and A.R. achieved the funding acquisition for this study, A.S. and C.M.D. made intellectual contributions, especially with the conceptualization and the methodology used. D.L. conceived and performed the experiments and wrote the original draft. All the authors have been also involved in drafting and revising the manuscript, so that everyone has given final approval of the current version to be published in Catalysts Journal.

Funding: This research received no external funding.

Acknowledgments: This work was supported by Regional Government of Madrid, project CARESOIL (S2018/EMT-4317), and from the Spanish Ministry of Economy, Industry and Competitiveness, projects CTM2016-77151-C2-1-R.

Conflicts of Interest: The authors declare no conflict of interest.

Notation

Acronyms

1,2 DCB = 1,2-dicholorbenzene
1,2,3 TCB = 1,2,3-tricholorbenzene
1,2,4 TCB = 1,2,4-tricholorbenzene
1,3 DCB = 1,3-dicholorbenzene
1,4 DCB = 1,4-dicholorbenzene
AOPs = Advanced oxidation processes
AES = atomic emission spectroscopy
B# = number of blank experiments
CB = monochlorobenzene
Cl^- = chloride
COCs = chlorinated organic compounds
CWPO = catalytic wet peroxide oxidation
FID = flame ionization detector
GC = gas chromatography
GOE = goethite
HA = hydroxylamine
HCHs = hexachlorocyclohexanes
IC = ion chromatography
MS = mass spectrometry detector
nCBS = chlorobenzenes (n = mono, di, tri, tetra)
POPs = persistent organic compounds
S# = number of experiments using spiked water
TetraCB-a = tetrachlorobenzenes-a
TetraCB-b = tetrachlorobenzenes-b

Symbols

C_j = concentration of compound j, mM
n = number of clorines in nCBS
P = pressure, bar
S_{BET} = specific surface area, $m^2 \cdot g^{-1}$
T = temperature °C
t = time, *min* or *h*
V = adsorbed volume, $cm^3 \cdot g^{-1}$
X_j = conversion compound j
Y_j = yield compound j

Subscripts

0 = initial
j = compounds $j = \{GOE, HA, H_2O_2, NO_3^-, Cl^-, nCBs\}$

Superscripts

St = stoichiometric

References

1. Weber, R.; Watson, A.; Forter, M.; Oliaei, F. Persistent organic pollutants and landfills—A review of past experiences and future challenges. *Waste Manag. Res.* **2011**, *29*, 107–121. [CrossRef] [PubMed]

2. Schulze, S.; Zahn, D.; Montes, R.; Rodil, R.; Quintana, J.B.; Knepper, T.P.; Reemtsma, T.; Berger, U. Occurrence of emerging persistent and mobile organic contaminants in European water samples. *Water Res.* **2019**, *153*, 80–90. [CrossRef]
3. Li, Z.J.; Jennings, A. Worldwide Regulations of Standard Values of Pesticides for Human Health Risk Control: A Review. *Int. J. Environ. Res. Public Health* **2017**, *14*, 826. [CrossRef] [PubMed]
4. Kumar, R.; Mukherji, S. Threat Posed by Persistent Organochlorine Pesticides and their Mobility in the Environment. *Curr. Org. Chem.* **2018**, *22*, 954–972. [CrossRef]
5. Weber, R.; Gaus, C.; Tysklind, M.; Johnston, P.; Forter, M.; Hollert, H.; Heinisch, E.; Holoubek, I.; Lloyd-Smith, M.; Masunaga, S.; et al. Dioxin- and POP-contaminated sites-contemporary and future relevance and challenges. *Enviro. Sci. Pollut. Res.* **2008**, *15*, 363–393. [CrossRef]
6. Moeck, C.; Radny, D.; Huggenberger, P.; Affolter, A.; Auckenthaler, A.; Hollender, J.; Berg, M.; Schirmer, M. Spatial distribution of anthropogenic inputs into groundwater: A case study. *Grundwasser* **2018**, *23*, 297–309. [CrossRef]
7. Pirsaheb, M.; Hossini, H.; Asadi, F.; Janjani, H. A systematic review on organochlorine and organophosphorus pesticides content in water resources. *Toxin Rev.* **2017**, *36*, 210–221. [CrossRef]
8. Wang, M.J.; Jones, K.C. Behavior and fate of chlorobenzenes (cbs) introduced into soil-plant systems by sewage-sludge application—A review. *Chemosphere* **1994**, *28*, 1325–1360. [CrossRef]
9. van Wijk, D.; Cohet, E.; Gard, A.; Caspers, N.; van Ginkel, C.; Thompson, R.; de Rooij, C.; Garny, V.; Lecloux, A. 1,2,4-trichlorobenzene marine risk assessment with special emphasis on the Osparcom region North Sea. *Chemosphere* **2006**, *62*, 1294–1310. [CrossRef]
10. Djohan, D.; Yu, Q.; Connell, D.W. Partition isotherms of chlorobenzenes in a sediment-water system. *Water Air Soil Pollut.* **2005**, *161*, 157–173. [CrossRef]
11. Lecloux, A.J. Scientific activities of Euro Chlor in monitoring and assessing naturally and man-made organohalogens. *Chemosphere* **2003**, *52*, 521–529. [CrossRef]
12. Van Wijk, D.; Thompson, R.S.; De Rooij, C.; Garny, V.; Lecloux, A.; Kanne, R. 1,2-Dichlorobenzene marine risk assessment with special reference to the OSPARCOM region: North Sea. *Environ. Monit. Assess.* **2004**, *97*, 87–102. [CrossRef] [PubMed]
13. Boutonnet, J.C.; Thompson, R.S.; De Rooij, C.; Garny, V.; Lecloux, A.; Van Wijk, D. 1,4-Dichlorobenzene marine risk assessment with special reference to the OSPARCOM region: North Sea. *Environ. Monit. Assess.* **2004**, *97*, 103–117. [CrossRef] [PubMed]
14. Van Wijk, D.; Thompson, R.S.; De Rooij, C.; Garny, V.; Lecloux, A.; Kanne, R. Monochlorobenzene Marine Risk Assessment with Special Reference to the Osparcom Region: North Sea. *Environ. Monit. Assess.* **2004**, *97*, 69–86. [CrossRef] [PubMed]
15. Schroll, R.; Brahushi, F.; Dörfler, U.; Kühn, S.; Fekete, J.; Munch, J.C. Biomineralisation of 1,2,4-trichlorobenzene in soils by an adapted microbial population. *Environ. Pollut.* **2004**, *127*, 395–401. [CrossRef]
16. Li, J.-H.; Sun, X.-F.; Yao, Z.-T.; Zhao, X.-Y. Remediation of 1,2,3-trichlorobenzene contaminated soil using a combined thermal desorption–molten salt oxidation reactor system. *Chemosphere* **2014**, *97*, 125–129. [CrossRef]
17. Zhang, T.; Li, X.; Min, X.; Fang, T.; Zhang, Z.; Yang, L.; Liu, P. Acute toxicity of chlorobenzenes in Tetrahymena: Estimated by microcalorimetry and mechanism. *Environ. Toxicol. Pharmacol.* **2012**, *33*, 377–385. [CrossRef]
18. Santos, A.; Fernandez, J.; Guadaño, J.; Lorenzo, D.; Romero, A. Chlorinated organic compounds in liquid wastes (DNAPL) from lindane production dumped in landfills in Sabiñanigo (Spain). *Environ. Pollut.* **2018**, *242*, 1616–1624. [CrossRef]
19. Mercado, D.F.; Weiss, R.G. Polydimethylsiloxane as a Matrix for the Stabilization and Immobilization of Zero-Valent Iron Nanoparticles. Applications to Dehalogenation of Environmentally Deleterious Molecules. *J. Braz. Chem. Soc.* **2018**, *29*, 1427–1439. [CrossRef]
20. Wan, X.; Liu, Y.; Chai, X.-S.; Li, Y.; Guo, C. A quint-wavelength UV spectroscopy for simultaneous determination of dichlorobenzene, chlorobenzene, and benzene in simulated water reduced by nanoscale zero-valent Fe/Ni bimetal. *Spectrochimica Acta Part a-Molecular and Biomolecular Spectroscopy* **2017**, *181*, 55–59. [CrossRef]
21. Zhang, X.; Wu, Y.Q. Application of coupled zero-valent iron/biochar system for degradation of chlorobenzene-contaminated groundwater. *Water Sci. Technol.* **2017**, *75*, 571–580. [CrossRef]

22. Dominguez, C.M.; Romero, A.; Fernandez, J.; Santos, A. In situ chemical reduction of chlorinated organic compounds from lindane production wastes by zero valent iron microparticles. *J. Water Process Eng.* **2018**, in press. [CrossRef]
23. Lhotský, O.; Krákorová, E.; Mašín, P.; Žebrák, R.; Linhartová, L.; Křesinová, Z.; Kašlík, J.; Steinová, J.; Rødsand, T.; Filipová, A.; et al. Pharmaceuticals, benzene, toluene and chlorobenzene removal from contaminated groundwater by combined UV/H2O2 photo-oxidation and aeration. *Water Res.* **2017**, *120*, 245–255. [CrossRef] [PubMed]
24. Saharan, V.K.; Pinjari, D.V.; Gogate, P.R.; Pandit, A.B. Chapter 3—Advanced Oxidation Technologies for Wastewater Treatment: An Overview. In *Industrial Wastewater Treatment, Recycling and Reuse*; Ranade, V.V., Bhandari, V.M., Eds.; Butterworth-Heinemann: Oxford, UK, 2014; pp. 141–191.
25. Dilmeghani, M.; Zahir, K.O. Kinetics and mechanism of chlorobenzene degradation in aqueous samples using advanced oxidation processes. *J. Environ. Qual.* **2001**, *30*, 2062–2070. [CrossRef] [PubMed]
26. Santos, A.; Fernandez, J.; Rodriguez, S.; Dominguez, C.M.; Lominchar, M.A.; Lorenzo, D.; Romero, A. Abatement of chlorinated compounds in groundwater contaminated by HCH wastes using ISCO with alkali activated persulfate. *Sci. Total Environ.* **2018**, *615*, 1070–1077. [CrossRef] [PubMed]
27. Sedlak, D.L.; Andren, A.W. Oxidation of chlorobenzene with Fenton's reagent. *Environ. Sci. Technol.* **1991**, *25*, 777–782. [CrossRef]
28. Pagano, M.; Volpe, A.; Lopez, A.; Mascolo, G.; Ciannarella, R. Degradation of chlorobenzene by Fenton-like processes using zero-valent iron in the presence of Fe_{3+} and Cu_{2+}. *Environ. Technol.* **2011**, *32*, 155–165. [CrossRef] [PubMed]
29. Kuang, Y.; Wang, Q.P.; Chen, Z.L.; Megharaj, M.; Naidu, R. Heterogeneous Fenton-like oxidation of monochlorobenzene using green synthesis of iron nanoparticles. *J. Colloid Interface Sci.* **2013**, *410*, 67–73. [CrossRef] [PubMed]
30. Santos, A.; Rodríguez, S.; Pardo, F.; Romero, A. Use of Fenton reagent combined with humic acids for the removal of PFOA from contaminated water. *Sci. Total Environ.* **2016**, *563–564*, 657–663. [CrossRef]
31. Karthikeyan, S.; Boopathy, R.; Gupta, V.K.; Sekaran, G. Preparation, characterizations and its application of heterogeneous Fenton catalyst for the treatment of synthetic phenol solution. *J. Mol. Liq.* **2013**, *177*, 402–408. [CrossRef]
32. Hou, X.; Huang, X.; Jia, F.; Ai, Z.; Zhao, J.; Zhang, L. Hydroxylamine Promoted Goethite Surface Fenton Degradation of Organic Pollutants. *Environ. Sci. Technol.* **2017**, *51*, 5118–5126. [CrossRef] [PubMed]
33. Neyens, E.; Baeyens, J. A review of classic Fenton's peroxidation as an advanced oxidation technique. *J. Hazard. Mater.* **2003**, *98*, 33–50. [CrossRef]
34. Munoz, M.; Domínguez, P.; de Pedro, Z.M.; Casas, J.A.; Rodriguez, J.J. Naturally-occurring iron minerals as inexpensive catalysts for CWPO. *Appl. Catal. B Environ.* **2017**, *203*, 166–173. [CrossRef]
35. Baloyi, J.; Ntho, T.; Moma, J. Synthesis and application of pillared clay heterogeneous catalysts for wastewater treatment: A review. *RSC Adv.* **2018**, *8*, 5197–5211. [CrossRef]
36. Ren, M.; Qian, X.F.; Fang, M.Y.; Yue, D.T.; Zhao, Y.X. Ferric (hydr)oxide/mesoporous carbon composites as Fenton-like catalysts for degradation of phenol. *Res. Chem. Intermed.* **2018**, *44*, 4103–4117. [CrossRef]
37. Wang, H.; Jiang, H.; Wang, S.; Shi, W.B.; He, J.C.; Liu, H.; Huang, Y.M. Fe_3O_4-MWCNT magnetic nanocomposites as efficient peroxidase mimic catalysts in a Fenton-like reaction for water purification without pH limitation. *RSC Adv.* **2014**, *4*, 45809–45815. [CrossRef]
38. Watts, R.J.; Jones, A.P.; Chen, P.H.; Kenny, A. Mineral-catalyzed Fenton-like oxidation of sorbed chlorobenzenes. *Water Environ. Res.* **1997**, *69*, 269–275. [CrossRef]
39. Kwan, W.P.; Voelker, B.M. Rates of hydroxyl radical generation and organic compound oxidation in mineral-catalyzed Fenton-like systems. *Environ. Sci. Technol.* **2003**, *37*, 1150–1158. [CrossRef]
40. Chen, L.W.; Huang, Y.M.; Zhang, J.; Wu, B.C.; Wang, P. Enhancement on Fenton system by N-substituted hydroxylamines. *Abstr. Pap. Am. Chem. Soc.* **2016**, *252*, 1155.
41. Chen, L.W.; Ma, J.; Li, X.C.; Guan, Y.H. Effect of common cations, anions and organics on the Fenton-hydroxylamine system. *Abstr. Pap. Am. Chem. Soc.* **2012**, *243*, 1155.
42. Chen, L.W.; Ma, J.; Li, X.C.; Zhang, J.; Fang, J.Y.; Guan, Y.H.; Xie, P.C. Strong Enhancement on Fenton Oxidation by Addition of Hydroxylamine to Accelerate the Ferric and Ferrous Iron Cycles. *Environ. Sci. Technol.* **2011**, *45*, 3925–3930. [CrossRef] [PubMed]

43. Fayazi, M.; Taher, M.A.; Afzali, D.; Mostafavi, A. Enhanced Fenton-like degradation of methylene blue by magnetically activated carbon/hydrogen peroxide with hydroxylamine as Fenton enhancer. *J. Mol. Liq.* **2016**, *216*, 781–787. [CrossRef]
44. Ding, Y.B.; Huang, W.; Ding, Z.Q.; Nie, G.; Tang, H.Q. Dramatically enhanced Fenton oxidation of carbamazepine with easily recyclable microscaled $CuFeO_2$ by hydroxylamine: Kinetic and mechanism study. *Sep. Purif. Technol.* **2016**, *168*, 223–231. [CrossRef]
45. Bengtsson, G.; Fronæus, S.; Bengtsson-Kloo, L. The kinetics and mechanism of oxidation of hydroxylamine by iron(iii). *J. Chem. Soc. Dalton Trans.* **2002**, 2548–2552. [CrossRef]
46. Dominguez, C.M.; Oturan, N.; Romero, A.; Santos, A.; Oturan, M.A. Lindane degradation by electrooxidation process: effect of electrode materials on oxidation and mineralization kinetics. *Water Res.* **2018**, *35*, 220–230. [CrossRef] [PubMed]
47. Han, D.H.; Wan, J.Q.; Ma, Y.W.; Wang, Y.; Huang, M.Z.; Chen, Y.M.; Li, D.Y.; Guan, Z.Y.; Li, Y. Enhanced decolorization of Orange G in a Fe(II)-EDDS activated persulfate process by accelerating the regeneration of ferrous iron with hydroxylamine. *Chem. Eng. J.* **2014**, *256*, 316–323. [CrossRef]
48. Dominguez, C.M.; Romero, A.; Santos, A. Improved Etherification of Glycerol with Tert-Butyl Alcohol by the Addition of Dibutyl Ether as Solvent. *Catalysts* **2019**, *9*, 378. [CrossRef]
49. Fernández, J.; Arjol, M.; Cacho, C. POP-contaminated sites from HCH production in Sabiñánigo, Spain. *Environ. Sci. Pollut. Res.* **2013**, *20*, 1937–1950. [CrossRef]
50. Cornell, R.M.; Schwertmann, U. *The Iron Oxides: Structure, Properties, Reactions, Occurrences and Uses*; John Wiley & Sons: Hoboken, NJ, USA, 2003.
51. Wang, Y.; Gao, Y.; Chen, L.; Zhang, H. Goethite as an efficient heterogeneous Fenton catalyst for the degradation of methyl orange. *Catal. Today* **2015**, *252*, 107–112. [CrossRef]
52. Liu, H.B.; Chen, T.H.; Frost, R.L. An overview of the role of goethite surfaces in the environment. *Chemosphere* **2014**, *103*, 1–11. [CrossRef]

© 2019 by the authors. Licensee MDPI, Basel, Switzerland. This article is an open access article distributed under the terms and conditions of the Creative Commons Attribution (CC BY) license (http://creativecommons.org/licenses/by/4.0/).

Article

Activation of Persulfate by Biochars from Valorized Olive Stones for the Degradation of Sulfamethoxazole

Elena Magioglou [1], Zacharias Frontistis [2], John Vakros [1,3], Ioannis D. Manariotis [4] and Dionissios Mantzavinos [1,3,*]

1. Department of Chemical Engineering, University of Patras, Caratheodory 1, University Campus, GR-26504 Patras, Greece; mag.elena14@hotmail.gr (E.M.); vakros@chemistry.upatras.gr (J.V.)
2. Department of Environmental Engineering, University of Western Macedonia, GR-50100 Kozani, Greece; zaxoys@gmail.com
3. INVALOR: Research Infrastructure for Waste Valorization and Sustainable Management, Caratheodory 1, University Campus, GR-26504 Patras, Greece
4. Department of Civil Engineering, Environmental Engineering Laboratory, University of Patras, University Campus, GR-26504 Patras, Greece; idman@upatras.gr
* Correspondence: mantzavinos@chemeng.upatras.gr; Tel.: +30-26-1099-6136

Received: 9 April 2019; Accepted: 24 April 2019; Published: 3 May 2019

Abstract: Biochars from spent olive stones were tested for the degradation of sulfamethoxazole (SMX) in water matrices. Batch degradation experiments were performed using sodium persulfate (SPS) as the source of radicals in the range 250–1500 mg/L, with biochar as the SPS activator in the range 100–300 mg/L and SMX as the model micro-pollutant in the range 250–2000 μg/L. Ultrapure water (UPW), bottled water (BW), and secondary treated wastewater (WW) were employed as the water matrix. Removal of SMX by adsorption only was moderate and favored at acidic conditions, while SPS alone did not practically oxidize SMX. At these conditions, biochar was capable of activating SPS and, consequently, of degrading SMX, with the pseudo-first order rate increasing with increasing biochar and oxidant concentration and decreasing SMX concentration. Experiments in BW or UPW spiked with various anions showed little or no effect on degradation. Similar experiments in WW resulted in a rate reduction of about 30%, and this was attributed to the competitive consumption of reactive radicals by non-target water constituents. Experiments with methanol and *t*-butanol at excessive concentrations resulted in partial but generally not complete inhibition of degradation; this indicates that, besides the liquid bulk, reactions may also occur close to or on the biochar surface.

Keywords: adsorption; antibiotics; emerging micro-pollutants; waste valorization; water matrix

---

1. Introduction

Over the past decades, the occurrence of harmful xenobiotic compounds in the environment has constantly been increasing. This causes a number of serious problems related to human health, the quality of surface- and groundwaters, and generally, ecosystem protection. Thus, much work has been performed for the remediation of sites contaminated by persistent pollutants [1,2].

Amongst different classes of pollutants, pharmaceuticals and especially antibiotics have attracted the interest of the scientific community due to adverse effects associated with their existence in water bodies [3]. In particular, many scientists alarmingly state that exposure to antibiotics can lead to increased antimicrobial resistance [4]. It is, therefore, not surprising that a large number of studies deal with the removal of common antibiotics, such as sulfamethoxazole (SMX) and amoxicillin, with different physicochemical processes, such as photocatalysis [5], ozonation [6], Fenton-like reaction [7], electrochemical oxidation [8], and sonochemistry [9].

In recent years, the use of persulfates as a source of the sulfate radical, $SO_4^{\bullet-}$, has become attractive for in situ oxidation, since it is more stable, easy to store, and less costly than other oxidants such as hydrogen peroxide [10]. The conversion of persulfates to sulfate radicals requires some kind of activation agent, including transition metals, high temperatures, UV irradiation, ultrasound irradiation, and microwaves, amongst others [11–15]. Of these, heterogeneous catalysis exhibits several advantages such as easy recovery of the catalysts and possible reuse, relatively low concentrations, and higher efficiency [14,15]. In recent years, there have been several reports concerning the use of carbonaqueous materials such as graphene, graphene oxide, carbon nanotubes, and activated carbons as persulfate activators [16–19].

Biochars, the solid residue produced from biomass thermal decomposition with no or little oxygen at moderate temperatures, are low-cost materials with high surface area and desirable physicochemical properties in terms of pore size distribution, the amount of functional groups (e.g., C–O, C=O, –COOH, –OH), and minerals (e.g., N, P, S, Ca, Mg, and K) that can be employed as adsorbents, catalysts or catalytic supports [20–24]. Moreover, they can be employed for carbon storage, thus avoiding the emission of 0.1–0.3 billion tons of CO_2 annually [25]. Although any kind of lignocellulosic biomass can be used for biochar production, its origin may have a significant effect on properties, such as the moisture content and the composition in terms of organic carbon, minerals, and ash content. Furthermore, the pyrolysis temperature can influence the surface area and the point of zero charge, pzc, of the produced biochars [26,27].

The ability of biochars to activate persulfates has been demonstrated in a few recent studies [28–30]. The use of a metal-free material originating from valorized biomass is a rather attractive concept since conventional catalysts can be replaced by a green material [28,29].

In a recent study of our group [30], we demonstrated the use of biochar from spent malt rootlets for persulfate activation and the subsequent oxidation of the antibiotic SMX. Effective persulfate activation took place on the biochar surface followed by considerable degradation of SMX in various environmental matrices, such as bottled water and secondary treated wastewater.

In this work, biochar from olive stones was prepared, exhaustively characterized, and eventually tested for the adsorption and oxidative degradation of SMX by means of activated persulfate.

2. Results and Discussion

2.1. Biochar Properties

The thermogravimetric analysis (TGA) curve of the prepared biochar is shown in Figure 1. As can be seen, there is a significant mass loss of about 10% starting at about 100 °C. This is due to water and moisture content that has been adsorbed in the biochar. Between 450 and 600 °C, the biochar is almost completely burned, while only a small amount (ca. 7%) corresponding to minerals remains. The carbonaqueous phase is almost homogeneous, as can be seen from the differential curve of the TGA results.

Figure 2 presents the titration curves for a blank solution and the biochar suspension. The point of zero charge coincides with the section point between these two curves, which is at around 3. The rather expected acidic character of the biochar has to do with the increased oxygen content of the starting biomass and the limited O_2 atmosphere in the pyrolysis step. Oxygen helps the formation of surface groups with acidic behavior, while the presence of minerals at low concentrations (see Figure 1) is also consistent with the reduced sample basicity. Interestingly, the biochar surface does not exchange significant amounts of H^+ at pH > 3, as can be seen from the titration curve in the 10 > pH > 4 region, as well as in the inset of Figure 2. The inset of Figure 2 represents the amount of H^+ consumed by the surface groups. As a result, the total negative charge of the surface is not expected to be high at pH > 3.

The specific surface area (SSA) of the sample was measured equal to 50 m^2/g using the BET equation. The microporous surface area, calculated with the t-plot method, was found equal to 44 m^2/g. The pore size distribution is presented in Figure 3. The microporous has a diameter of 2 nm, while there are some mesoporous-macroporous with an average diameter greater than 160 nm. Almost no mesoporous was observed in the range of 5–100 nm.

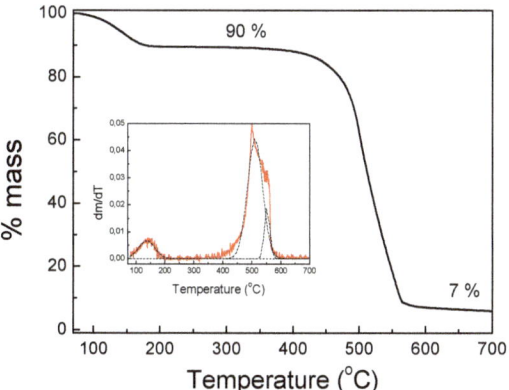

Figure 1. TGA curve obtained at a heating rate of 10 °C/min under 20 mL/min air flow. Inset: Differential TGA graph.

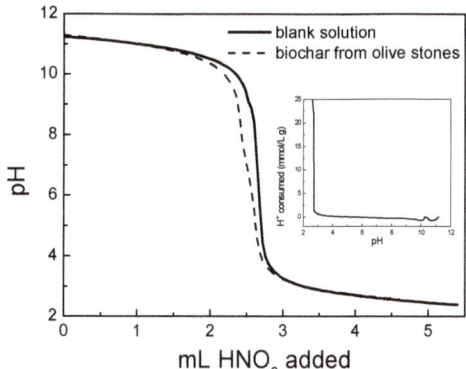

Figure 2. PMT curves for the blank solution (solid line) and the biochar suspension (dotted line). Inset: Consumption of H^+ ions by the surface as a function of suspension pH.

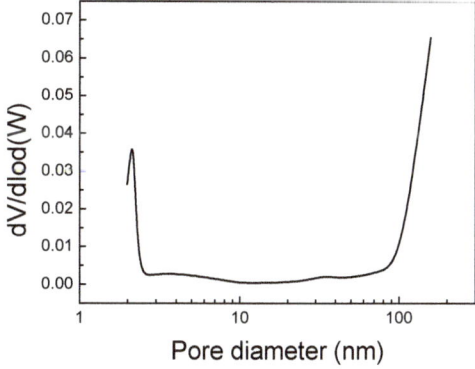

Figure 3. Pore size distribution of the biochar.

Representative SEM images in Figure 4a,b shows that the surface is covered with deposits. Due to the moderate SSA and the absence of mesoporosity, the content of these deposits is not expected to be significant since it is limited to the external surface area in accordance with the TGA results. EDS analysis (Figure 4c) shows that C and O, respectively, comprise almost 90% and 6% in atomic ratio, while the two elements account for almost 90% of the total mass. Other elements detected are K, Ca, Na, Mg, Si, and Cl, with Cl (1.53% atomic ratio), Na (0.9% atomic ratio), and K (0.6% atomic ratio) existing in higher amounts.

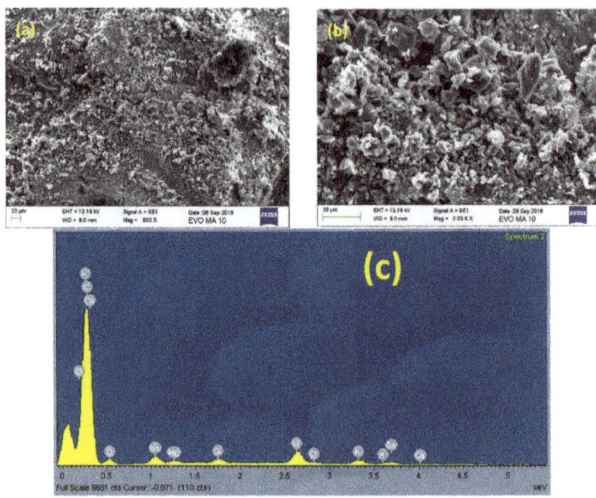

Figure 4. SEM images (**a**) and (**b**) and EDS analysis (**c**) of the biochar.

The FTIR spectrum of the biochar (Figure 5) exhibits a wide band at 3436 cm^{-1} and a peak at 1045 cm^{-1} mainly due to H$_2$O content, in accordance with TGA results and surface C–OH groups. The aromatic character is not significant as evidenced by the absence of peaks above 3000 cm^{-1} (C–H in aromatic compounds). The low intensity of the peaks in the region 1000–1800 cm^{-1} is characteristic of the heterogeneity of the biochar. The peak at 1580 cm^{-1} can be attributed to C=C with conjugation of π electrons [31] by functional groups with high electronegativity. Finally, the peak at 1742 cm^{-1} can be assigned to C=O groups, while the peaks at 2922 and 2846 cm^{-1} are due to C–H bonds [32,33].

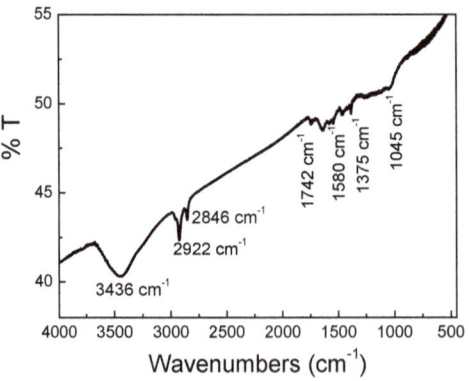

Figure 5. The FTIR spectrum of the biochar.

The XRD pattern of the biochar (Figure 6) clearly shows the amorphous carbon phase with a wide peak at 25°. This peak is typical of a carbonaceous material with a less ordered structure due to pyrolysis. The intense sharp peaks are identified as halite, and this is supported by the EDS analysis shown in Figure 4c.

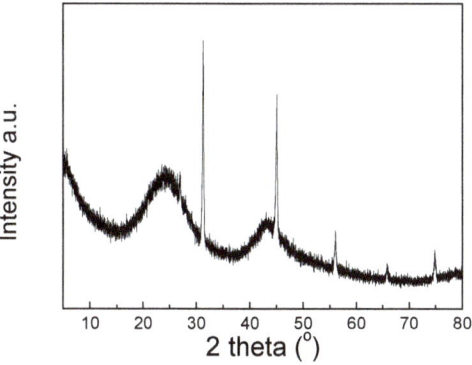

Figure 6. The XRD pattern of the biochar.

2.2. Adsorption Capacity

The adsorption isotherm of SMX at pH = 6.5 and ambient temperature is shown in Figure 7.

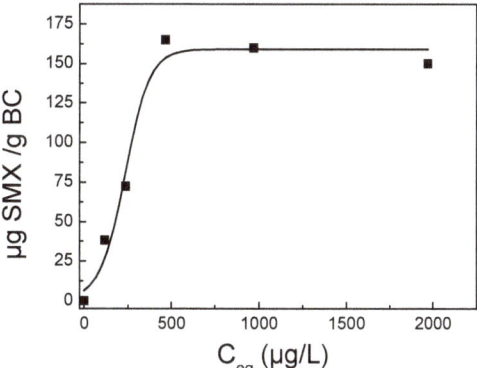

Figure 7. Adsorption isotherm of sulfamethoxazole (SMX) at pH = 6.5, ambient temperature, and 200 mg/L biochar.

As can be seen, SMX concentrations up to 500 µg/L can saturate the surface with the adsorption maximum being about 150 µg/g of the biochar (BC). The amount of adsorbed SMX on biochar in the range 100–300 mg/L is almost constant and equal to 8%, leaving more than 90% of SMX in the solution (data not shown). The starting solution pH was 5.8 and slightly increased after sorption to 6.5–7. This pH shift is typical of an anion sorption behavior, implying that the adsorption of SMX is closely related to the negative charge of SMX. The pK_a value of the sulfonamide group of SMX is about 5.7 and the isoelectric point 4.5. Generally, at pH values <1.4, SMX exists as a cation with the terminal –NH_2 group being protonated, while at a pH >5.8, SMX is negatively charged with deprotonation of –NH in the sulfonoamido group. On the other hand, SMX is characterized by a low positive charge and can be considered as neutral at 1.4 < pH < 5.8 [34,35].

Briefly, when an anion concentration increases in the interfacial region, the surface releases OH^- into the solution to partially neutralize the negative charge of the surface. The increment of pH results in a more negatively charged surface as well as more negatively charged SMX species and thus a lower deposition of SMX onto the biochar surface. Even if SMX interacts with the C=C bonds, as has been reported in the literature [36], the negative charge increases and should be partially neutralized.

The influence of the solution pH on the adsorption is significant. As can be seen in Figure 8, the deposition of SMX is rather low at high pH (i.e., final pH > 7), while it is significant at low pH in accordance with the literature [37,38]. Since the pzc of the biochar is 3 and the isoelectric point of SMX 4.5, both the surface and SMX are negatively charged at high pH values, so adsorption is not favored. Two experiments were performed at an initial pH value of 5.7: one was buffered at this value, while the other was left uncontrolled reaching a final value of 7.2. In the latter case, SMX adsorption is evidently slower than in the former, thus pointing out the significance of pH. Fast adsorption occurs at pH = 3, where the biochar has almost zero charge and has the ability to consume a significant amount of H^+ ions without altering the solution pH, while SMX is slightly positively charged and therefore can approach the surface and become adsorbed. The ability of biochar to consume H^+ in this pH region helps the sorption process, which can be either electrostatic, due to the different charged surface sites and SMX molecules, or due to the reaction of SMX with the surface –OH groups.

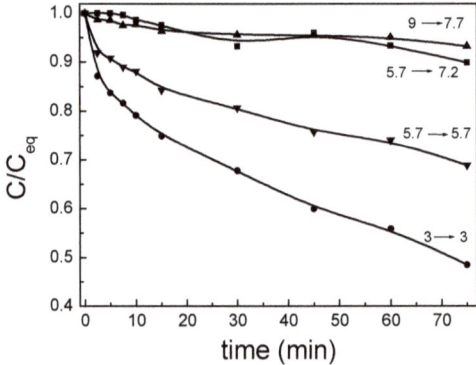

Figure 8. Kinetics of 500 µg/L SMX adsorption as a function of solution pH with 200 mg/L biochar. For each experiment, the starting and final pH values are indicated.

On the other hand, the biochar surface exhibits small differences at 4 < pH < 10, showing a small degree of H^+ consumption. At higher than pzc, the surface is charged negatively releasing H^+ in the solution and thus lowering the solution pH and making SMX less negative, both of which help the deposition of SMX onto the biochar surface. At pH = 5.7, the total negative surface charge of biochar is less than it is at pH = 9, and the SMX is also less negative. Therefore, it is easier for SMX to approach the surface and become sorbed. In the 4 < pH < 10 region, deposition is not electrostatic but probably involves surface reactions between surface groups and SMX or hydrogen bonding due to high –O content of the biochar. As pH increases, SMX deposition becomes more difficult, and equilibrium is reached more slowly. This explains why the adsorption/deposition of SMX onto the biochar surface is almost the same at pH 7–7.7 regardless of the starting pH. In a buffer system, there are no changes in pH, while the speciation of SMX and the surface of biochar are not altered during deposition. The pH value remains low and the deposition is higher, as can be seen in Figure 8.

2.3. Oxidative Degradation of SMX

2.3.1. Effect of Biochar, Sodium Persulfate, and SMX Concentration

Figure 9 shows the effect of biochar concentration on SMX degradation in the presence of sodium persulfate (SPS). The concentration profiles are normalized against the equilibrium concentration of SMX, C_{eq}, after adsorption for 15 min.

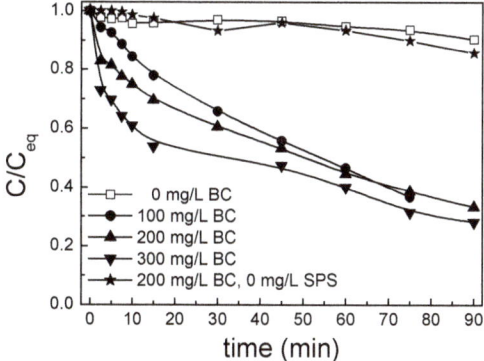

Figure 9. Effect of biochar concentration on 500 µg/L SMX degradation with 1000 mg/L sodium persulfate (SPS) in ultrapure water (UPW) and inherent pH.

Persulfate alone, although a mild oxidant itself, is not capable of degrading SMX to an appreciable extent, leading to ca. 10% removal after 75 min of reaction. Similarly, the extent of adsorption onto 300 mg/L biochar without oxidant is not considerable, and this is consistent with the adsorption data shown in Figure 8 for experiments at neutral conditions (i.e., the pH changed from 5.7 to 7.3 during the adsorption experiment shown in Figure 9). The simultaneous use of biochar and oxidant is evidently beneficial for SMX degradation, leading to 65–70% conversion after 75 min at either of the three biochar concentrations tested; this said, it appears that the effect of concentration is more pronounced during the early stages of the reaction (i.e., 15 min) as conversion takes values of 22%, 30%, and 47% at 100, 200, and 300 mg/L of biochar, respectively.

Figure 10 shows the effect of SPS concentration in the range of 0–1500 mg/L on SMX degradation with 200 mg/L biochar.

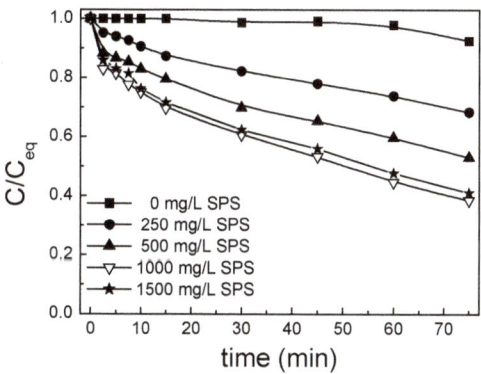

Figure 10. Effect of SPS concentration on 500 µg/L SMX degradation with 200 mg/L biochar in UPW and inherent pH.

Adsorption alone is insignificant, but the extent of degradation increases with increasing oxidant concentration up to 1000 mg/L and remains practically constant thereafter. On the assumption that SMX degradation follows a pseudo-first order kinetic expression, the logarithm of the normalized concentration profiles can be plotted against time to compute the respective apparent rate constants; these are equal to 5.4×10^{-3}, 9.4×10^{-3}, 13.3×10^{-3}, and 12.2×10^{-3} min^{-1} for the runs at 250, 500, 1000, and 1500 mg/L SPS, respectively. Although increased oxidant concentrations will expectedly generate more radicals, i.e., $SO_4^{\bullet -}$ and $OH \bullet$, this may be counterbalanced by a stronger competitive adsorption between SMX and SPS for the biochar's active sites. Furthermore, radicals in excess may suffer partial scavenging and be converted to less reactive species, such as $S_2O_8\bullet^-$ and O_2 [39].

The effect of initial SMX concentration in the range 250–2000 µg/L on its degradation is depicted in Figure 11.

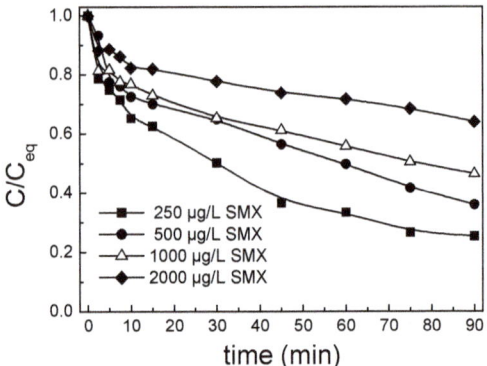

Figure 11. Effect of initial SMX concentration on its degradation with 200 mg/L biochar and 1000 mg/L SPS in UPW and inherent pH.

Degradation is clearly retarded at higher substrate concentrations and the apparent rate constants are computed equal to 17.9×10^{-3}, 13.3×10^{-3}, 9.7×10^{-3}, and 5.6×10^{-3} min^{-1} for the runs at 250, 500, 1000, and 2000 µg/L SMX, respectively. These results clearly show that the reaction is not truly first order (although data fitting is still satisfactory) since the rate constant depends on the initial concentration. The fact that the constant decreases with increasing SMX concentration indicates that the reaction shifts towards orders lower than first, reaching eventually the zeroth order. The rationale behind this is associated with the concentration of reactive species relative to the substrate. At a fixed set of operating conditions (i.e., the concentration of biochar and oxidant, pH, water matrix), the concentration of the generated radicals is expected to be nearly constant (this is particularly true at the early stages of the reaction), so the critical factor determining kinetics would be the substrate concentration. As the latter increases, lower rates will occur [30].

2.3.2. The Water Matrix Effect

All the experiments described so far were performed in UPW, thus ignoring the possible interactions of the inorganic and organic species typically found in real water matrices. In this respect, additional experiments were performed spiking bottled water (BW) and secondary treated wastewater (WW) with 500 µg/L SMX to study its degradation in the presence of 200 mg/L biochar and 1000 mg/L SPS. Interestingly, SMX degradation in BW was as fast as it was in UPW, with the respective rate constants being identical ($13.4 \pm 0.1 \times 10^{-3}$ min^{-1}), but it decreased by about 30% in WW with a rate constant of 9.3×10^{-3} min^{-1} (constants were computed from the respective concentration–time profiles, which are not shown for brevity. The same also happens for the experiments with various anions and HA spiked in UPW).

Since bicarbonate is the dominant, in terms of concentration, anion in waters, experiments were performed in UPW adding bicarbonate in the range 50–250 mg/L; the rate constant decreased from 13.3×10^{-3} in UPW to 11.8×10^{-3}, 8.9×10^{-3}, and 8.4×10^{-3} min^{-1} at 50, 100, and 250 mg/L bicarbonate, respectively. Similarly, experiments were performed adding chloride or nitrate in UPW; the addition of chloride at 50–250 mg/L slightly retarded the SMX degradation with a rate constant equal to $11.5 \pm 0.3 \times 10^{-3}$ min^{-1}, while the addition of nitrate at 50–250 mg/L slightly enhanced degradation with a rate constant equal to $15.1 \pm 1 \times 10^{-3}$ min^{-1}. These results are in agreement with those in BW that contains, amongst others, bicarbonate, chloride, and nitrate anions; although the radicals formed through SPS activation may partly be scavenged by either of these anions, the formation of secondary oxidizing species (e.g., carbonate or chloride radicals) cannot be disregarded, thus compensating for the undesired radical consumption.

An additional experiment was performed adding 10 mg/L humic acid in UPW to simulate the organic content of the WW sample. The addition of HA had a mild negative effect on SMX degradation with a rate constant equal to 10.9×10^{-3} min^{-1}, in line with the results in WW. Non-target organic species in water may either react with the relatively non-selective radicals or/and compete with the substrate for the surface active sites.

Figure 12a shows temporal SMX profiles normalized against initial (solid lines) and equilibrium (dotted lines) SMX concentrations for UPW and WW. In the case of WW, the two lines practically overlap, implying that SMX adsorption is very low. On the other hand, adsorption partially contributes to SMX removal in UPW, and this contribution is more pronounced during the early stages of the reaction. This implies that non-target matrix components in WW compete with SMX for the active sites. Figure 12b shows the temporal profile of the logarithm of the SMX concentration ratio in UPW and WW. The dependence of the above parameter (the logarithm arises from the fact that the reaction is modeled as pseudo-first order) on time is linear, indicating that the mechanism of degradation is common for the two water matrices. Therefore, the lower degree of degradation in WW can be attributed to the interactions of SPS with non-target matrix components that have competitively been adsorbed on the biochar surface and the subsequent consumption of the generated radicals.

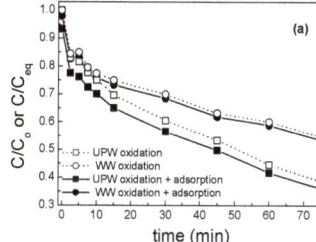
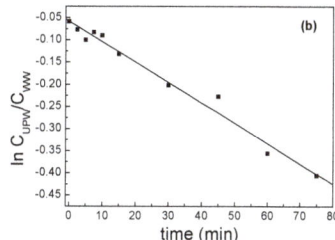

Figure 12. (a) Comparison between the actual oxidation and adsorption process and oxidation alone for SMX removal in UPW and wastewater (WW) with 200 mg/L biochar and 1000 mg/L SPS. (b) The linear dependence of the logarithm of SMX concentration ratio in UPW and WW.

2.3.3. The Role of Radical Scavengers

In an attempt to evaluate indirectly the role of sulfate and hydroxyl radicals on SMX degradation, experiments were conducted in UPW, adding methanol or t-butanol in excess (i.e., 10 and 100 g/L). Both alcohols act as radical scavengers; methanol reacts with both sulfate and hydroxyl radicals at comparable rates, while t-butanol reacts preferentially with hydroxyl radicals [40,41]. As seen in Figure 13a, SMX degradation is only partly impeded in the presence of methanol, with the rate constant decreasing by about 25% and 40% at 10 and 100 g/L alcohol, respectively. The detrimental effect of t-butanol (Figure 13b) is more pronounced with the rate constant decreasing by about 45% at 10 g/L, while the reaction is completely hindered at 100 g/L. These findings may imply that (i) hydroxyl radicals dominate over sulfate radicals, and/or (ii) reactions occur not only in the liquid bulk but also in the vicinity of the biochar surface, where adsorption becomes important; this is consistent with the fact that SMX adsorption is not affected in the presence of methanol, but it is completely hindered in the presence of t-butanol.

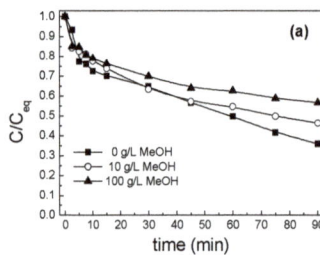 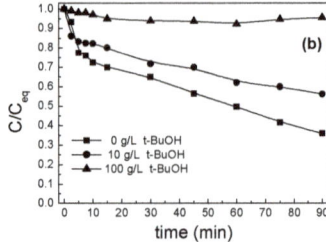

Figure 13. Effect of methanol (**a**) and *t*-butanol (**b**) on 500 µg/L SMX degradation with 200 mg/L biochar and 1000 mg/L SPS in UPW and inherent pH.

3. Materials and Methods

3.1. Preparation of Biochar

The biochar used in this study was prepared from olive stones, the main by-product of the olive-oil- and table-olive-producing companies. They were dried overnight at 50 °C and then heated up to 850 °C in a limited amount of air. The yield to biochar was 22% due to the high calcination temperature [27].

3.2. Chemicals

Sulfamethoxazole (SMX, $C_{10}H_{11}N_3O_3S$, analytical standard, CAS number: 723-46-6) and sodium persulfate (SPS, $Na_2S_2O_8$, 99+%, CAS number: 7775-27-1) were purchased from Sigma-Aldrich, St. Louis, MO, USA. Acetonitrile (99.9 wt %), humic acid (HA, technical grade), sodium hydroxide (98 wt %), and sulfuric acid (95 wt %) were also obtained from Sigma Aldrich. Methanol (99.9%) and *t*-butanol (99%) were purchased from Fluka, Loughborough, England.

3.3. Experimental Procedures

Experiments were conducted in a cylindrical glass reaction vessel of a 250 mL capacity that was open to the atmosphere (open air equilibrium). A stock SMX solution was prepared in ultrapure water (UPW), and a measured volume was mixed with the water matrix to achieve the desired SMX concentration (from 250 to 2000 µg/L). The range of SMX concentrations tested was chosen so that at least 90% of SMX conversion could be measured. The solution was then supplemented with SPS to achieve the desired concentration of up to 1500 mg/L biochar to start the reaction under magnetic stirring (400 rpm) and at ambient temperature. Samples of 1.2 mL were periodically withdrawn from the vessel (every 2.5 min in the first 10 min of reaction and every 15 min from 15 to 120 min of reaction), filtered and analyzed by liquid chromatography. In most experiments, UPW (pH = 6; 0.056 µS/cm conductivity) was the water matrix, while some experiments were also performed in bottled water (BW, pH = 7.5; 396 µS/cm conductivity; 211 mg/L bicarbonates; 15 mg/L sulfates; 9.8 mg/L chlorides) and secondary treated wastewater (WW, pH = 8; 7 mg/L total organic carbon; 4.5 mg/L total suspended solids; 21 mg/L chemical oxygen demand; 311 µS/cm conductivity; 30 mg/L sulfates; 0.44 mg/L chlorides). Unless otherwise stated, experiments were performed at the matrix's inherent pH; in some experiments, the initial pH was adjusted or buffered. Most of the experiments were performed in duplicate, and mean values (<5% difference) are quoted as results.

The general kinetic model for a *n*-th order reaction rate is as follows:

$$\text{Rate} = -\frac{dC}{dt} = k_{app}C^n \quad (1)$$

where C is the concentration of SMX, k_{app} the apparent rate constant, and n is the reaction order. The value of k_{app} can be calculated from the slope of the integrated form of Equation (1). For the case of zeroth order reaction ($n = 0$), the integrated form of Equation (1) becomes

$$C = C_o - k_{app}\, t \qquad (2)$$

where C_o is the initial concentration of SMX. For the case of a first order reaction ($n = 1$), the integrated form of Equation (1) becomes

$$lnC = lnC_o - k_{app}\, t. \qquad (3)$$

3.4. High Performance Liquid Chromatography (HPLC)

SMX was analyzed by chromatography using an Alliance HPLC system equipped with a photodiode array detector (Waters 2996 Milford, PA, USA), a gradient pump (Waters 2695, Milford, PA, USA) for solvent delivery (0.35 mL/min), and a Kinetex C18 100A column (150 × 3 mm; 2.6 μm particle size) maintained at 45 °C. The isocratic mobile phase consisted of UPW (68%) and acetonitrile (32%). The SMX absorbance peaked at 270 nm, as determined from the corresponding UV-vis absorbance spectrum. The limits of SMX detection and quantitation were 6.9 and 20.7 μg/L, respectively [42].

3.5. Physicochemical Characterization

The biochar was characterized by several techniques as follows: (i) Nitrogen adsorption isotherms were used at liquid N_2 temperature (Tristar 3000 porosimeter) for the determination of specific surface area (SSA), micropore surface area, and pore size distribution. (ii) X-ray diffraction (XRD) patterns were recorded in a Bruker D8 (Billerica, MA, USA) Advance diffractometer equipped with a nickel-filtered CuKa (1.5418 Å) radiation source. (iii) Surface topography images were obtained using scanning electron microscopy (SEM), EVO MA10 (ZEISS, Oberkochen, Germany), with high vacuum mode in which the chamber vacuum was 1×10^{-5} mbar. The samples before SEM measurements were sputtered with an 18 nm Au thin film in order to avoid electron beam charging phenomena observed in non-conductive samples. For the highest resolution images, the lowest possible beam current used, which was 50 pA. (iv) Fourier transform infrared (FTIR) spectroscopy was performed using a Perkin Elmer Spectrum RX FTIR system. The measurement range was 4000–400 cm^{-1}. (v) Thermogravimetric analysis (TGA) was performed in a TGA Perkin Elmer system (Waltham, MA, USA) under air atmosphere, with a heating rate of 10 °C/min in the range 50–900 °C. (vi) To study the acid–base behavior of the biochar, potentiometric mass titration (PMT) was applied [43]. According to PMT, the pzc value of the biochar is the common intersection point of the titration curves of suspensions with different amounts of biochar and the titration curve of a blank solution. The latter contains exactly the same amounts of inert electrolyte and base solution without biochar. Applying the mass balance equation for the H$^+$ ions for each titration curve [44], the H$^+$ consumption on the biochar surface was determined. More details about the techniques used can be found elsewhere [30].

4. Conclusions

In this work, biochars from spent olive stones were synthesized, characterized, and tested for the activation of persulfate to reactive radicals and the subsequent degradation of a model antibiotic micro-pollutant in water matrices. The main conclusions of this study are as follows:

(1) Biochars are characterized by low pzc and mineral content and a moderate specific surface area; the latter is associated with a moderate capacity for SMX adsorption.
(2) Adsorption is strongly affected by solution pH and is enhanced at acidic environments.
(3) Biochars are capable of activating persulfate, thus inducing SMX oxidative degradation; the single use of either biochar (without oxidant) or oxidants (without biochar) does not practically contribute to SMX removal.

(4) Degradation rates depend on factors, such as biochar, oxidant, and substrate concentration. An increase in the concentration of the latter retards degradation, while the opposite occurs for the other two factors.
(5) Environmental matrices such as bottled water and wastewater have no or a moderately detrimental effect on degradation; this is encouraging since the process could be applied in real life applications.
(6) It appears that reactions occur both in the liquid bulk and in the vicinity of the biochar surface; this has indirectly been evidenced by the partial inhibition of SMX degradation in the presence of excessive amounts of alcohols in the liquid bulk (i.e., at concentrations 20,000–200,000 times greater than SMX).
(7) Biochars from different biomass sources should be tested for persulfate activation and subsequent organic pollutant degradation to gain a more thorough understanding of the oxidation mechanism and tailor their properties according to specific environmental applications.

Author Contributions: Investigation, E.M., Z.F., J.V., I.D.M. and D.M.; Methodology, E.M., Z.F., J.V., I.D.M. and D.M. All authors have contributed equally.

Funding: J.V. and D.M. acknowledge support of this work by the project "INVALOR: Research Infrastructure for Waste Valorization and Sustainable Management" (MIS 5002495), which is implemented under the Action "Reinforcement of the Research and Innovation Infrastructure", funded by the Operational Programme "Competitiveness, Entrepreneurship and Innovation" (NSRF 2014–2020) and co-financed by Greece and the European Union (European Regional Development Fund).

Acknowledgments: Authors are thankful to Vagelis Karoutsos, University of Patras, Department of Materials Science for SEM measurements.

Conflicts of Interest: The authors declare no conflict of interest.

References

1. Liu, X.; Zhou, Y.; Zhang, J.; Luo, L.; Yang, Y.; Huang, H.; Peng, H.; Tang, L.; Mu, Y. Insight into electro-Fenton and photo-Fenton for the degradation of antibiotics: Mechanism study and research gaps. *Chem. Eng. J.* **2018**, *347*, 379–397. [CrossRef]
2. Ribeiro, A.B.; Dias-Ferreira, C.; Gomes, H.I. Overview of in situ and ex situ remediation technologies for PCB-contaminated soils and sediments and obstacles for full-scale application. *Sci. Total Environ.* **2013**, *445–446*, 237–260. [CrossRef]
3. Michael, I.; Frontistis, Z.; Fatta-Kassinos, D. Removal of Pharmaceuticals from Environmentally Relevant Matrices by Advanced Oxidation Processes (AOPs). *Compr. Anal. Chem.* **2013**, *62*, 345–407. [CrossRef]
4. Barancheshme, F.; Munir, M. Strategies to Combat Antibiotic Resistance in the Wastewater Treatment Plants. *Front. Microbiol.* **2018**, *8*, 2603. [CrossRef] [PubMed]
5. Ioannidou, E.; Frontistis, Z.; Antonopoulou, M.; Venieri, D.; Konstantinou, I.; Kondarides, D.I. Solar photocatalytic degradation of sulfamethoxazole over tungsten—Modified TiO_2. *Chem. Eng. J.* **2017**, *318*, 143–152. [CrossRef]
6. Dantas, R.F.; Contreras, S.; Sans, C.; Esplugas, S. Sulfamethoxazole abatement by means of ozonation. *J. Hazard. Mater.* **2008**, *150*, 790–794. [CrossRef] [PubMed]
7. Ribeiro, R.S.; Frontistis, Z.; Mantzavinos, D.; Venieri, D.; Antonopoulou, M.; Konstantinou, I.; Silva, A.M.T.; Faria, J.L.; Gomes, H.T. Magnetic carbon xerogels for the catalytic wet peroxide oxidation of sulfamethoxazole in environmentally relevant water matrices. *Appl. Catal. B Environ.* **2016**, *199*, 170–186. [CrossRef]
8. Hussain, S.; Gul, S.; Steter, J.R.; Miwa, D.W.; Motheo, A.J. Route of electrochemical oxidation of the antibiotic sulfamethoxazole on a mixed oxide anode. *Environ. Sci. Pollut. Res.* **2015**, *22*, 15004–15015. [CrossRef]
9. Guo, W.-Q.; Yin, R.-L.; Zhou, X.-J.; Du, J.-S.; Cao, H.-O.; Yang, S.-S.; Ren, N.-Q. Sulfamethoxazole degradation by ultrasound/ozone oxidation process in water: Kinetics, mechanisms, and pathways. *Ultrason. Sonochem.* **2015**, *22*, 182–187. [CrossRef]
10. Lutze, H.V.; Bircher, S.; Rapp, I.; Kerlin, N.; Bakkour, R.; Geisler, M.; von Sonntag, C.; Schmidt, T.C. Degradation of Chlorotriazine Pesticides by Sulfate Radicals and the Influence of Organic Matter. *Environ. Sci. Technol.* **2015**, *49*, 1673–1680. [CrossRef]

11. Ioannidi, A.; Frontistis, Z.; Mantzavinos, D. Destruction of propyl paraben by persulfate activated with UV-A light emitting diodes. *J. Environ. Chem. Eng.* **2018**, *6*, 2992–2997. [CrossRef]
12. Wang, J.; Wang, S. Activation of persulfate (PS) and peroxymonosulfate (PMS) and application for the degradation of emerging contaminants. *Chem. Eng. J.* **2018**, *334*, 1502–1517. [CrossRef]
13. Chenju, L.; Bruell, C.J. Thermally Activated Persulfate Oxidation of Trichloroethylene: Experimental Investigation of Reaction Orders. *Ind. Eng. Chem. Res.* **2008**, *47*, 2912–2918. [CrossRef]
14. Anipsitakis, G.P.; Dionysiou, D.D. Radical Generation by the Interaction of Transition Metals with Common Oxidants. *Environ. Sci. Technol.* **2004**, *38*, 3705–3712. [CrossRef]
15. Graça, C.A.L.; de Velosa, A.C.; Teixeira, A.C.S.C. Amicarbazone degradation by UVA-activated persulfate in the presence of hydrogen peroxide or Fe^{2+}. *Catal. Today* **2017**, *280*, 80–85. [CrossRef]
16. Shao, P.; Tian, J.; Yang, F.; Duan, X.; Gao, S.; Shi, W.; Luo, X.; Cui, F.; Luo, S.; Wang, S. Identification and Regulation of Active Sites on Nanodiamonds: Establishing a Highly Efficient Catalytic System for Oxidation of Organic Contaminants. *Adv. Funct. Mater.* **2018**, *28*, 1705295. [CrossRef]
17. Cheng, X.; Guo, H.; Zhang, Y.; Wu, X.; Liu, Y. Non-photochemical production of singlet oxygen via activation of persulfate by carbon nanotubes. *Water Res.* **2017**, *113*, 80–88. [CrossRef]
18. Tang, L.; Liu, Y.; Wang, J.; Zeng, G.; Deng, Y.; Dong, H.; Feng, H.; Wang, J.; Peng, B. Enhanced activation process of persulfate by mesoporous carbon for degradation of aqueous organic pollutants: Electron transfer mechanism. *Appl. Catal. B Environ.* **2018**, *231*, 1–10. [CrossRef]
19. Dimitriadou, S.; Frontistis, Z.; Petala, A.; Bampos, G.; Mantzavinos, D. Carbocatalytic activation of persulfate for the removal of drug diclofenac from aqueous matrices. *Catal. Today* **2019**. [CrossRef]
20. Abbas, T.; Rizwan, M.; Ali, S.; Zia-Ur-Rehman, M.; Farooq Qayyum, M.; Abbas, F.; Hannan, F.; Rinklebe, J.; Ok, Y.S. Effect of biochar on cadmium bioavailability and uptake in wheat (*Triticum aestivum* L.) grown in a soil with aged contamination. *Ecotoxicol. Environ. Saf.* **2017**, *140*, 37–47. [CrossRef]
21. Ahmad, M.; Rajapaksha, A.U.; Lim, J.E.; Zhang, M.; Bolan, N.; Mohan, D.; Vithanage, M.; Lee, S.S.; Ok, Y.S. Biochar as a sorbent for contaminant management in soil and water: A review. *Chemosphere* **2014**, *99*, 19–33. [CrossRef]
22. Ali, S.; Rizwan, M.; Noureen, S.; Anwar, S.; Ali, B.; Naveed, M.; Abd_Allah, E.F.; Alqarawi, A.A.; Ahmad, P. Combined use of biochar and zinc oxide nanoparticle foliar spray improved the plant growth and decreased the cadmium accumulation in rice (*Oryza sativa* L.) plant. *Environ. Sci. Pollut. Res.* **2019**, *26*, 11288–11299. [CrossRef]
23. Liu, W.-J.; Jiang, H.; Yu, H.-Q. Development of Biochar-Based Functional Materials: Toward a Sustainable Platform Carbon Material. *Chem. Rev.* **2015**, *115*, 12251–12285. [CrossRef]
24. Vakros, J. Biochars and Their Use as Transesterification Catalysts for Biodiesel Production: A Short Review. *Catalysts* **2018**, *8*, 562. [CrossRef]
25. Williams, S.; Higashi, C.; Phothisantikul, P.; Van Wesenbeeck, S.; Antal, M.J. The fundamentals of biocarbon formation at elevated pressure: From 1851 to the 21st century. *J. Anal. Appl. Pyrolysis* **2015**, *113*, 225–230. [CrossRef]
26. Zhao, L.; Cao, X.; Mašek, O.; Zimmerman, A. Heterogeneity of biochar properties as a function of feedstock sources and production temperatures. *J. Hazard. Mater.* **2013**, *256–257*, 1–9. [CrossRef]
27. Manariotis, I.D.; Fotopoulou, K.N.; Karapanagioti, H.K. Preparation and Characterization of Biochar Sorbents Produced from Malt Spent Rootlets. *Ind. Eng. Chem. Res.* **2015**, *54*, 9577–9584. [CrossRef]
28. Duan, X.; Sun, H.; Wang, S. Metal-Free Carbocatalysis in Advanced Oxidation Reactions. *Acc. Chem. Res.* **2018**, *51*, 678–687. [CrossRef]
29. Huang, B.-C.; Jiang, J.; Huang, G.-X.; Yu, H.-Q. Sludge biochar-based catalysts for improved pollutant degradation by activating peroxymonosulfate. *J. Mater. Chem. A* **2018**, *6*, 8978–8985. [CrossRef]
30. Kemmou, L.; Frontistis, Z.; Vakros, J.; Manariotis, I.D.; Mantzavinos, D. Degradation of antibiotic sulfamethoxazole by biochar-activated persulfate: Factors affecting the activation and degradation processes. *Catal. Today* **2018**, *313*, 128–133. [CrossRef]
31. Kaczmarczyk, B. FTIR study of conjugation in selected aromatic polyazomethines. *J. Mol. Struct.* **2013**, *1048*, 179–184. [CrossRef]
32. Zeng, M.; Shah, S.A.; Huang, D.; Parviz, D.; Yu, Y.H.; Xuezhen Wang, X.; Micah, J.; Green, M.J.; Cheng, Z. Aqueous Exfoliation of Graphite into Graphene Assisted by Sulfonyl Graphene Quantum Dots for Photonic Crystal Applications. *ACS Appl. Mater. Interfaces* **2017**, *9*, 30797–30804. [CrossRef]

33. Luo, J.; Zeng, M.; Peng, B.; Tang, Y.; Zhang, L.; Wang, P.; He, L.; Huang, D.; Wang, L.; Wang, X.; et al. Electrostatic-driven Dynamic Jamming of 2D Nanoparticles at Interfaces for Controlled Molecular Diffusion. *Angew. Chem. Int. Ed.* **2018**, *57*, 11752–11757. [CrossRef]
34. Avisar, D.; Primor, O.; Gozlan, I.; Mamane, H. Sorption of Sulfonamides and Tetracyclines to Montmorillonite Clay. *Water Air Soil Pollut.* **2010**, *209*, 439–450. [CrossRef]
35. Schott, H.; Astigarrabia, E. Isoelectric Points of Some Sulfonamides: Determination by Microelectrophoresis and by Calculations Involving Acid–Base Strength. *J. Pharm. Sci.* **1988**, *77*, 918–920. [CrossRef]
36. Heo, J.; Yoon, Y.; Lee, G.; Kim, Y.; Han, J.; Park, C.M. Enhanced adsorption of bisphenol A and sulfamethoxazole by a novel magnetic $CuZnFe_2O_4$–biochar composite. *Bioresour. Technol.* **2019**, *281*, 179–187. [CrossRef]
37. Thiele, S. Adsorption of the antibiotic pharmaceutical compound sulfapyridine by a long-term differently fertilized loess Chernozem. *J. Plant Nutr. Soil Sci.* **2000**, *163*, 589–594. [CrossRef]
38. Boxall, A.B.A.; Blackwell, P.; Cavallo, R.; Kay, P.; Tolls, J. The sorption and transport of a sulphonamide antibiotic in soil systems. *Toxicol. Lett.* **2002**, *131*, 19–28. [CrossRef]
39. Fang, G.; Wu, W.; Liu, C.; Dionysiou, D.D.; Deng, Y.; Zhou, D. Activation of persulfate with vanadium species for PCBs degradation: A mechanistic study. *Appl. Catal. B Environ.* **2017**, *202*, 1–11. [CrossRef]
40. Yin, R.; Guo, W.; Wang, H.; Du, J.; Wu, Q.; Chang, J.-S.; Ren, N. Singlet oxygen-dominated peroxydisulfate activation by sludge-derived biochar for sulfamethoxazole degradation through a nonradical oxidation pathway: Performance and mechanism. *Chem. Eng. J.* **2019**, *357*, 589–599. [CrossRef]
41. Yu, J.; Tang, L.; Pang, Y.; Zeng, G.; Wang, J.; Deng, Y.; Liu, Y.; Feng, H.; Chen, S.; Ren, X. Magnetic nitrogen-doped sludge-derived biochar catalysts for persulfate activation: Internal electron transfer mechanism. *Chem. Eng. J.* **2019**, *364*, 146–159. [CrossRef]
42. Özkal, C.B.; Frontistis, Z.; Antonopoulou, M.; Konstantinou, I.; Mantzavinos, D.; Meriç, S. Removal of antibiotics in a parallel-plate thin-film-photocatalytic reactor: Process modeling and evolution of transformation by-products and toxicity. *J. Environ. Sci.* **2017**, *60*, 114–122. [CrossRef] [PubMed]
43. Bourikas, K.; Vakros, J.; Kordulis, C.; Lycourghiotis, A. Potentiometric Mass Titrations: Experimental and Theoretical Establishment of a New Technique for Determining the Point of Zero Charge (PZC) of Metal (Hydr)Oxides. *J. Phys. Chem. B* **2003**, *107*, 9441–9451. [CrossRef]
44. Sfaelou, S.; Vakros, J.; Manariotis, I.D.; Karapanagioti, H.K. The use of Potentiometric Mass Titration technique for determining the acid-base behavior and surface charge of activated sludge. *Glob. NEST J.* **2015**, *17*, 397–405.

© 2019 by the authors. Licensee MDPI, Basel, Switzerland. This article is an open access article distributed under the terms and conditions of the Creative Commons Attribution (CC BY) license (http://creativecommons.org/licenses/by/4.0/).

Article

Enhanced Degradation of Phenol by a Fenton-Like System (Fe/EDTA/H₂O₂) at Circumneutral pH

Selamawit Ashagre Messele [1,2], Christophe Bengoa [1], Frank Erich Stüber [1], Jaume Giralt [1], Agustí Fortuny [3], Azael Fabregat [1] and Josep Font [1,*]

1. Departament d'Enginyeria Química, Universitat Rovira i Virgili, Av. Països Catalans 26, 43007 Tarragona, Catalunya, Spain; messele@ualberta.ca (S.A.M.); christophe.bengoa@urv.cat (C.B.); frankerich.stuber@urv.cat (F.E.S.); jaume.giralt@urv.cat (J.G.); azael.fabregat@urv.cat (A.F.)
2. Department of Civil & Environmental Engineering, University of Alberta, Edmonton, AB T6G 1H9, Canada
3. Departament d'Enginyeria Química, Universitat Politècnica de Catalunya, EUPVG, Av. Víctor Balaguer, s/n, 08800 Vilanova i la Geltrú, Catalunya, Spain; agustin.fortuny@upc.edu
* Correspondence: jose.font@urv.cat; Tel.: +34-9775-59646; Fax: +34-9775-59621

Received: 30 April 2019; Accepted: 19 May 2019; Published: 22 May 2019

Abstract: This work deals with the degradation of phenol based on the classical Fenton process, which is enhanced by the presence of chelating agents. Several iron-chelating agents such as ethylenediaminetetraacetic acid (EDTA), nitrilotriacetic acid (NTA), diethylenetriamine pentaacetic acid (DTPA), and ethylenediamine-N,N'-diacetic acid (EDDA) were explored, although particular attention was given to EDTA. The effect of the molar ligand to iron ratio, EDTA:Fe, initial pH, and temperature on the oxidation process was studied. The results demonstrate that the proposed alternative approach allows the capacity for degrading phenol to be extended from the usual acidic pH (around 3.0) to circumneutral pH range (6.5–7.5). The overall feasibility of the process depends on the concentration of the chelating agent and the initial pH of the solution. The maximum phenol conversion, over 95%, is achieved using a 0.3 to 1 molar ratio of EDTA:Fe, stoichiometric ratio of H₂O₂ at an initial pH of 7.0, and a temperature of 30 °C after 2 hours of reaction, whereas only 10% of phenol conversion is obtained without EDTA. However, in excess of ligand (EDTA:Fe > 1), the generation of radicals seems to be strongly suppressed. Improvement of the phenol removal efficiency at neutral pH also occurs for the other chelating agents tested.

Keywords: phenol degradation; EDTA; circumneutral pH; Fenton system; ligands

---

1. Introduction

The treatment of wastewater has increasingly become a challenge for a number of industries. In many cases, biological treatment is sufficient and the most economical solution for this problem. Nevertheless, many industrial and some urban effluents contain refractory and/or biotoxic compounds, which need a specific chemical treatment in order to eliminate or partly reduce the concentration of contaminants to the required level allowing for direct discharge to conventional sewage plants [1,2]. Phenols are the major organic constituents found in effluents of petroleum refineries, phenolic resin manufacturing, herbicide manufacturing, and petrochemicals [3,4]. Phenol and its derivatives are a major source of environmental pollutants.

Most of the applied technologies to treat refractory compounds are based on expensive chemical oxidation, either because of the drastic operating conditions in catalytic wet air oxidation (CWAO), costly equipment (H_2O_2/UV), or dedicated oxidants (O_3) [5]. In this regard, the well-known Fenton reagent (Fe^{2+}/H_2O_2) has shown interesting results and some significant advantages: (i) iron is a widely available and a non-toxic element, (ii) H_2O_2 is easy to handle and its decomposition leads to harmless products [6], and (iii) the process can be applied at room conditions and with simple equipment [7–9].

However, its application to the treatment of real wastewater has so far been limited mainly due to the requirements of chemicals to acidify the wastewater, which results in major operational costs [10,11]. Classical Fenton-based process is restricted to an acidic pH range, the optimum pH being 3.0, due to the inability of the homogeneous iron catalyst to remain in solution at a pH beyond 4.0. This means that, as the pH increases from strongly acidic to neutral, the application of Fenton's reagent to wastewater remediation is hampered by the formation of amorphous iron oxide precipitates [12–14]. This reduces the efficiency of reagents during the reaction and generates a high amount of chemical sludge too [15].

Many studies have been devoted to applying modifications in order to circumvent the drawbacks associated with conventional Fenton and Fenton-like processes. Among many others, use of different sources of oxyradicals, photo irradiation, ultrasound, and electrochemical methods have been tested for their intensified performance [16–19]. In this sense, addition of chelating agents in the reaction system could prevent precipitation of iron in the solution at higher pH (6.0-7.0) by forming stable chelates with iron ions and promoting its availability for hydroxyl radical generation from peroxide in a wider pH range [13].

Some studies have reported a quick entire destruction of chlorophenols using hydrogen peroxide activated with iron catalysts complexed with tetraamidomacrocyclic ligand (TAML). This provides an efficiency of one order of magnitude higher than the classic Fenton [20,21]. The route has also been tested for destruction of phenolic structures [22] and oxidation of the colorant Orange II [23]. Moreover, it has been demonstrated that other ligands like EDTA have been widely used to enhance the efficiency of Fenton's reaction due to their strong complexing ability with multivalent cation [24,25] and also their capability to activate the formation of hydroxyl radical when they are added in Fenton systems [26]. In addition, the capacity of the chelating ligands to activate the decomposition of the peroxides and to intensify the generation of radicals has been widely confirmed [27]. In recent studies [28], ligand enhanced Fenton reaction was successfully used for the oxidation of As(III) to As(V), but As(III) oxidation was inhibited in the presence of excess EDTA at acidic and neutral pH. Rastogi et al. [29] also reported the effect of inorganic, synthetic, and naturally occurring chelating agents on Fe(II) mediated advanced oxidation of chlorophenols. Even though there are different studies that deal with how to overcome the drawbacks of the Fenton reaction, there is a lack of systematic investigation about the effect of different operational parameters and the possibilities of the addition of different ligands on the oxidation of phenol with Fenton process. Therefore, a comparison study in the oxidation of phenol under acidic and basic initial pH conditions was accomplished using different EDTA:Fe molar ratios. Furthermore, decomposition rate of H_2O_2 during the reaction, effect of other iron-chelating agents, effect of iron concentration, and temperature were also investigated.

2. Results and Discussion

2.1. Effect of Ligand to Metal (L:M) Molar Ratio

The effect of EDTA:Fe molar ratio was investigated in the range from 0 to 2 in order to determine the optimum conditions for best phenol removal at initial phenol concentration of 1000 mg/L and 7 mg/L of Fe^{2+} for 2 hours at pH 3.0. The results are illustrated in Figure 1a. Similar phenol conversion (95%) was obtained for the free iron catalyst and for the Fe^{2+}-EDTA complexes in the range from 0 to 0.5 of L:M molar ratio after 2 hours of reaction. This implies that no improvement was obtained in acidic conditions due to the addition of EDTA. From the result, it is important to note that L:M molar ratios beyond 1:1 inhibited the oxidation of phenol. This fact can be related with the degradation of EDTA with H_2O_2 in the presence of different catalysts [30–34]. Thus, the presence of excess EDTA does not improve the catalytic behavior; rather, it may inhibit the generation of radicals. It has been reported elsewhere that the ratio of ligand to metal is of significant importance since the generation of radicals can be reduced in the presence of excess ligand [35–37].

Figure 1b shows the consumption of hydrogen peroxide for different L:M molar ratio at pH 3.0 in the absence of phenol. At higher ligand concentration, hydrogen peroxide decomposition rate was negligible as the H₂O₂ decomposition is inhibited with the presence of excess EDTA, as observed elsewhere [35]. Therefore, it shows the same tendency as the phenol removal efficiency, which depends on the L:M ratio.

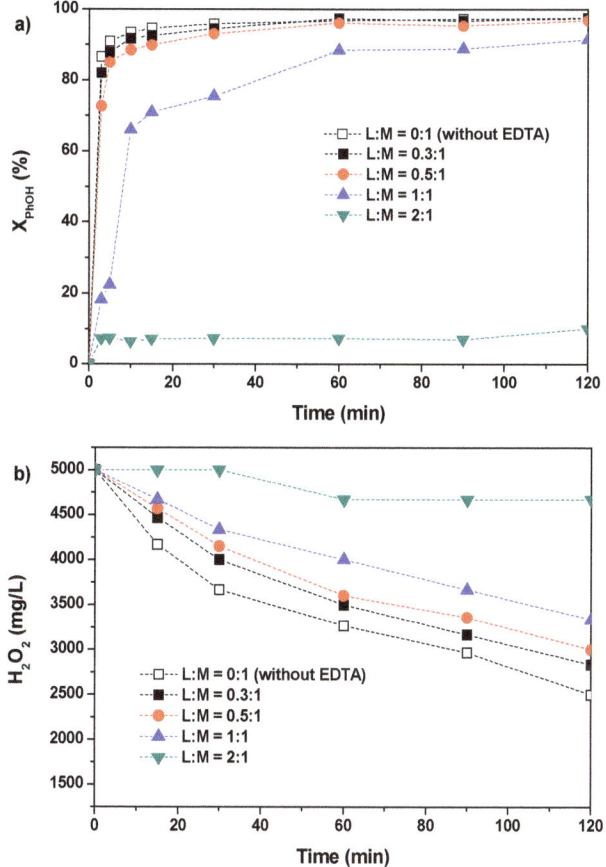

Figure 1. Phenol (**a**) and H$_2$O$_2$ (**b**) conversion versus time for different EDTA:Fe molar ratios. [Phenol] = 1000 mg/L, [H$_2$O$_2$] = 5000 mg/L, [Fe^{2+}]$_0$ = 7 mg/L, T = 30 °C, [pH]$_0$ = 3.0, and t = 120 min

2.2. Speciation of Fe^{2+} and Fe^{3+} in the Presence of EDTA

It is well-known that Fe and EDTA in solution form a diversity of species, whose distribution depends on the pH to a great extent. However, this speciation not only depends on the pH, but also on the complex formation kinetics [36,37] and the probability of EDTA degradation when exposed to H$_2$O$_2$ [38]. The speciation diagrams of both Fe^{2+} and Fe^{3+} with EDTA as a function of pH were obtained using the thermodynamic data from the MINTEQA2 [39] database and are shown in Figure 2 for an overall iron concentration range from 0 to 0.125 mM. As can be seen from Figure 2a, Fe^{2+} precipitates in the form of hydroxides at pH above 4.0. However, both Fe^{2+} and Fe^{3+} form stable complexes in the presence of strong complex forming agents like EDTA in a wide pH range, up to pH 11 [9]. DTPA, EDDA, and NTA give similar behavior (diagrams not shown) too.

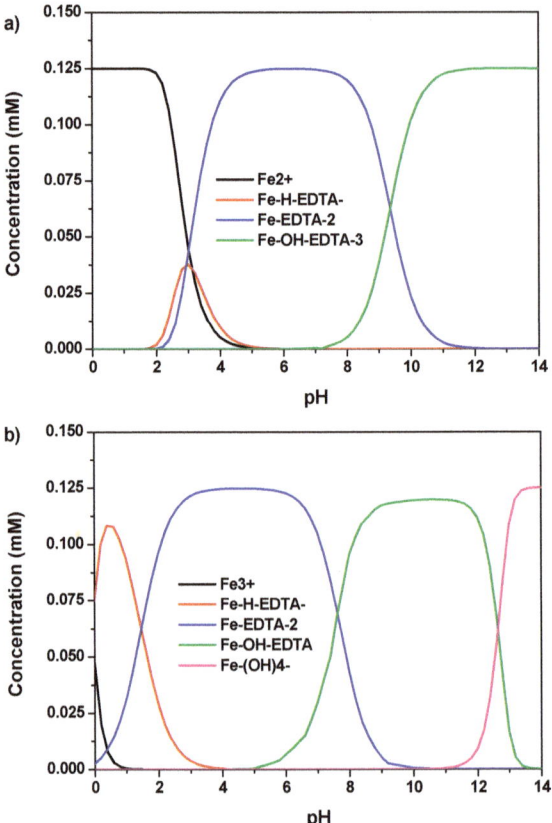

Figure 2. Equilibrium distribution of [Fe-EDTA] species as a function of pH for (**a**) Fe^{2+} and (**b**) Fe^{3+}, assuming 0.125 mM concentration of Fe and EDTA [40].

Furthermore, this distribution diagram shows that Fe^{3+}-EDTA is the predominant species at a pH range of 2.0–7.0, whereas Fe^{2+}-EDTA is the major species in the range of 3.0–9.0. Based on the speciation distribution in Figure 2b, pH range was divided into three regions, i.e., low pH (pH < 3.0, Region I), mid pH (3.0 < pH < 7.0, Region II), and high pH (pH > 7.0, Region III). The speciation in the low pH range contains protonated $Fe^{2+/3+}$-EDTA complexes and free Fe^{2+}, whereas the high pH range contains the hydroxyl complexes. The above experiments were conducted at pH 3.0, so there was already highly active free Fe^{2+} and no possibility for Fe^{3+} to form insoluble precipitate. Therefore, the presence of EDTA does not modify the conditions of the classical Fenton, hence the nil effect of the EDTA. However, at higher pH, where Fe^{3+} precipitation in the form of hydroxide occurs, the addition of EDTA should play a relevant role. Thus, mid pH, where the Fe^{2+}-EDTA and Fe^{3+}-EDTA are predominant species, was chosen for the subsequent experiments.

2.3. Effect of Initial pH

Considering the Fe-EDTA species diagrams, different initial pHs were evaluated in order to explore the effect of pH over the Fenton process in the presence of EDTA. As can be seen in Figure 3b, only small performance changes were noted again for acidic conditions (pH 3.0 and 5.0) in the presence of EDTA (0.3:1 L:M ratio at 7 mg/L Fe^{2+}, i.e., 0.125 mM) compared to the equivalent free Fe^{2+} catalyst (Figure 3a), reaching a final 95% phenol conversion with both Fe^{2+} and Fe^{2+}-EDTA.

However, under circumneutral conditions (pH 6.5–7.0), the addition of EDTA brings to phenol conversion over 95% (Figure 3b). It is important to note that the conversion at circumneutral pH without EDTA only reaches 10% of phenol conversion (Figure 3a). As expected, at pH above 4–5 without EDTA, the reaction does not proceed because the iron precipitation yields inactive iron. Moreover, Figure 3a,b respectively show that at higher pH, e.g., 8.0, the conversion drops again both in the absence and in presence of EDTA. For instance, under the same operating conditions, only 10% was achieved in the absence of EDTA and 35% in the presence of EDTA. The decrease in the reaction rates at pH 8.0 can be due to changes in the speciation of Fe^{3+} towards hydroxide complex species, which are probably not active for the generation of radicals and suppress the catalytic properties of iron. In this case, the presence of EDTA only partially prevents the formation of such less active iron species. In addition, hydrogen peroxide stability is also strongly affected by pH conditions [36]. Thus, a further increase of the pH above neutral values results in favoring the H_2O_2 decomposition into water and molecular oxygen [41].

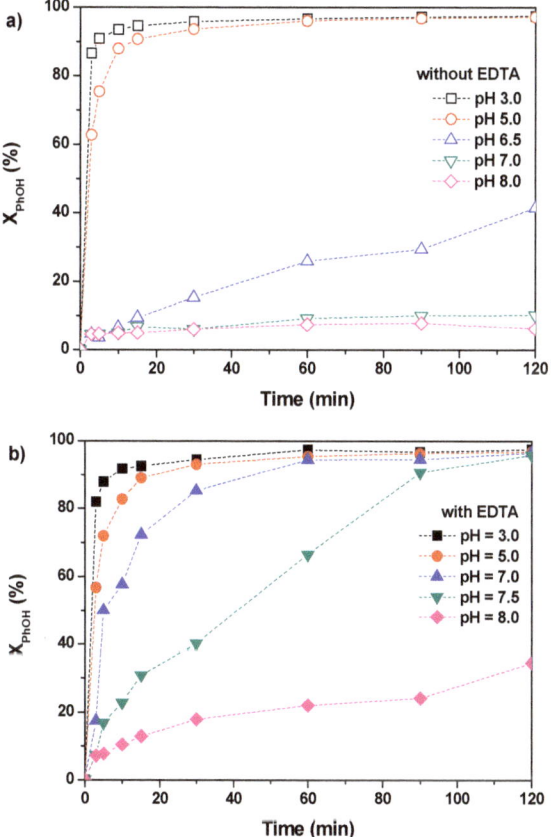

Figure 3. Phenol conversion versus time for different initial pH solution: (**a**) without EDTA (L:M = 0:1), (**b**) with EDTA (L:M = 0.3:1). [Phenol] = 1000 mg/L, [H_2O_2] = 5000 mg/L, [Fe^{2+}]$_0$ = 7 mg/L, T = 30 °C, and t = 120 min.

Figure 4 shows the pH evolution, phenol, and TOC conversion against time for different L:M ratios under the same experimental conditions but at initial pH solution of 7.0. The result illustrates that the feasibility of the process mainly depends on both the chelating agent concentration used to promote the reaction and the initial pH of the solution. This means that the presence of EDTA resulted in increased iron solubility at higher pH. Consequently, the Fenton reaction can be conducted over a broader range of pH when iron is complexed with EDTA.

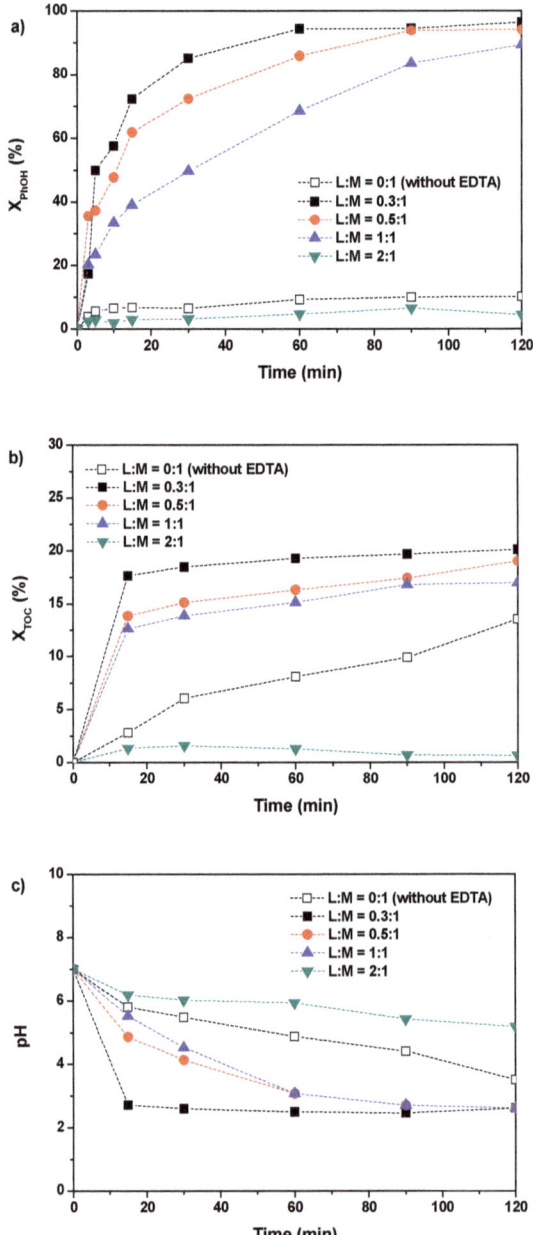

Figure 4. Phenol conversion (**a**), TOC conversion (**b**), and pH evolution (**c**) versus time for different EDTA:Fe molar ratio. [Phenol] = 1000 mg/L, [H_2O_2] = 5000 mg/L, [Fe^{2+}]$_0$ = 7 mg/L, T = 30 °C, [pH]$_0$ = 7.0, and t = 120 min.

Figure 4a shows the degradation rate of phenol for each EDTA:Fe ratio, which follows the order 0.3:1 > 0.5:1 > 1:1. However, in excess of ligand (EDTA:Fe >1), only 10% of phenol conversion was achieved, because the presence of excess EDTA inhibited the generation of radicals, and this strongly negatively impacts the efficiency of phenol removal. Overall, the presence of EDTA highly increased the pH range where the classical Fenton system is feasible, which results in major cost savings as this eliminates the need for initial acidification.

Enhanced TOC removal was also observed during the oxidation. As can be seen in Figure 4b, the TOC reduction was 4.7%, 20.1%, 19.0%, 16.6%, and 0.6% for L:M ratio of 0:1, 0.3:1, 0.5:1, 1:1, and 2:1, respectively, after 2 hours of reaction. From the results, it can be seen that the amount of ligand also has a great effect on phenol total mineralization. Thus, the phenol and TOC conversion at initial pH of 7.0, in the absence (0:1) and excess (2:1) of EDTA is insignificant. However, the addition of small amounts of EDTA in the range of 0.3:1 to 1:1 EDTA:Fe ratio shows a significant improvement on the degradation.

Figure 4c illustrates the pH variation of the solution during the reaction. After starting the reaction, the pH of the solution decreases dramatically in the first 15 min from original pH 7.0 to pH 5.8, 2.7, 4.9, 5.5, and 6.2 for L:M ratio of 0:1, 0.3:1, 0.5:1, 1:1, and 2:1, respectively. During the reaction, the decrease of the pH of the solution is very significant for the EDTA:Fe ratio of 0.3:1, 0.5:1, 1:1, which agrees with the formation of low molecular weight acid intermediates. For the EDTA:Fe ratio of 0.3:1, the solution pH plateaus after 30 min of reaction. In this experiment, the best results were obtained using 0.3:1 EDTA:Fe molar ratio (Figure 4), reaching over 95 % of phenol and 20.1% TOC conversion under neutral pH. Subsequently, the EDTA:Fe ratio was set at 0.3:1 in the following experiments.

2.4. Effect of Temperature and Iron Concentration

The influence of two temperatures (30 °C and 60 °C) and two iron concentrations (7 mg/L and 28 mg/L) on the oxidation of phenol was studied under the optimal EDTA:Fe molar ratio previously found, i.e., 0.3:1, and without EDTA (0:1); in all cases, the initial pH of the solution was 7.0. Figure 5a clearly demonstrates that an increase in iron concentration and temperature favors phenol degradation. The phenol degradation profile obtained in the presence of EDTA at 30 °C and 7 mg/L of iron showed that, even at the lower temperature, 96% of phenol disappears after 2 hours of reaction. As expected, a higher temperature and iron concentration led to the increase of the reaction rate. At 60 °C with 7 mg/L of iron, the same conversion of 96% was found after only 60 min and the conversion was complete in just 30 min when 28 mg/L of iron was applied at 30 °C. Interestingly, without EDTA (L:M = 0:1), the results obtained for the phenol removal indicated a maximum phenol conversion of 10% and 36% for 7 mg/L of iron at 30 and 60 °C, respectively, whereas, at 30 °C and 28 mg/L of iron, the conversion was 58%. This confirms that, at neutral pH, the presence of EDTA plays a major role. Thus, the addition of EDTA at one-third of the molar stoichiometric ratio with respect to the iron is able to increase the phenol conversion from 10% to 96% at 30 °C with 7 mg/L of iron; this is almost ten times higher.

Figure 5b shows the hydrogen peroxide consumption at the two different temperatures and iron concentrations in the presence of EDTA using an EDTA:Fe molar ratio of 0.3:1 and without EDTA (0:1); in all the cases, the initial pH of the solution was again 7.0. At a higher temperature and iron concentration, the hydrogen peroxide decomposition rate also improved. Accordingly, phenol removal efficiency shows the same tendency as the peroxide decomposition rate, so small values of hydrogen peroxide conversion were encountered for the systems without EDTA at pH 7.0, even at a higher temperature and iron concentration.

This trend has already been reported in the literature for different cases. Walling [42] and Oakes and Smith [43] confirmed that while Fenton's reaction is effective in many cases, Fe^{2+} catalyzed hydrogen peroxide decomposition rate takes feasible values only in a narrow pH range (3–4) where the activity is significant. However, it has also been reported that complexed forms of iron are active for hydrogen peroxide decomposition over a much wider pH range [44].

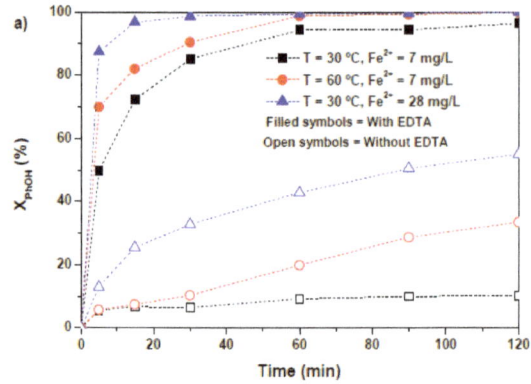

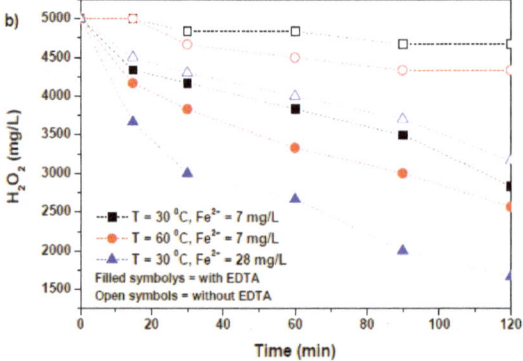

Figure 5. Effect of temperature and Fe concentration on (**a**) phenol and (**b**) H_2O_2 conversion. [Phenol] = 1000 mg/L, $[H_2O_2]_0$ = 5000 mg/L, $[pH]_0$ = 7.0, L:M = 0.3:1, and t = 120 min.

There are several studies on the reaction between hydrogen peroxide and free or complexed iron ions in aqueous solution. Two different reaction mechanisms have been proposed. The first reaction pathway considers the generation of radicals through a classical set of reactions as proposed by Haber and Weiss [45].

$$Fe^{2+} + H_2O_2 \rightarrow Fe^{3+} + HO^\bullet + OH^- \quad (1)$$

$$Fe^{2+} + HO^\bullet \rightarrow Fe^{3+} + OH^- \quad (2)$$

$$Fe^{3+} + H_2O_2 \rightarrow FeOOH^{2+} + H^+ \quad (3)$$

$$FeOOH^{2+} \rightarrow Fe^{2+} + HO_2^\bullet \quad (4)$$

$$Fe^{2+} + HO_2^\bullet \rightarrow Fe^{3+} + HO_2^- \quad (5)$$

$$Fe^{3+} + HO_2^\bullet \rightarrow Fe^{2+} + O_2 + H^+ \quad (6)$$

$$HO^\bullet + H_2O_2 \rightarrow H_2O + HO_2^\bullet \quad (7)$$

In the presence of a radical scavenger, e.g., some organic compound, the radicals can attack it and this alternative pathway can compete, often favorably, with the self-decomposition of the hydrogen peroxide into water and molecular oxygen [46]. If Fe^{3+} is removed from the system, Fe^{2+} is progressively exhausted and the reaction stops, which occurs when the pH is not acidic enough for preventing the formation of inactive, insoluble Fe(OH)$_3$. Hence, the main role of the EDTA seems to be maintaining Fe^{3+} in solution even at neutral pH, without negative impact on the rest of steps.

Although other researchers have alternatively suggested the intermediate generation of highly reactive ferryl ion (Fe^{4+}) [46–48], in spite of their fundamental differences, the two schemes are surprisingly hard to distinguish. Rahhal and Ritcher [48] suggested that the pH of the system is the determining factor as to whether hydroxyl radicals or ferryl ions are generated.

Therefore, from our results, the addition of EDTA in the Fenton system indeed leads to a more efficient consumption of H$_2$O$_2$, which indicates an enhanced iron-catalyzed H$_2$O$_2$ decomposition into radicals and, in accordance, an improved phenol removal rate. In conclusion, this can be considered as an intensification of the conventional Fenton process.

2.5. Effect of the Chelating Agent

To study the effect of the chelating agent over this oxidation process, four different chelating species were tested (EDTA, EDDA, DTPA, and NTA). These compounds were used in this study because they are commercially available, similar in structure, and represent a potentially useful class of Fe^{2+}/Fe^{3+} chelate catalysts. Table 1 shows the structures of these compounds.

Table 1. Formula and structure of the different chelating agents tested.

Compound	Structure	Molecular Formula	Molecular Weight (g/mol)
Ethylenediaminetetraacetic acid (EDTA)		C$_{10}$H$_{16}$N$_2$O$_8$	292.24
Ethylenediamine-N,N'-diacetic acid (EDDA)		C$_6$H$_{12}$N$_2$O$_4$	176.17
Diethylenetriaminepenta-acetic acid (DTPA)		C$_{14}$H$_{23}$N$_3$O$_{10}$	393.35
Nitrilotriacetic acid (NTA)		C$_6$H$_9$NO$_6$	191.14

The tests were carried out for each chelating agent with an L:M ratio of 0.3:1 (except NTA 0.6:1, as NTA is tridentated) [13], 7 mg/L of iron concentration, an initial concentration of phenol of 1000 mg/L, a stoichiometric ratio of H$_2$O$_2$, at initial pH solution of 7.0, and a temperature of 30 °C.

As can be seen in Figure 6, the phenol degradation markedly depends on the chelating agent used for enhancing the oxidation. At 30 min of reaction, in terms of phenol conversion, the reactivity order observed was EDDA (93%) > EDTA (85%) > DTPA (81%) > NTA (63%). However, quite similar conversions (96%) were obtained for all chelating agents after 60 min of reaction. These results could be due to the fact that the stability of Fe^{2+} complex formed with each chelating agent could follow the speciation of Fe^{2+}, thus avoiding precipitation of iron at higher pH as suggested for EDTA. The pH evolution (data not shown) exhibits almost the same values, a minimum value (2.5) at approximately 120 min for all chelating agents, again suggesting the formation of organic acids. Although the identification of the partial oxidation products was not a main objective, we occasionally conducted a complete analysis of the treated samples. In such cases, the partial oxidation intermediates did not differ from those identified in previous studies [49] conducted with powdered zero valent iron and air, which indeed included several short-chain organic acids present as later intermediates or end products.

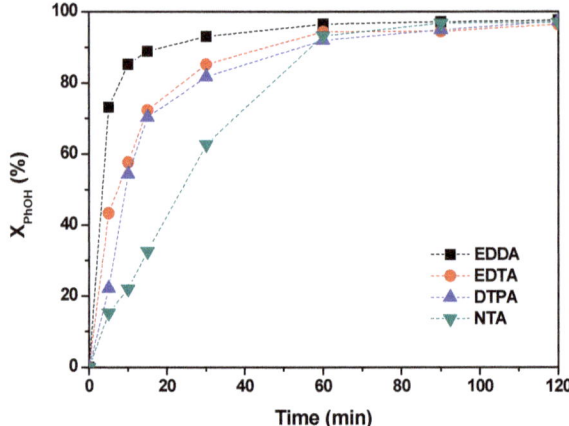

Figure 6. Phenol conversion for different chelating agent. [Phenol] = 1000 mg/L, [H_2O_2] = 5000 mg/L, [Fe^{2+}]$_0$ = 7 mg/L, T = 30 °C, [pH]$_0$ = 7.0, L:M = 0.3:1, and t = 120 min.

Although the main role of the chelating agent is believed to be preventing the formation of Fe(OH)$_3$, the complex formed must somehow modify the reaction steps where Fe^{3+} participates, particularly in Eq. 3 where it is reduced back to Fe^{2+}, allowing the generation of oxyradicals from the hydrogen peroxide to continue. This reaction is the slowest and often is the rate-controlling step for the overall process. Therefore, a too stable Fe(III)-chelant complex may slow down this step. From this point of view, EDDA seems to possess a structure less favorable to forming a very stable complex so the reduction of the Fe^{3+} is facilitated, yet it is able to keep it in solution.

3. Materials and Methods

3.1. Chemicals

Phenol (PhOH), used as model compound, was purchased from Panreac (>99% purity). Fenton reagents, hydrogen peroxide (H_2O_2 30% w/v solution), and Iron(II) sulphate heptahydrated (>98% purity) were also purchased from Panreac (Barcelona, Spain). The chelating agents used in this study were: ethylenediaminetetraacetic acid disodium salt dihydrate (98% purity) purchased from Panreac; nitrilotriacetic acid trisodium salt (98% purity) supplied by Sigma-Aldrich (St. Louis, MS, USA); diethylentriamine pentaacetic acid (≥98% purity); and ethylenediamine diacetic acid (≥98% purity) obtained from Fluka (Seelze, Germany). Sulphuric acid (95-97% purity) and sodium hydroxide (98% purity) were also purchased from Sigma-Aldrich; these reagents were used to adjust the initial pH values. Deionized water was used to prepare all the aqueous solutions.

3.2. Experimental Set-Up and Procedure

A magnetically stirred jacketed reactor was used for all oxidation reactions. The reactor has a 200 mL capacity. The reaction temperature was set and controlled by circulating deionized water from a thermostatic bath through a jacket.

The reactor was filled with 100 mL of solution containing 1000 mg/L of phenol, 7 mg/L of Fe^{2+}, and a variable concentration of EDTA, selected to give the desired EDTA:Fe ratio (0:1, 0.3:1, 0.5:1, 1:1, and 2:1). The pH was adjusted by adding NaOH or H_2SO_4. Once the desired temperature was reached (30 °C), a small volume (1.7 mL) of concentrated hydrogen peroxide was added to provide the stoichiometric amount of H_2O_2 (5000 mg/L) and, thus, to start the reaction. All the experiments were carried out at a stirring rate of 300 rpm and for 2 hours.

During the reaction, 1 mL samples were withdrawn at 0, 3, 5, 10, 15, 30, 60, 90, and 120 min. Each sample was immediately quenched by using 40 µL of NaOH 6 N to stop the Fenton reaction. Then, it was filtered with a syringe filter of 0.45 µm nylon (Teknokroma, ref.TR-200101) and placed in a glass vial (Agilent) for immediate analysis. Some experiments were conducted three times to check the reproducibility of the results. The experimental error was within ±4%.

The main parameter used to compare the results in the discussion section is the conversion of phenol, X_{PhOH}, defined as:

$$X_{PhOH}(\%) = \frac{[PhOH]_0 - [PhOH]_t}{[PhOH]_0}, \qquad (8)$$

where $[PhOH]_0$ is the initial concentration and $[PhOH]_t$ is the concentration at time t.

3.3. Analytical Methods

The concentration of phenol was determined by HPLC (Agilent Technologies, model 1220 Infinity LC, Santa Clara, CA, USA) equipped with a C18 reverse phase column (Hypersil ODS, 5µm, 25 × 0.4 cm from Agilent technologies, Santa Clara, CA, USA). The analyses were performed with a mobile phase of a 40/60 mixture (volume %) of methanol and ultrapure water (Milli-Q water) at a flow rate of 1 mL/min. The pH of the water was adjusted at 1.41 with sulphuric acid (H_2SO_4). The detection was performed using UV absorbance at a wavelength of 254 nm. The automatic injection volume was 20 µL of sample.

The total organic carbon (TOC) was measured in a TC Multi Analyzer 2100 N/C, equipment from Analytik Jena AG, with a non-diffractive IR detector (Jena, Germany). The non-purgeable organic carbon (NPOC) combustion infrared standard method 5310B [50] was used. TOC determination was performed using chemical oxidation of the sample in a high temperature furnace (800 °C) in presence of a platinum catalyst. The carbon dioxide produced during the oxidation was quantitatively determined by means of an infrared spectrophotometer detector. Sample acidification and aeration was carried out prior to analysis to eliminate inorganic carbon.

The TOC conversion, X_{TOC}, was defined as:

$$X_{TOC}(\%) = \frac{[TOC]_0 - [TOC]_t}{[TOC]_0} \text{ and} \qquad (9)$$

$$[TOC]_0 = [TOC]_{(PhOH)0} + [TOC]_{(EDTA)0}, \qquad (10)$$

where $[TOC]_0$ is the initial total TOC according to Equation (10); $[TOC]_{(PhOH)0}$, the initial TOC from phenol; $[TOC]_{(EDTA)0}$, the initial TOC from EDTA; and $[TOC]_t$, the total TOC at time t.

Finally, hydrogen peroxide concentration was determined by iodometric titration.

4. Conclusions

Aqueous phase oxidation of phenol solutions (1000 mg/L) was conducted from acidic up to circumneutral pH using a classical Fenton system (Fe^{2+}/H_2O_2) with or without the addition of a chelating agent in order to enhance oxidation performance. EDTA was selected for most of the tests under the same Fe^{2+} catalyst load (7 mg/L) and H_2O_2 dose (stoichiometric with respect to phenol).

In acidic conditions, close to the optimal pH (3–4), the presence of EDTA does not improve the phenol conversion achieved under classical Fenton conditions while, in EDTA excess, the phenol conversion becomes insignificant.

On the contrary, the phenol removal efficiency and peroxide decomposition rate significantly improved in the presence of EDTA at near circumneutral pH. Over 95% of phenol conversion was obtained using an EDTA:Fe ratio of 0.3:1 at pH 7.0, which is almost tenfold that obtained in the absence of EDTA. Among the different EDTA:Fe molar ratios tested, the ratio 0.3:1 was found to be the optimum.

Other chelating agents, like EDDA, DTPA, and NTA, were also tested in this study. They all enhanced the oxidation ability of the Fenton system at neutral pH, although EDDA provided the best oxidation performance.

Overall, the presence of a chelating agent in small quantities greatly broadens the pH range where the Fenton-like system is feasible, up to circumneutral pH. Thus, pH adjustment would not be required or would be just limited to caustic real wastewaters, which could result in major savings of operational costs.

Author Contributions: Conceptualization, S.A.M. and J.F.; investigation, S.A.M.; methodology, S.A.M. and J.F.; resources, C.B., F.E.S., J.G., Agustí Fortuny, and Azael Fabregat; supervision, J.F.; validation, C.B., F.E.S., J.G., Agustí Fortuny, and Azael Fabregat; writing—original draft, S.A.M.; writing—review & editing, J.F. All authors discussed the results and commented on the manuscript in addition to their funding acquisition involvement.

Funding: Financial support for this research was provided by the Spanish Ministerio de Ciencia e Innovación, the Ministerio de Economía, Industria y Competitividad, the Ministerio de Ciencia, Innovación y Universidades and the European Regional Development Fund (ERDF), projects CTM2011-23069, CTM2015-67970-P, and RTI2018-096467-B-I00. The Ministerio de Ciencia e Innovación is also thanked for providing a doctoral scholarship (Programme FPI, BES-2009-017016) to carry out this research work. The authors research group is recognized by the Comissionat per a Universitats i Recerca, DIUE de la Generalitat de Catalunya (2017 SGR 396), and supported by the Universitat Rovira i Virgili (2018PFR-URV-B2-33 and 2019OPEN).

Conflicts of Interest: The authors declare no conflicts of interest. The funders had no role in the design of the study; in the collection, analyses, or interpretation of data; in the writing of the manuscript, or in the decision to publish the results.

References

1. Matatov-Meytal, Y.I.; Sheintuch, M. Catalytic abatement of water pollutants. *Ind. Eng. Chem. Res.* **1998**, *37*, 309–326. [CrossRef]
2. Levec, J.; Pintar, A. Catalytic wet-air oxidation processes. *Catal. Today* **2007**, *124*, 172–184. [CrossRef]
3. Fortuny, A.; Bengoa, C.; Font, J.; Castells, F.; Fabregat, A. Water pollution abatement by catalytic wet air oxidation in a trickle bed reactor. *Catal. Today* **1999**, *53*, 107–114. [CrossRef]
4. Santos, A.; Yustos, P.; Quintanilla, A.; Rodríguez, S.; García-Ochoa, F. Route of the catalytic oxidation of phenol in aqueous phase. *Appl. Catal. B* **2002**, *39*, 97–113. [CrossRef]
5. Pera-Titus, M.; García-Molina, V.; Baños, M.A.; Giménez, J.; Esplugas, S. Degradation of chlorophenols by means of advanced oxidation processes: A general review. *Appl. Catal. B* **2004**, *47*, 219–256. [CrossRef]
6. Jones, C.W. *Applications of Hydrogen Peroxide and Derivates*; Royal Society of Chemistry: London, UK, 1999.
7. Azbar, N.; Yonar, T.; Kestioglu, K. Comparison of various advanced oxidation processes and chemical treatment methods for COD and color removal from a polyester and acetate fiber dyeing effluent. *Chemosphere* **2004**, *55*, 35–43. [CrossRef] [PubMed]
8. Esplugas, S.; Giménez, J.; Contreras, S.; Pascual, E.; Rodríguez, M. Comparison of different advanced oxidation processes for phenol degradation. *Water Res.* **2002**, *36*, 1034–1042. [CrossRef]
9. Pignatello, J.J.; Oliveros, E.E.; Mackay, A. Advanced oxidation processes for organic contaminant destruction based on the Fenton reaction and related chemistry. *Crit. Rev. Environ. Sci. Technol.* **2006**, *36*, 1–84. [CrossRef]
10. Villegas-Guzman, P.; Giannakis, S.; Rtimi, S.; Grandjean, D.; Bensimon, M.; de Alencastro, L.F.; Torres-Palma, R.; Pulgarin, C. A green solar photo-Fenton process for the elimination of bacteria and micropollutants in municipal wastewater treatment using mineral iron and natural organic acids. *Appl. Catal. B Environ.* **2017**, *219*, 538–549. [CrossRef]
11. Villegas-Guzman, P.; Giannakis, S.; Torres-Palma, R.A.; Pulgarin, C. Remarkable enhancement of bacterial inactivation in wastewater through promotion of solar photo-Fenton at near-neutral pH by natural organic acids. *Appl. Catal. B Environ.* **2017**, *205*, 219–227. [CrossRef]
12. Duesterberg, C.K.; Mylon, S.E.; Waite, T.D. pH effects on Iron-catalyzed oxidation using Fenton's reagent. *Environ. Sci. Technol.* **2008**, *42*, 8522–8527. [CrossRef]
13. Tachiev, G.; Roth, J.A.; Bowers, A.R. Kinetics of hydrogen peroxide decomposition with complexed and "free" iron catalysts. *Int. J. Chem. Kinet.* **2000**, *32*, 24–35. [CrossRef]
14. Usman, M.; Hanna, K.; Haderlein, S. Fenton oxidation to remediate PAHs in contaminated soils: A critical review of major limitations and counter-strategies. *Sci. Total Environ.* **2016**, *569–570*, 179–190. [CrossRef]

15. Barros, W.R.P.; Steter, J.R.; Lanza, M.R.V.; Tavares, A.C. Catalytic activity of $Fe_{3-x}Cu_xO_4$ ($0 \leq x \leq 0.25$) nanoparticles for the degradation of Amaranth food dye by heterogeneous electro-Fenton process. *Appl. Catal. B Environ.* **2016**, *180*, 434–441. [CrossRef]
16. Neyens, E.; Baeyens, J. A review of classic Fenton's peroxidation as an advanced oxidation technique. *J. Hazard. Mater.* **2003**, *98*, 33–50. [CrossRef]
17. Yu, Y.; Li, S.; Peng, X.; Yang, S.; Zhu, Y.; Chen, L.; Wu, F.; Mailhot, G. Efficient oxidation of bisphenol A with oxysulfur radicals generated by iron-catalyzed autoxidation of sulfite at circumneutral pH under UV irradiation. *Environ. Chem. Lett.* **2016**, *14*, 527–532. [CrossRef]
18. Salazar, L.M.; Grisales, C.M.; Garcia, D.P. How does intensification influence the operational and environmental performance of photo-Fenton processes at acidic and circumneutral pH. *Environ. Sci. Pollut. Res.* **2019**, *26*, 4367–4380. [CrossRef] [PubMed]
19. Bello, M.M.; Raman, A.A.A.; Asghar, A. A review on approaches for addressing the limitations of Fenton oxidation for recalcitrant wastewater treatment. *Process Saf. Environ.* **2019**, *126*, 119–140. [CrossRef]
20. Gupta, S.S.; Stadler, M.; Noser, C.A.; Ghosh, A.; Steinhoff, B.; Lenoir, D.; Horwitz, C.P.; Schramm, K.W.; Collins, T.J. Rapid total destruction of chlorophenols by activated hydrogen peroxide. *Science* **2002**, *296*, 326–328. [CrossRef]
21. Collins, T.J. TAML oxidant activators: A new approach to the activation of hydrogen peroxide for environmentally significant problems. *Acc. Chem. Res.* **2002**, *35*, 782–790. [CrossRef]
22. Wingate, K.G.; Stuthridge, T.R.; Wright, L.J.; Horwitz, C.P.; Collins, T.J. Application of TAML catalysts to remove colour from pulp and paper mill effluents. *Water Sci. Technol.* **2004**, *49*, 255–260. [CrossRef] [PubMed]
23. Chahbane, N.; Popescu, D.; Mitchell, D.A.; Chanda, A.; Lenoir, D.; Ryabov, A.D.; Schramm, K.; Collins, T.J. Fe^{III}–TAML-catalyzed green oxidative degradation of the azo dye Orange II by H_2O_2 and organic peroxides: Products, toxicity, kinetics, and mechanisms. *Green Chem.* **2007**, *9*, 49–57. [CrossRef]
24. Li, L.; Abea, Y.; Kanagawa, K.; Shoji, T.; Mashino, T.; Mochizuki, M.; Tanaka, M.; Miyata, N. Iron-chelating agents never suppress Fenton reaction but participate in quenching spin-trapped radicals. *Anal. Chim. Acta* **2007**, *599*, 315–319. [CrossRef]
25. Sun, Y.; Pignatello, J.J. Chemical treatment of pesticide wastes: Evaluation of Fe(III) chelates for catalytic hydrogen peroxide oxidation of 2,4-D at circumneutral pH. *J. Agric. Food Chem.* **1992**, *40*, 322–327. [CrossRef]
26. Wink, D.A.; Nims, R.W.; Desrosiers, M.F.; Ford, P.C.; Keefer, L.K. A kinetic investigation of intermediates formed during the Fenton reagent mediated degradation of N-Nitrosodimethylamine: Evidence for an oxidative pathway not involving hydroxyl radical. *Chem. Res. Toxicol.* **1991**, *4*, 510–512. [CrossRef] [PubMed]
27. Bull, C.; McClune, G.J.; Fee, J.A. The mechanism of Fe-EDTA catalyzed superoxide dismutation. *J. Am. Chem. Soc.* **1983**, *105*, 5290–5300. [CrossRef]
28. Wang, Z.; Bush, R.T.; Liu, J. Arsenic(III) and iron(II) co-oxidation by oxygen and hydrogen peroxide: Divergent reactions in the presence of organic ligands. *Chemosphere* **2013**, *93*, 1936–1941. [CrossRef] [PubMed]
29. Rastogi, A.; Al-abed, S.R.; Dionysiou, D.D. Effect of inorganic, synthetic and naturally occurring chelating agents on Fe(II) mediated advanced oxidation of chlorophenols. *Water Res.* **2009**, *43*, 684–694. [CrossRef]
30. Sillanpää, M.; Pirkanniemi, K.; Sorokin, A. Oxidation of EDTA with H_2O_2 catalysed by metallophthalocyanines. *Environ. Technol.* **2009**, *30*, 1593–1600. [CrossRef] [PubMed]
31. Kunz, A.; Peralta-Zamora, P.; Durán, N. Hydrogen peroxide assisted photochemical degradation of ethylenediaminetetraacetic acid. *Adv. Environ. Res.* **2002**, *7*, 197–202. [CrossRef]
32. Engelmann, M.D.; Bobier, R.T.; Hiatt, T.; Cheng, I.F. Variability of the Fenton reaction characteristics of the EDTA, DTPA, and citrate complexes of iron. *BioMetals* **2003**, *16*, 519–527. [CrossRef] [PubMed]
33. Sharma, V.K.; Millero, F.J.; Homonnay, Z. The kinetics of the complex formation between iron(III)–ethylenediaminetetraacetate and hydrogen peroxide in aqueous solution. *Inorg. Chim. Acta* **2004**, *357*, 3583–3587. [CrossRef]
34. Rämö, J.; Sillanpää, M. Degradation of EDTA by hydrogen peroxide in alkaline conditions. *J. Clean. Prod.* **2001**, *9*, 191–195. [CrossRef]
35. Chiriţă, P. Hydrogen Peroxide Decomposition by Pyrite in the Presence of Fe(III)-ligands. *Chem. Biochem. Eng. Q.* **2009**, *23*, 259–265.
36. Walling, C.; Kurz, M.; Schugar, H.J. The Iron(III)-Ethylenediaminetetraacetic Acid-Peroxide system. *Inorg. Chem.* **1970**, *9*, 931–937. [CrossRef]

37. Nowack, B.; Sigg, L. Dissolution of Fe(III)(hydr)oxides by metal-EDTA complexes. *Geochim. Cosmochim. Acta* **1997**, *61*, 951–963. [CrossRef]
38. Pierce, E.M.; Wellman, D.M.; Lodge, A.M.; Rodriguez, E.A. Experimental determination of the dissolution kinetics of zero-valent iron in the presence of organic complexants. *Environ. Chem.* **2007**, *4*, 260–270. [CrossRef]
39. Allison, J.D.; Brown, D.S.; Novo-Gradac, K.J. *MINTEQA2/PRODEFA2-A Geochemical Model for Environmental Systems: Version 3.0 User's Manual*; EPA/600/3-91/021; US Environmental Protection Agency, Environmental Research Laboratory: Washington, DC, USA, 1991.
40. Szpyrkowicz, L.; Juzzolino, C.; Kaul, S.N. A comparative study on oxidation of disperse dyes by electrochemical process, ozone, hypochlorite and Fenton reagent. *Water Res.* **2001**, *35*, 2129–2136. [CrossRef]
41. Lin, S.H.; Chen, M.L. Treatment of textile wastewater by chemical methods for reuse. *Water Res.* **1997**, *31*, 868–876. [CrossRef]
42. Walling, C. Fenton's reagent revisited. *Acc. Chem. Res.* **1975**, *8*, 125–131. [CrossRef]
43. Oakes, J.; Smith, E.G. Nuclear magnetic resonance studies of transition-metal complexes of Ethylenedi-amine-NNN'N'-tetramethylphosphonate in aqueous solution. *J. Chem. Soc. Dalton Trans.* **1983**, 601–605. [CrossRef]
44. Francis, K.C.; Cummins, D.; Oakes, J. Kinetic and structural investigations of [FeIII(edta)] − [edta = Ethylenediaminetetra-acetate(4-)] catalysed decomposition of hydrogen peroxide. *J. Chem. Soc. Dalton Trans.* **1985**, 493–501. [CrossRef]
45. Haber, F.; Weiss, J. The catalytic decomposition of hydrogen peroxide by iron salts. *Proc. R. Soc. Lond. Ser. A* **1934**, *147*, 332–351.
46. Barb, W.G.; Baxendale, J.H.; George, P.; Hargrave, K.R. Reactions of ferrous and ferric ions with hydrogen peroxide. *Trans. Faraday Soc.* **1951**, *47*, 462–500. [CrossRef]
47. Rush, J.D.; Maskos, Z.; Koppenol, W.H. Distinction between hydroxyl radical and ferryl species. *Methods Enzymol.* **1990**, *186*, 148–156.
48. Rahhal, S.; Richter, H.W. Reduction of hydrogen peroxide by the ferrous iron chelate of diethylenetriamine-N,N,N',N'',N''-pentaacetate. *J. Am. Chem. Soc.* **1988**, *110*, 3126–3133. [CrossRef]
49. Sanchez, I.; Stüber, F.; Font, J.; Fortuny, A.; Fabregat, A.; Bengoa, C. Elimination of phenol and aromatic compounds by zero valent iron and EDTA at low temperature and atmospheric pressure. *Chemosphere* **2007**, *68*, 338–344. [CrossRef]
50. American Public Health Association; American Water Works Association. Method 5310 B. In *Standard Methods for the Examination of Water and Wastewater*, 17th ed.; Clesceri, L.S., Greenberg, A.E., Trusel, R.R., Franson, M.A., Eds.; American Public Health Association: Washington, DC, USA, 1989.

© 2019 by the authors. Licensee MDPI, Basel, Switzerland. This article is an open access article distributed under the terms and conditions of the Creative Commons Attribution (CC BY) license (http://creativecommons.org/licenses/by/4.0/).

MDPI
St. Alban-Anlage 66
4052 Basel
Switzerland
Tel. +41 61 683 77 34
Fax +41 61 302 89 18
www.mdpi.com

Catalysts Editorial Office
E-mail: catalysts@mdpi.com
www.mdpi.com/journal/catalysts